孙玉

1962年毕业于清华大学，后被分配到中国电子科技集团第54研究所工作至今。其间，从事军事通信设备研制和通信系统总体工程设计；领导创建了电信网络专业和数字家庭专业；出版电信科技著作13部。1995年当选中国工程院院士。现任，国防电信网络重点实验室科技委主任；兼任，中央军委科技委顾问。

·孙玉院士技术全集·
中国工程院院士文集

PDH for
Telecommunication Network

◎ 孙 玉 编著

人民邮电出版社
北京

图书在版编目（ＣＩＰ）数据

电信网络中的PDH技术 ＝ PDH for Telecommunication Network：英文／孙玉编著．－－北京：人民邮电出版社，2017.9
（孙玉院士技术全集）
ISBN 978-7-115-44675-6

Ⅰ．①电… Ⅱ．①孙… Ⅲ．①数字通信系统－英文 Ⅳ．①TN914.3

中国版本图书馆CIP数据核字(2017)第232813号

Summary

This book is a basic work specialized on modern digital communication, which consists of plesiochronous digital hierarchy, digital multiplexing principle, essential realization technology, relevant recommendations of ITU, CCITT/TSS and the application design of typical engineering project. It is intended for communication engineers and technicians as well as teachers and students of the universities and colleges.

◆ 编　著　孙　玉
　责任编辑　杨　凌
　责任印制　彭志环

◆ 人民邮电出版社出版发行　北京市丰台区成寿寺路11号
邮编 100164　电子邮件 315@ptpress.com.cn
网址 http://www.ptpress.com.cn
北京圣夫亚美印刷有限公司印刷

◆ 开本：700×1000　1/16
印张：25.25　彩插：1
字数：444千字　2017年9月第1版
　　　　　　　　2017年9月北京第1次印刷

定价：198.00元

读者服务热线：(010)81055488　印装质量热线：(010)81055316
反盗版热线：(010)81055315

《中国工程院院士文集》总序

二〇一二年暮秋,中国工程院开始组织并陆续出版《中国工程院院士文集》系列丛书。《中国工程院院士文集》收录了院士的传略、学术论著、中外论文及其目录、讲话文稿与科普作品等。其中,既有早年初涉工程科技领域的学术论文,亦有成为学科领军人物后,学术观点日趋成熟的思想硕果。卷卷《文集》在手,众多院士数十载辛勤耕耘的学术人生跃然纸上,透过严谨的工程科技论文,院士笑谈宏论的生动形象历历在目。

中国工程院是中国工程科学技术界的最高荣誉性、咨询性学术机构,由院士组成,致力于促进工程科学技术事业的发展。作为工程科学技术方面的领军人物,院士们在各自的研究领域具有极高的学术造诣,为我国工程科技事业发展做出了重大的、创造性的成就和贡献。《中国工程院院士文集》既是院士们一生事业成果的凝练,也是他们高尚人格情操的写照。工程院出版史上能够留下这样丰富深刻的一笔,余有荣焉。

我向来以为,为中国工程院院士们组织出版《院士文集》之意义,贵在"真善美"三字。他们脚踏实地,放眼未来,自朴实的工程技术升华至引领学术前沿的至高境界,此谓其"真";他们热爱祖国,提携后进,具有坚定的理想信念和高尚的人格魅力,此谓其"善";他们治学严谨,著作等身,求真务实,科学创新,此谓其"美"。《院士文集》集真善美于一体,辩而不华,质而不俚,既有"居高声自远"之澹泊意蕴,又有"大济于苍生"之战略胸怀,斯人斯事,斯情斯志,令人阅后难忘。

读一本文集,犹如阅读一段院士的"攀登"高峰的人生。让我们翻开《中国工程院院士文集》，进入院士们的学术世界。愿后之览者，亦有感于斯文，体味院士们的学术历程。

徐匡迪

二〇一二年

全集序言

20世纪70年代后期，我国的通信网开始模/数转换，当时国内自行研制的 PCM 基群设备和二次群数字复接设备先于国外引进的产品在国内试验并应用，打破了国外的技术封锁。我与孙院士相识也是从那时开始，孙院士在这之前就成功主持了我国第一代散射数字传输系统和第一套 PDH 数字复接设备的研制，我当时负责 PCM 基群复用设备的研制和试验。PCM 基群与 PDH 数字复接设备分属一次群与二次群，在网络上是上下游的关系，我们连续几年一起参加国际电信联盟（ITU）数字网研究组的标准化会议，后来在各自的工作中又有不少的联系，从中了解了他的学识，也学习了他的做人准则。他在通信工程方面有非常丰富的经验，他对通信网的理解、对通信标准的掌握和治学精神的严谨一直为我所敬佩，他勤于思考和积极探索，善于总结和举一反三，乐于诲人和提携后进，与他共事受益不浅。在这之后他又相继研制成功数字用户程控交换机、ISDN 交换机、B-ISDN 交换机及相应的试验网，还主持研制成功接入网和用户驻地网网络平台，并将上述成果应用到专用通信网和民用通信工程中，很多研发工作都是国内首次完成。

孙玉院士将研发体会写成著作交由人民邮电出版社出版，他的著作如同他的科技成果一样丰硕，从20世纪80年代初的《数字复接技术》一书开始，陆续出版了《数字网传输损伤》、*PDH for Telecommunications Network*、《数字网专用技术》《电信网络总体概念讨论》《电信网络安全总体防卫讨论》《应急通信技术总体框架讨论》《数字家庭网络总体技术》《电信网络中的数字方法》和《孙玉院士技术报告文集》，其中《数字复接技术》与《数字网传输损伤》两本书还都出了修订本。这些论著所涉及的领域或视角在当时为国内首次出版。他鼓励我将科研成果也写成书出版，既可将宝贵的经验与同行共享，也是自身对专业认识的深化过程。我写过一本书，深感要写出自己满意且读者认可的书非要下苦功不可。孙玉院士难能可贵的是笔耕三十年，著作十余本，网聚新技术，敢为世人先。这一系列专著覆盖了电信网的诸多方面，每一本既独立成书但又彼此关联，虽然时间跨度几十年，但就像一气呵成那样连贯，这些著作体现了他的一贯风格，概念清晰准确，思路层次分明，理论与实践结合，解读深入浅出。这些论著在写作上以电信网系统工程为主线，突出了总体设计思想和方法，既有严格的电信标准规范，又有创新性的解决方案，学术思想寓于工程应用中，兼具知识性与实用性，不论是对电信工程师还是相关专业的高校师生都不无裨益，在我国电信网的建设中发挥了重要作用。电信网技术演进很快，但这一系列著作所论述的设计思想及方法论对今后网络发展的认识仍有很好的指导意义，人民邮电出版社提议出版孙玉院士著作全集，更便于广大读者对电信网全局和系统性的了解，这是电信界的一件好事，并得到了中国工程院院士文集出版工作的大力支持，我期待这一全集的隆重问世。

中国工程院院士

2017 年 6 月于北京

全集出版前言

1962—1995 年期间，我在科研生产第一线，有幸参加了我国电信技术数字化的全过程。其间根据科研工作进程的需要，也是创建电信网络专业的需要，我逐年编写并出版了一些著作。

1. 专著《数字复接技术》，人民邮电出版社出版，1983 年第一版；1991 年修订版；1994 年翻译版 *PDH for Telecommunication Network*，IPC.Graphics.U.S.A。这是我 1970—1980 年期间，从事复接技术研究的工作总结。其中提出了准同步数字体系（PDH）数字复用设备的国际通用工程设计方法。令我欣慰的是，这本书居然存活了十余年，创造并保持着人民邮电出版社科技专著销量纪录，让我在我国电信技术界建立了广泛的友谊。

2. 编著《数字网传输损伤》，人民邮电出版社出版，1985 年第一版；1991 年修订版。这是我 1970—1980 年期间，出于电信网络总体工程设计需要，参考国际电信联盟（ITU）文献，编写的工具书。为了便于应用，其中澄清了一些有关传输损伤的基本概念。

3. 编著《数字网专用技术》，人民邮电出版社 1988 年出版。这是为我的硕士研究生们编写的专业科普图书，介绍了一些当时出现不久的技术概念和原理。显然，无技术水平可言。

1995 年之后，我退居科研生产第二线，转入技术支持工作。其间，根据当时的技术问题，以及培育学生和理论研究的需要，我逐年编写并出版了一些著作。

4. 编著《数字家庭网络总体技术》，电子工业出版社 2007 年出版。这是我 2006—2009 年期间，受聘国家数字家庭应用示范产业基地（广州）技术顾问，为广州基地编写的培训教材。其中提出了数字家庭第二代产业目标——家庭网络平台和多业务系统，被基地和工信部接受。

5. 专著《电信网络总体概念讨论》，人民邮电出版社 2008 年出版。这是我 2005—2008 年期间，从事电信网络机理研究的总结。在我从事电信科研 30 多年之后发现，电信网络技术作为已经存在 160 多年、支撑着遍布全球电信网络的基础技术，居然尚未澄清电信网络机理分类，而且充满了概念混淆。我试图讨论这些问题。其中，澄清了电信网络的形成背景；电信网络技术分类；电信网络机理分类及其属性分析。但是，当我得出电信网络资源利用效率的数学结论时，竟然与我的物理常识大相径庭。为此，我在全国知名电信学府和研究院所做了 50 多场讲座，主要目的是请同行指点我的理论是否有误。这是我的代表著作，令我遗憾的是，这是一本未竟之作。书名称为"讨论"，是期盼后生能够接着讨论这个问题。

6. 编著《电信网络安全总体防卫讨论》，人民邮电出版社 2008 年出版。这是 2004—2005 年期间，我在国务院信息办参加解决"非法插播和电话骚扰问题"时编写的总结报告，经批准出版。其中提出了网络安全的概念；建议主管部门不要再利用通信卫星广播电视信号；建议国家发射广播卫星；建议国家建设信源定位系统。这本书曾经令同行误认为我懂得网络安全。其实，我仅仅经历了半年时间，参与解决上述特定问题。

7. 编著《应急通信技术总体框架讨论》，人民邮电出版社 2009 年出版。这是 2008—2009 年期间，在汶川地震前后，我参加国家应急通信技术研究时编写的技术报告。希望澄清应急通信总体概念，然后开展科研工作。可惜，我未能参与后续的工作。

8. 编著《电信网络技术中的数学方法》，人民邮电出版社 2017 年出版。我国电信界普遍认为，在电信技术中应用数学方法非常困难，同时，也看到一旦利用数学方法解决了问题，就会取得明显的工程效果。2009 年我曾建议人民邮电出版社出版《电信技术中的数学方法丛书》。所幸，一经提出就得到了人民邮电出版社和电信同仁的广泛支持。本书作为这套丛书的"靶书"，仅供同行讨论，以寻求编写这套丛书的规范。我认为数学方法对于电信技术的发展和人才的培养具有特殊的意义，我期待着这套丛书出版。

9. 编著《孙玉院士技术报告文集》，人民邮电出版社 2017 年出版。这是我历年技术报告的代表性文本，其中，主要是近年来关于研制和推广应用物联网的相关报告。这些报告多数属于科普报告，主要反映了我对于我国国民经济信息化的期望。

上述著作，出版时间跨越整整 34 年，电信科技内容覆盖了我 50 多年的科研历程。可见，这几本书基本上是一叠陈年旧账。然而，人民邮电出版社决定出版这套全集，也许，他们认为，这套全集大体上能够从电信技术出版业角度，反映出我国电信技术的发展历程；反映出我们这一代电信工程师的工作经历；同时，也反映了与我们同代的电信科技书刊编辑们的奉献。也许，他们认为，作为高技术中的基础学科，电信技术的某些理论和技术成就仍然起着支撑和指导作用。如实而言，不难发现，在我国现实、大量信息系统工程设计中，涉及信息基础设施（电信网络）设计，普遍存在概念性、技术性、机理性甚至常识性错误。我们国家已经走过生存、发展历程，正在走向强大。在我国电信领域，不仅需要加强技术研究（如 "863" 计划），而且需要加强理论研究（如 "973" 计划）。期待我国年轻的电信科技精英们，特别是年轻有为的院士们，能够编撰出更好、更多的电信科技著作。

2017 年 6 月于中国电子科技集团公司第 54 研究所

This book is given in commemoration of Mr. Robert A. Brooks' visit to Communication, Telemetry and Telecontrol Research Institute (CTI) in July 1993.

Preface to English Edition

The basic contents of the book is taken from "Digital Multiplexing Technology" published by PTPRESS, China in 1991. In the course of translation, the contents concerning the plesiochronous digital hierarchy (PDH) taking 1544kbit/s as primary digital group have been augmented. Now, its name is changed to "Plesiochronous Digital Hierarchy (PDH) in Communication Network", corresponding to the synchronous digital hierarchy (SDH) recommended by CCITT in 1988.

For the publishing of the English edition, I am grateful to Mr. Robert A. Brooks for his financial aid, and also to PTPRESS for the support; I give my special thanks to Mr. Wang Feng and Mr. Lu Chengzhao who translated this book together and also to Mr. Lung Hsiungchang who reviewed the translation.

December, 1993
CTI, P.R.China

Table of contents

Chapter 1 Introduction ··· 1
  1.1   DIGITAL MULTIPLEX CONCEPT ··· 1
  1.2   DIGITAL MULTIPLEX HIERARCHY ··· 3
  1.3   CCITT RECOMMENDATIONS ·· 5
  1.4   CONTENT OF THE BOOK ·· 6

Chapter 2 Synchronous Multiplexing ··· 7
  2.1   SYNCHRONOUS MULTIPLEXING EQUIPMENT COMPOSITION ······· 7
  2.2   FRAME STRUCTURE ·· 9
  2.3   FRAME ALIGNMENT ··· 11
  2.4   SYNCHRONIZATION ACQUISITION METHOD ·························· 13
  2.5   THE CODE TYPE OF FRAME ALIGNMENT SIGNALS ·················· 15
  2.6   AVERAGE ACQUISITION TIME ··· 16
  2.7   OPTIMUM LENGTH OF FRAME ALIGNMENT SIGNAL ··············· 20
  2.8   SYNCHROMIZATION STATUS PROTECTION ··························· 24
  2.9   ACQUISITION PROCEDURE CHECKING ································· 27
  2.10 FRAME ALIGNMENT PROTECTION PARAMETERS ·················· 31
  2.11 AVERAGE OUT-OF-FRAME TIME AND AVERAGE SYNCHRONIZATION TIME ·· 34
  2.12 THE BIT ERRORS OF SYNCHRONIZATION OUT-OF-FRAME ············ 35
  2.13 FRAME ALIGNMENT ACQUISITION/ MAINTENANCE LOGIC ······ 37
  2.14 GUARANTEEING THE SYNCHRONIZATION CIRCUMSTANCE ······· 41
  2.15 CCITT RECOMMENDATIONS ·· 44

Chapter 3 Positive Justification ··· 49
  3.1   PLESIOCHRONOUS MULTIPLEX ··· 49
  3.2   PRINCIPLE OF POSITIVE JUSTIFICATION ······························ 50
  3.3   BASIC FORMULA OF POSITIVE JUSTIFICATION ····················· 52
  3.4   JUSTIFICATION DESIGN ·· 54

3.5 JUSTIFICATION TRANSITION PROCESS 57
3.6 THE CHANGE RANGE OF THE READ/WRITE TIME DIFFERENCE IN A STABLE JUSTIFICATION PROCESS 61
3.7 CLASSIFICATION OF JUSTIFICATION TRANSITION PROCESS 64
3.8 BIT RATE RECOVERY DESIGN 66
3.9 PHASE-LOCK PARAMETERS DESIGN 69
3.10 VOLTAGE CONTROLLED OSCILLATOR DESIGN 73
3.11 BUFFER SIZE 76
3.12 CCITT Recommendations 78

Chapter 4 Impairment of Positive Justification 82
4.1 STUFFING JITTER 82
    4.1.1 Physical Concept of Stuffing Jitter 82
    4.1.2 Signal justification process justifying only q times in p frames 84
    4.1.3 Justifying q times in p frames with a residue 88
    4.1.4 The relationship between the number of code justification detectors and stuffing jitter 95
    4.1.5 Distribution of stuffing jitter 99
4.2 STUFFING ERROR 104
    4.2.1 Physical Concept of Stuffing Error 104
    4.2.2 Calculation of stuffing error 104
    4.2.3 Suppression of stuffing errors 109

Chapter 5 Positive/Negative Justification 111
5.1 THE PRINCIPLE OF POSITIVE/NEGATIVE JUSTIFICATION 111
5.2 JUSTIFICATION BY FIX ED DECISION CONTROL 113
5.3 ADAPTIVE JUSTIFICATION CONTROL 116
5.4 POSITIVE/NEGATIVE JUSTIFICATION CONTROL CIRCUIT 124
5.5 TRANSITION PROCESS OF JUSTIFICATION 126
5.6 PARAMETER DESIGN OF JUSTIFICATION 128
5.7 ENVIRONMENT DESIGN OF JUSTIFICATION 131
5.8 THE TECHNICAL APPLICATION OF THE POSITIVE/NEGATIVE JUSTIFICATION 135
5.9 EXAMPLE OF POSITIVE/NEGATIVE JUSTIFICATION DESIGN 136
5.10 CHARACTERISTIC COMPAISON OF JUSTIFICATIONS 142
5.11 RECOMMENDATION OF CCITT 143

Chapter 6 Positive/0/Negative Justification ··· 145
- 6.1 CONCEPT OF POSITIVE/0/NEGATIVE JUSTIFICATION ················ 145
- 6.2 THE PRINCIPLES OF DELTA CONTROLLED POSITIVE/0/NEGATIVE JUSTIFICATION ··· 146
- 6.3 DIGITAL SMOOTH POSITIVE/0/NEGATIVE JUSTIFICATION PRINCIPLES ·· 147
- 6.4 FRAME STRUCTURE ·· 149
- 6.5 JUSTIFICATION AND RECOVERY CONTROL ······························ 153
- 6.6 JUSTIFICATION TRANSITION PROCESS ······································ 156
- 6.7 DIGITAL SMOOTH ·· 162
- 6.8 PULSE SMOOTHLY PRINCIPLE OF WEST GERMAN PCM30D ······· 163
- 6.9 AUSTRALIA 2/8Mbit/s MULTIPLEXER PULSE SMOOTH PRINCIPLE ··· 166
- 6.10 ENGINEERING APPLICATIONS OF POSITIVE/0/NEGATIVE JUSTIFICATION ··· 169

Chapter 7 Measurement Techniques of Justification ································ 170
- 7.1 MEASUREMENT OF STUFFING JITTER ······································ 170
  - 7.1.1 Characteristics of stuffing jitter ·· 170
  - 7.1.2 Measurement range and precision requirement ······················· 172
  - 7.1.3 Measurement equipment ·· 173
  - 7.1.4 Measurement method ·· 176
  - 7.1.5 Example of measurement ·· 181
- 7.2 MEASUREMENT OF MULTIPLEX CODE ERROR ······················· 183
  - 7.2.1 Characteristics of multiplex code error ···································· 183
  - 7.2.2 Requirements of code error meter ·· 184
  - 7.2.3 Design of Special code error meter ·· 186
  - 7.2.4 Design of analog channel ·· 189
  - 7.2.5 Example of measurement ·· 190
- 7.3 MEASUREMENT CHARACTERISTICS OF THE PHASE-LOCKED LOOP OF THE CODE RECOVERY ·· 192
  - 7.3.1 Tracing error measurement ··· 192
  - 7.3.2 Measurement of jitter suppression characteristics ··················· 194

Chapter 8 Frame Adjustment Principle ··· 195
- 8.1 GENERAL ·· 195

8.2　FRAME REGULATION CATEGORIES ················· 198
8.3　THE WORKING PRINCIPLE OF PRE-BUFFER FRAME REGULATOR ················· 199
8.4　2048kbit/s FRAME REGULATOR DESIGN ················· 203
8.5　THE ADDITIONAL FUNCTION OF FRAME REGULATOR ············ 210

Chapter 9　PDH/SDH Interface ················· 211
9.1　GENERAL ················· 211
9.2　HIGHER ORDER GROUP SDH ················· 211
9.3　THE GENERAL ARRANGEMENT OF PDH/SDH INTERFACE ················· 214
9.4　THE ARRANGEMENT DESIGN FOR C-4 TO ENTER STM-1 ············ 217
9.5　THE ARRANGEMENT DESIGN FOR C-3 ENTERING STM-1 ···· 221
9.6　THE ARRANGEMENT DESIGN FOR TUG ENTERING HIGH ORDER VC ················· 224
9.7　BASIC CONTAINER ENTERING TUG ················· 227
9.8　STM-N MULTIPLEXING ················· 231
9.9　PAYLOAD CONTAINER INTERFACE IN CCITT REC ················· 234
9.10　THE IMPROVED DESIGN OF PAYLOAD CONTAINER INTERFACE ················· 240

Chapter 10　Anti Fading Frame Synchronization ················· 248
10.1　THE PRINCIPLES OF ANTI FADING FRAME SYNCHRONIZATION ···· 248
10.2　FRAME SYNCHRONIZATION AVERAGE KEEPING TIME ········ 253
　　10.2.1　Scheme (1) ················· 253
　　10.2.2　Scheme (2) ················· 255
　　10.2.3　Scheme (3) ················· 257
　　10.2.4　Scheme (4) ················· 261
　　10.2.5　Scheme (5) ················· 265
　　10.2.6　Scheme (6) ················· 269
10.3　FRAME LOSS AVERAGE KEEPING TIME ················· 273
　　10.3.1　Frame Loss deciding Time ················· 273
　　10.3.2　Frame Synchronization Search Time ················· 275
　　10.3.3　Frame Synchronization Deciding Time ················· 278
　　10.3.4　Frame Synchronization Reset Delay ················· 285
　　10.3.5　Frame Loss Keeping Time ················· 286
10.4　COMPARISON OF FRAME SYNCHRONIZATION SCHEMES ········ 286

	10.4.1	Frame Synchronization Average Keeping Time ······················ 286
	10.4.2	Frame Loss Average Keeping Time ····································· 289
	10.4.3	The Synthetical Criterion of the Frame Synchronization System ············· 293

Chapter 11　Engineering Application Design ································ 296

11.1　GENERAL PLESIOCHRONOUS GROUP MULTIPLEX DESIGN ······ 296
- 11.1.1　Elementary Parameters Design ··· 296
- 11.1.2　Transition Process Design ··· 300
- 11.1.3　Frame Loss Probability and Search Characteristics Design ···················· 300
- 11.1.4　Partition of the Basic Units ·· 301
- 11.1.5　Design Examples ··· 302

11.2　STANDARD/NON-STANDARD RATES TOLERANCE DESIGN ······ 307
- 11.2.1　Introduction ·· 307
- 11.2.2　Transmission of Non-Standard Binary Digits through Standard Channels ···· 308
- 11.2.3　Design Examples ··· 309
- 11.2.4　Compatible Multiplex of Different Tributary Rates ·························· 310

11.3　PLESIOCHRONOUS/SYNCHRONOUS COMPATIBLE DESIGN ······· 313
- 11.3.1　Introduction ·· 313
- 11.3.2　Instruction Justification Compatible method ······························· 313
- 11.3.3　Multiframe Fixed Control Compatible Method ····························· 316

11.4　2/34Mbit/s MULTIPLEX DESIGN ······································· 320
- 11.4.1　2/34Mbit/s Multiplex Scheme Comparison ································ 320
- 11.4.2　Model B 2/34Mbit/s Multiplexer Design ·································· 323
- 11.4.3　CCITT Recommendations ··· 329

11.5　DESIGN OF BRANCH IN GROUP TRUNK TRANSMISSION ············ 329
- 11.5.1　Introduction ·· 329
- 11.5.2　Trunk Branch Simplified Scheme One ··································· 331
- 11.5.3　Trunk Branch Simplified Scheme Two ··································· 331
- 11.5.4　Trunk Branch Simplified Scheme Three ·································· 332
- 11.5.5　Design of Branch Control Signals ······································ 335
- 11.5.6　Function Supplement ··· 337
- 11.5.7　CCITT Recommendations ··· 339

11.6　INTERNETWORK MULTIPLEX DESIGN ······························· 341
- 11.6.1　2048-6312kbit/s Interconnection Multiplex Design ························· 341
- 11.6.2　44 736-139 264kbit/s Interconnection Multiplex Design ······················ 343

- 11.7 SUBGROUP MULTIPLEX DESIGN ... 346
  - 11.7.1 Subgroup Rate Selection ... 346
  - 11.7.2 Subframe Structure Arrangement ... 349
  - 11.7.3 Examples of Subgroup Frame Structure Design ... 355
- 11.8 ISDN USERS/NETWORK INTERFACE MULTIPLEX DESIGN ... 360
  - 11.8.1 2B+D Interface Multiplex Design ... 360
  - 11.8.2 2048kbit/s Primary Interface Multiplex Design ... 362
  - 11.8.3 Multiplex Design of Entering 64kbit/s Channel ... 363
  - 11.8.4 Multiplex Design of X_1 Rate Entering 8/16kbit/s ... 363
  - 11.8.5 Design of V Rate Multiplexed into Middle Rate ... 366
- 11.9 DESIGN OF MULTIPLEX SYSTEM MAINTENANCE ... 371
  - 11.9.1 Introduction ... 371
  - 11.9.2 Maintenance Principles ... 371
  - 11.9.3 Multiplexer Maintenance Requirements ... 373
  - 11.9.4 System Consideration of Maintenance Design ... 375
  - 11.9.5 Examples of Maintenance Design ... 376

Bibliography ... 382

出集出版后记 ... 386

Chapter 1   Introduction

1.1   DIGITAL MULTIPLEX CONCEPT

In a digital communication network, in order to expand transmission capacity and increase transmission efficiency, it is often necessary to combine a number of low rate digital signals into a high rate digital signal which will be transmitted over a high rate channel. The digital multiplex is the special technique for implementation of the digital signals combination. The digital multiplex equipment in a digital communication network corresponds to a carrier communication equipment in an analog communication network. But the position of digital multiplex is much more important in a digital communication network than that of the carrier communication equipment in an analog communication network. In a digital communication network, digital multiplex is not only a special technique juxtaposed with information source coding, digital transmission and digital switching, but also the fundament for techniques such as frame alignment in network synchronization, line multiplex in line concentrators and time division access in digital switching. It is clear that the digital multiplex is a fundamental technique in a digital communication network.

A digital multiplex system is composed of a digital multiplexer and a digital demultiplexer. As shown in Figure 1-1, the digital multiplexer is such an equipment to combine two or more tributary digital signals into a single combined digital signal in accordance with time-division multiplexing; the digital demultiplexer is an equipment to divide the combined digital signal into original tributaries' digital signals. Usually the digital multiplexer and the digital demultiplexer are installed together to form an equipment called multiplexer-demultiplexer (abbreviated to muldex). which is called digital multiplex equipment in short.

A digital multiplexer consists of timing, justification and multiplex units while a digital demultiplexer consists of synchronization, timing, demultiplex and recovery units. The timing unit provides an unified reference time signal to the digital multiplex equipment. There is an internal clock in the multiplier's timing unit which can be also driven by an external clock. The timing unit in the demultiplexer can only be driven by the received clock, and by means of control of a synchronization unit, which makes the reference time signals of the demultiplexer hold a correct phase relation with that of multiplexer. i.e. holding synchronization. The justification unit is corresponding to the recovery unit; the multiplex unit; the multiplex unit is corresponding to the demultioplex unit. The function of the justification unit is to conduct necessary frequency or phase adjustment to digital signals from input tributaries, thus forming the digital signals synchronous with the local timing signal before the multiplex unit combines them into a combined digital signal by using time-division multiplexing. The functions of the demultiplex unit is to time-divide the combined digital signal into synchronous tributary digital signals, which shall be restored to original tributary digital signals via the recovery unit.

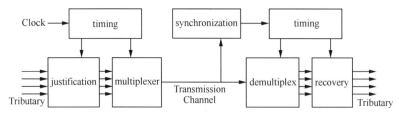

Figure 1-1 Simplified block diagram of a digital multiplex equipment

It can be seen from the principle of time-division multichannel communication that all the tributary digital signals at the multiplex unit input must be synchronous, i.e. a correct phase relation must be maintained between their significant instants and corresponding local timing signals, but there is no requirement at the justification unit inputs, i.e. the multiplex inputs. If the tributary digital signals at the multiplexer input are synchronous to the local timing signal, the justification unit is only to adjust the phase, at times even the phase need not to be adjusted, this type of multiplexer is called synchronous one. If the tributary input digital signals are asynchronous to the local timing signal, i.e. their corresponding significant instants occur probably at different rate, the justification unit needs adjusting rate and phase for each tributary digital signal so as to make it become a synchronous digital signal, this type of multiplexer is called heterochronous multiplexer. If the significant instants of the input tributary digital signals occur at the same nominal reat relative to the local corresponding timing signal, but any variation of the rate is limited to a range of specified tolerance, this type of multiplexer is called plesiochronous multiplexer. In recent years CCITT calls the a foresaid hetero-chronous muldex and plesiochronous muldex to be an asynchronous muldex.

1.2 DIGITAL MULTIPLEX HIERARCHY

Digital multiplex hierarchy means a digital multiplex series with different levels according to multiplex possibility, in which the multiplex at certain level means that the fixed number of digital signals with lower specified rate are combined into a digital signal with higher specified rate. In next higher level digital multiplex, the digital signal with higher rate[1] and other digital signals with the same rate are further combined. Therefore, before the digital multiplex technique is discussed in detail the digital multiplex hierarchy and the digital rate series must be defined.

Definition of the digital rate series and planning of the digital multiplex hierarchy are associated with overall digital network in all aspects. They depend upon digital transmission, digital multiplex, source coding and network development, etc., of which each aspect involves many factors again. For example, digital multiplex involves many factors in cluding primary PCM group structure, frame structure, network synchronization system and digital switching system, etc. The requirement for frame structure, in turn. involve many more detailed factors, such as frame alignment, justification, signaling and service digit. The causalities between the digital rate series and these aspects or factors is given in Figure 1-2[2]. For these causalities, some can be described by analytical expressions, some can not do so. Generally speaking, this is only an induction after the event because between these aspects or factors mutual restrictions have been formed historically in a hundred and one way during the long term development and evolution of digital communication network technique. For these restrictions, some are proper, some are not so proper, even some are mutually contradictory, Even if it has been found afterwards that some aspects or factors were unsuitable, but generally it is difficult to change the existing fact, because these problems are not only the technical matter being reasonable or not, but also the economic matter involving every a spects at the high cost.

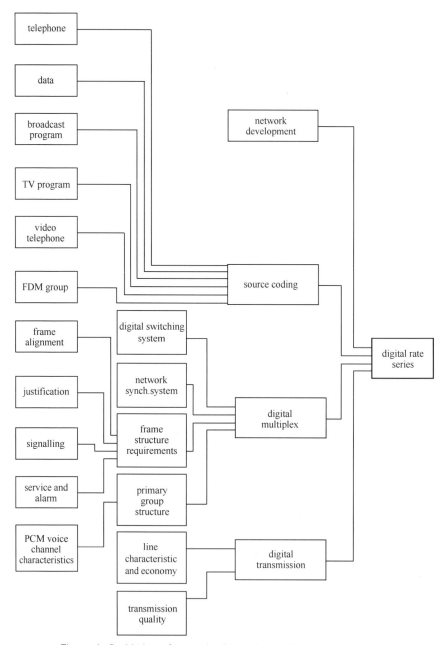

Figure 1-2 Various factors in determining digital rate hierarchy

1.3 CCITT RECOMMENDATIONS

On the basis of numerous theoretic and experimental research in different countries, and through full discussion and repeated trade-off, CCITT now has recommended two types of digital rate serieses and digital multiplex hierarchy, see Figure 1-3[3] and Figure 1-4. North America and Japan adopt a digital rate series taking 1544kbit/s as the first level rate (called primary group), Europe and ex-USSR adopt another digital rate series 2048kbit/s as the first level rate. For the latter, although the digital rate series is unified, division of the digital multiplex hierarchy is different. Level-by-level combination, such as the combination of 2Mbit/s into 8Mbit/s and in turn 8Mbit/s into 34Mbit/s etc. is called a digital multiplex hierarchy using $n-(n+1)$ mode; level-intervallic combination, for example the combination of 2Mbit/s into 34Mbit/s or 8Mbit/s into 140Mbit/s, is called a digital multiplex hierarchy using $n-(n+2)$ mode. For example, previously France did not adopt the digital multiplex hierarchy from 34Mbit/s to 140Mbit/s. In 1988 CCITT researched and adopted a new digital rate series, i.e. high order group synchronous digital rate series. whose specific rates are 155 520kbit/s and 622 080kbit/s etc. These digital rates are unified in the world.

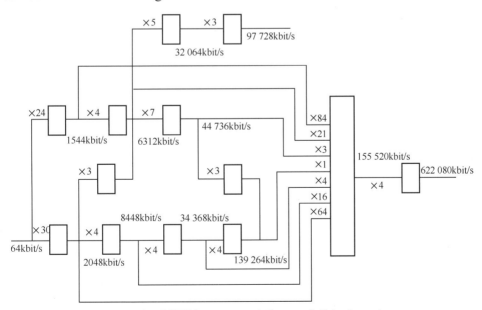

Figure 1-3 CCITT Recommendation on digital rate series

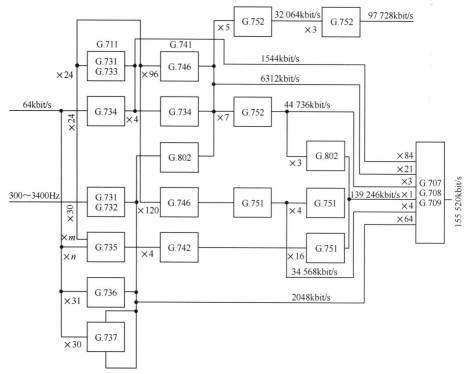

Figure 1-4 CCITT Recommendation on digital multiplex hierarchy

1.4 CONTENT OF THE BOOK

The book is composed of four parts. The first part is synchronous digital multiplex, which is the basis of digital multiplex technique. In this part emphasis is put on discussion of frame structure and frame synchronization. The second part is plesiochronous (or called heterochronous) digital multiplex which is main content of digital multiplex technique. In this part emphasis is put on discussion of various justification methods. The third part is SDH/PDH interface technique, which was recommended by CCITT in 1988 and basically it has comprehensive application of synchronous and plesiochronous multiplex techniques. The fourth part is engineering application desig. In this part specitic design methods is discussed for various typical engineering applications. For digital multiplex hierarchy and their technical specification CCITT has laid down a full set of interrelated recommendations with strict structure, which are authoritative international standards and valuable international experience, so the content in this book will join up closely with these recommendations.

Chapter 2 Synchronous Multiplexing

2.1 SYNCHRONOUS MULTIPLEXING EQUIPMENT COMPOSITION

As stated in the introduction, if the digital signals from input tributaries are synchronous to the local timing signal and only phase adjustment is required (sometimes no adjustment is required) in the digital multiplexing, it is synchronous multiplexing. As shown in Figure 2-1, when the adjustment unit and recovery unit are deleted it becomes the principle schematics of the synchronous multiplexing equipment.

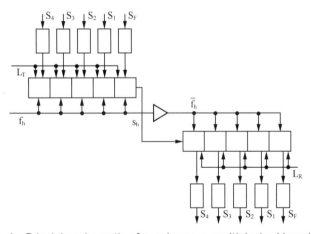

Figure 2-1 Principle schematic of synchronous multiplexing/demultiplexing

The basic function of synchronous multiplexer/demultiplexer are to combine the synchronous tributary digital signals into the combined digital signal and in turn separate the combined digital signal into the original tributary digital signals. It is clear from the Figure 2-1 that the digital multiplexing/demultiplexing equipment can be divided into a timing system and a multiplexing/demultiplexing system. The timing system is composed of multiplexing timing unit, demultiplexing timing unit and synchronous unit. The multiplexing timing unit provides the multiplexer with all kinds of timing signals while the demultiplexing timing unit supplies the demultiplexer with all kinds of timing signals. The function of the synchronous unit is to make timing signals produced by the demultiplexing timing unit set up and maintain a correct phase relation relative to the timing signals produced by the multiplexing timing unit, i.e. establishing and keeping the synchronous status, which is the prerequisite for the correct implementation of synchronous demultiplexing, and will be discussed hereinafter in detail. The synchronous multiplexing/demultiplexing system consists of the synchronous multiplexing unit and synchronous demultiplexing unit. The synchronous multiplexing unit combines the synchronous tributary digital signals into a combined digital signal while the synchronous demultiplexing unit separates the combined digital signal into the tributary digital signals.

In the timing system, if the multiplexing timing unit only transmit the clock signal to the demultiplexing unit, it is unable to establish and maintain overall synchronous relationship; in the multiplexing/demultiplexing system, if the digital signals are multiplexed only in terms of circulating and interleaving the combined digital signal is difficulty to be demultiplexed correctly. Therefore, for which a characteristic signal are required to ensure the establishment of overall synchronous relation for the timing system while the correct implementation of the demultiplexing is guaranteed for the demultiplexer. Such signal is named as frame alignment signal or frame synchronization signal, which is stuffed into the combined signal regularly and the interval between two stuffing is defined as frame circle. The time relations of multiplexing or demultiplexing timing signals are called as frame status. The frame alignment signal and frame structure will be discussed hereinafter in detail.

In the synchronous multiplexing, the numbers of the symbols stuffed per tributary signal by way of interleaving every time may be one or several ones. Each time one symbol being stuffed per tributary is called as bit multiplexing and each time one code word being stuffed per tributary as code multiplexing. Different multiplexing mode is suitable for different application cases. For instance, the tributary multiplexing together with PCM coding is required to preserve complete word structure and the encoder has the buffering functions, so it is suitable to adopt the word multiplexing. The normal digital group multiplexing is not required to maintain the word structure, it is considered to make the equipment as simple as possible, the bit multiplexing is common used.

Chapter 2 Synchronous Multiplexing

When the bit multiplexing is used, the principle schematic of synchronous multiplexing/demultiplexing is shown in Figure 2-1; corresponding timing in Figure 2-2. The synchronous multiplexer consists of tributary buffers and combined shift register. Assuming that the tributary signals have been written into their own buffers at t_0; and the contents in tributary buffers are written in parallel into the combined shift register at t_1 under the control of timing signal L_T; which will begin to be read out in serial at t_2, thus the synchronous multiplexing is finished. If the transmission time delay is not considered, the combined digital signal will be transmitted to the input of synchronous demultiplexer immediately. The synchronous demultiplexer is composed of a combined shift register and tributary buffers. Under the control of receiving clock (f_h), the received combined digital signal begins to be written in serial into the shift buffer at t_3. Under the control of timing signal L_R at t_4, the tributary signals will be written into their own buffer, thus the synchronous demultiplexing is completed.

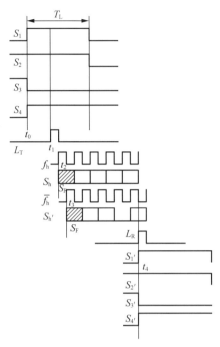

Figure 2-2 The timing of synchronous multiplexing/demultiplexing

It is clear from relative statement of synchronous multiplexing aforesaid that the synchronous multiplexing/demultiplexing technology will involve some concrete technical issues, which include the frame structure, frame alignment and how to guarantee synchronous environment etc. All these issues will be discussed one by one in the chapter of synchronous multiplexing.

2.2 FRAME STRUCTURE

As mentioned above, in order to make it possible for the frame status of demultiplexer to obtain and maintain the correct phase relation relative to that of multiplexer and implement the demultiplexing correctly, the frame alignment signals must be circularly stuffed into the combined digital signal. Thus, a strucure taking the frame as unit exists in the combined digital signal. The strict definition of the frame means such kind of a group of adjacent digital time slots, among which, the positions of various digital time slots can be identified in accordance with frame alignment signals. If required, sometimes a few frames form one multiframe or one frame is divided into a few subframes. Generally, one frame includes the following:

(1) Frame alignment signal, or also named as frame synchronization signal, as described in the foregoing paragraph, which is used to guarantee the flag signals to be obtained and held for the frame synchronization. Depending on their positions distributed in one frame, the frame alignment signal in code block can be categorized into bunched frame alignment signal and distributed frame alignment signal. The former means its code blocks occupying adjacent time slots and the latter means its code blocks seizing non-adjacent time slots. There is no substantial difference in mechanism for both of them, but a little difference in frame synchronization performance, Both of them are applicable in different frame structure respectively. For instance,CCITT Recommendation G.732 (2048kbit/s) adopts the bunched frame alignment signal while the Rec. G.733 (1544kbit/s) employees the distributed frame alignment signal.

(2) message bits: These are main contents transmitted in the frame, which occupy most of the time slots in each frame. For example, In Rec. G.732, 94% time slots is used for transmitting the message bits and in Rec. G.733, 87% tim slots for message bits. As mentioned above, the message bits from each tributary are stuffed into the combined digital signal by means of the circulating and interleaving. Each time one symbol being stuffed is called bit multiplexing while one word being stuffed each time is named word multiplexing. For instance, in the Rec. G.732 and G.733, the word multiplexing is adopted.

(3) Signaling: The signaling is the message bits which are related to the access being established and controlled in the communication network and the network management. The signaling usually occupies the defined time slot in the primary group. For example, in the Rec. G.732, it is stipulated that signaling occupies the 16th time slot, if not enough, uses the time slot beginning from 31st time slot, in the Rec. G.733, it is stipulated that signaling occupies the 8th bit in each channel time slot. The signaling is always used together with the voice channel. If the

signaling time slot is divided into a few sub-signaling time slots and each of them will be assigned to a certain fixed channel correspondingly it is named channel associated signaling; if all the channels share this one signaling path, it is named common-channel signaling.

(4) service digit: also named housekeeping digit: generally, it means some digital signals such as alarm signals, justification service signals and other indicating and controlling signals which guarantee the equipment working normally and provide all kinds of conveniences.

When the signaling mentioned above are properly arranged and standardized the general-purpose frame structure is formed. The last paragraph of the chapter is related to CCITT Recommendations concerning the frame structures of synchronous channel multiplexing, in the following chapters, CCITT Recommendations on the flame structures of plesiochronous group multiplexing are described.

2.3　FRAME ALIGNMENT

As said above, in the synchronous multiplexer/demultiplexer equipment, the prerequisite to correctly implement demultiplexing is that the frame status of the demultiplexer must hold a correct phase relation with that of multiplexer, i.e. it is must to hold frame synchronization. Frame alignment means the procedure in which the frame status of the demultiplexer is aligned to have correct phase relation relative to that of multiplexer and hold such phase relation. The type of alignment procedure is generally called synchronization acquisition.

The principle of frame alignment system is shown in Figure 2-3. It consists of clock source, multiplexing timing unit, demultiplexing timing unit and synchronization acquisition/holding unit. The multiplexing timing unit provides the multiplexer with all kinds of timing signal and the demultiplexing timing unit supplies the demultiplexer with all kinds of timing signals. The frame structures produced by these two are similar, and the frame length produced by these two are equal. In principle, the frame alignment codes are same too, which represent the beginning and end (or other special parts) of their own frame respectively, i.e. representing their own frame status. The multiplexer needs transmitting clock (f_h) and the signal of frame alignment (S_F) to the demultiplexer. Usually the timing unit of demultiplexer also needs to be driven by the received clock signal (f_h), which produces its own frame alignment signal (S'_F). The relative location between SF and S'_F is compared in the synchronization acquisition/holding unit. If it is not compliance with correct phase relation, a certain control measurement will be taken to make the demultiplexing frame status produce a relative time delay corresponding to the multiplexing frame status until the correct phase relation between those two being reached, i.e. the synchronization being realized then the operation of such synchronous status will be held. This is the whole procedure of the frame alignment. Of course, this is an explanation in principle, the concrete realization may be different.

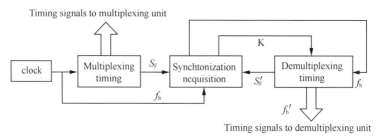

Figure 2-3　Simplified diagram of frame alignment system

The frame alignment system is the most important part of the synchronous multiplexing equipment, which involves many issues in the engineering design; and has a bigger effect on the features of the integrated multiplexing equipment. The following concrete design issues needs to be settled in the frame alignment system design:

(1) synchronization acquisition mode;

(2) code type design of frame alignment;

(3) determination of frame length;

(4) code length selection of frame alignment code;

(5) protection method of frame alignment;

(6) selection of protection parameters of frame alignment.

After the frame alignment system design is completed, the following technical performances will be defined for the synchronous multiplexer.

(1) average synchronization acquisition time;

(2) average time to find the out-of-frame;

(3) average time to acknowledge frame synchronization;

(4) average time interval to occur the out-of-frame;

(5) average duration of out-of-frame;

(6) average duration of frame synchronization;

(7) average bit error rate due to out-of-frame.

The design items mentioned above and the relation between the items and the technical features will be discussed hereinafter.

2.4 SYNCHRONIZATION ACQUISITION METHOD

There are two types of traditional synchronization acquisition: bit-by-bit alignment method and preset initiation method.

The working procedure of bit-by-bit alignment method is as follows. First the frame status of receiving equipment stops for a beat, i.e. the frame status is delayed for one bit relative to the received signal, then the frame status of the receiving equipment will be compared with that of received signal in one checking circle. If it is not compliance with the correct phase relation, the frame status of receiving equipment is delayed for one bit again, the procedure said above will be repeated; if it is compliance with the correct phase relation, such kind of relation will be maintained and the acquisition procedure will be terminated. It can be seen from procedure said above that in case the frame alignment code will not occur in any section of message code among the receiving code stream and the bit error will not occur in the frame alignment code, the time consumed by the bit-by-bit alignment and checking seems to be a little longer. As a matter of factor, the code type of frame alignment may occur in any section of message code of receiving code stream (i.e. the virtual alarm phenomenon exists) and the bit errors may occur in the frame alignment code (i.e. virtual missing phenomenon exists), now the time consumed by such kind of bit-by-bit alignment and checking becomes necessary. and only if the checking circle is selected to be long enough the effect of virtual missing probability will be better decreased, at last the acquisition procedure will be made to be as short as possible.

When the preset initiation acquisition method is adopted, in the out-of-frame duration, the time sequencing generator of the receiving equipment is preset at a specific waiting state, i.e. the frame status of the receiving equipment is at a specific preset state; the receiving code stream enters into frame alignment signal checking circuit, once all n-bit code among the stream are same as the defined code type of frame alignment signal a control signal will be output to initiate the time sequencing generator of the receiving equipment and at same time the received clock signal will be used to drive it. After that, it will be checked and decided during a checking circle. If the correct phase relation established, the procedure mentioned above will be repeated; if the correct phase relation is established in deed, which will be maintained and the acquisition procedure terminated. The procedure said above shows if the frame alignment code type does no occur in any section of message code before synchronization in the receiving code stream and the bit error does not occur in the flame alignment code, then only one complete frame alignment code block is met, that is enough to establish the synchronization. It is thus clear that the acquisition procedure is completed a little fast. Actually, the phenomena of virtual alarm and virtual missing exist, i.e. some section of message code may be formed into a frame alignment code type, while the real frame alignment code may not be found because of bit errors (i.e. without control signal output), thus, the acquisition procedure will be enlarged.

Making a comparison between the two acquisition methods, it is thus evident that at non-synchronization location, each alignment requires a checking for the bit-by-bit alignment mode but for preset initiation method a checking is only required at the time of virtual alarm phenomenon, so, we can say that the latter save more time. At synchronization location, the bit-by-bit alignment method always have a checking no matter if the bit errors occur in the frame alignment code or not, and sometimes it determines that the phase relation is correct even if there are some bit errors to a certain extent; but for preset initiation method only if the bit error exists in the frame alignment code, the opportunity to set up the synchronization will be missed definitively, it is thus evident that the latter will expand the acquisition time.

Synthesizing the two aspects mentioned above. shows that the lower the bit error rate in the combined digital signals is, the shorter the average acquisition time of the preset initiation method is; the higher the bit error rate in the combined digital signal is, the shorter the average acquisition time of the bit-by-bit alignment is. The bit error rate of a digital channel is quite low according to the international Recommendations and the inapplicable threshold of bit error rate is only 10^{-3}, therefore, generally it is proper to use the preset initiation method, which is common used in the design of digital multiplexer. The method to be discussed in the book will always be the preset initiation one unless otherwise stated.

2.5 THE CODE TYPE OF FRAME ALIGNMENT SIGNALS

If the received combined digital signal is able to meet the following two conditions simultaneously: the first, the probabilities to occur "1" or "0" of any message bit is 1/2 and the bit codes are not interrelated each other; the second, any two tests are not interrelated each other when checking the frame alignment signal, that case, the performance of the frame alignment system is not interrelated with the code type of frame alignment. Actually, the second condition can be met thoroughly. In the two synchronization acquisition methods discussed in previous chapter, the test is done step-by-step in sequence and the tests altogether are interrelated each other. For instance, assuming that the out-of-frame has occurred (refer to Figure 2-4) but the delay(τ) of the frame status of the demultiplexing relative to that of the multiplexing is smaller than the time length of frame alignment signal (nT_h—in which, n is the number of bits of a frame alignment code and T_h is the time slot width of the combined signal), i.e. located in the overlapped section, there will be different virtual alarm probability (P_y) if the code type of the frame alignment is different. Three typical cases are listed in Figure 2-4: their code length of frame alignment (n) are all 6 digital, in scheme (a) at $\tau=1T_h$, $P_y=1/2$; in scheme (b) at $\tau=2T_h$, $P_y=(1/2)^2$; and in scheme (c) at $\tau=1T_h$ until $\tau=(n-1)T_h$, $P_y=0$ exists. It is thus clear that the code type of frame alignment has an obvious effect on the probability value of virtual alarm in the overlapped section.

So, when selecting the code type of frame alignment signal, first of all, the virtual alarm probability in the overlapped section [$\tau<(n-1)T_h$] should be guaranteed as low as possible. The code type of frame alignment signal for the group multiplexing code stream is as follows:

CCITT Rec	Bit rate(kbit/s)	Code type of frame alignment signal
G.742	8448	1111010000
G.751	34 368	1111010000
G.751	139 264	111110100000
G.922	564 992	111110100000

It is quite clear that the virtual alarm probabilities in the overlapped section are all equal to zero when these code types of frame alignment signals are selected.

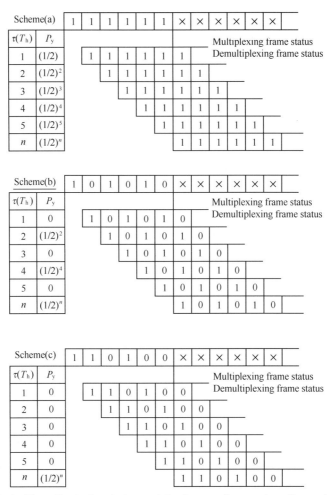

Figure 2-4 The effect of code type of the frame alignment on the probability of virtual alarm in the overlapped section

2.6 AVERAGE ACQUISITION TIME

(1) The average retention time on the non-synchronization bit positions.

The average acquisition time means the time duration which begins at the moment when the acquisition starts upon finding the out-of-synchronization and ends at the moment when the synchronization has been acknowledged upon establishing the synchronization. It should be explained that no any protection action for synchronization acquisition has been considered to be taken.

Assuming that the code type of frame alignment will be designed in accordance with the method stated in previous chapter, in such case, it is possible for (L_s-n) non-synchronization bit positions in one frame to have virtual alarm phenomena with the probability of P_y (See Figure 2-5). The average retention time remaining on the non-synchronization bit positions during the operation of acquisition is:

$$\begin{aligned}\Delta t_a' &= (1-P_y)T_h + P_y(1-P_y)(T_h+T_s) \\ &\quad + P_y^2(1-P_y)(T_h+2T_s) + L \\ &= (1-P_y)T_h(1+P_y+P_y^2+L) \\ &\quad + P_y(1-P_y)T_s(1+2P_y+3P_y^2+L) \\ &= (1-P_y)T_h\sum_{i=0}^{\infty}P_y^i + P_y(1-P_y)T_s\sum_{j=1}^{\infty}jP_y^{j-1} \\ &= T_h + \frac{P_y}{1-P_y}\cdot T_s \\ \Delta t_a' &= T_h + \frac{P_y}{1-P_y}\cdot T_s\end{aligned} \quad (2\text{-}1)$$

The formula (2-1) is an expression for average retention time remaining on the nonsynchronization bits positions during the operation of acquisition when the phenomena of virtual alarm may occur, in which, T_h is the width of time slot of the combined signal; T_s is the frame circle of the combined signal; L_s is the frame length of the combined signal.

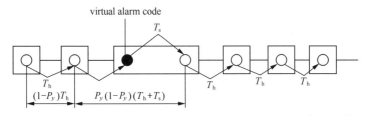

Figure 2-5 The average retention time on non-synchronization bits

When the code type of the frame alignment signal is selected according to the method discussed in the previous chapter, the virtual alarm phenomenon can not occur on $(n-1)$ bit positions of the non-synchronization code in the overlapped section, i.e. the probability of virtual alarm is zero. Therefore, the average retention time of the acquisition operation on such non-synchronization bit position is:

$$\Delta t'' = T_h \tag{2-2}$$

(2) The time duration from non-synchronization bit positions shifting to synchronization bit position

Considering that there are (L_S-1) non-synchronization bit positions in each frame in total, among them, the virtual alarm phenomena may occur on (L_s-n) non-synchronization bit positions; but the virtual alarm phenomena may not occur on $(n-1)$ non-synchronization bit positions. The maximum time (t'_{smax}), minimum time (t'_{smin}) and average time (t'_s) from non-synchronization bit position shifting to synchronization bit position can be derived directly by bring the expressions (2-1) and (2-2):

$$\begin{aligned}
t'_{smax} &= (L_s - n)\Delta t'_a + (n-1)\Delta t''_a \\
&= (L_s - n)\left(T_h + \frac{P_y T_s}{1-P_y}\right) + (n-1)T_h \\
&= \left[1 + (L_s - n)\frac{P_y}{1-P_y}\right]T_s - T_h
\end{aligned} \tag{2-3}$$

$$t'_{smin} = T_h \tag{2-4}$$

$$\begin{aligned}
t'_s &= \frac{1}{2}(t'_{smax} + t'_{smin}) \\
&= \frac{T_s}{2}\left[1 + (L_s - n)\frac{P_y}{1-P_y}\right]
\end{aligned} \tag{2-5}$$

The expressions (2-3), (2-4) and (2-5) are the formulas to calculate the maximum. minimum and average times from nonsynchronization bit position shifting to synchronization bit position.

(3) Average acquisition time

At the synchronization bits position, any bit or more bits of the frame alignment signal code block having bit errors are able to cause the opportunity to be lost for the establishment of synchronization. i.e. the virtual missing events happen. The probability to have such events is:

$$P_1 = \sum_{x=1}^{n} C_n^x (1 - P_e^{n-x}) \cdot P_e^x \approx n P_e \tag{2-6}$$

in which, P_e is the bit error rate for the frame alignment code block. When the error occurring on each bit of the combined digital signals is random and of equal probability, P_e is equal to the average bit error ratio of the bit code of the combined digital signals.

The time duration from non-synchronization bit position shifting to synchronization and in which the correct judgment being made is the average acquisition time:

$$t_a = (1-P_1)t'_s + P_1(1-P_1)(t'_s + t'_{s\max})$$
$$+ P_1(1-P_1)(t'_s + 2t'_{s\max}) + \cdots$$
$$= (1-P_1)t'_s(1 + P_1 + P_1^2 + \cdots)$$
$$+ P_1(1-P_1)t'_{s\max}(1 + 2P_1 + 3P_1^2 + \cdots)$$
$$= (1-P_1)t'_s \sum_{i=0}^{\infty} P_1^i + P_1(1-P_1)t'_{s\max} \sum_{j=1}^{\infty} jP_1^{j-1}$$
$$= t'_s + \frac{P_1 t'_{s\max}}{1-P_1}$$

$$t'_{s\max} \approx 2t'_s$$

$$\therefore t_a \approx \frac{1+P_1}{1-P_1} \cdot t'_s$$
$$= \frac{1+P_1}{1-P_1} \cdot \frac{T_s}{2} \left[1 + (L_s - n) \cdot \frac{P_y}{1-P_y}\right]$$
$$= \frac{(1+P_1)[1+(L_s-n-1)P_y]}{2(1-P_1)(1-P_y)} \cdot T_s$$
$$\approx (1+P_1)[1+(L_s-n-1)P_y](1+P_1+P_y) \cdot \frac{T_s}{2} \quad (2\text{-}7)$$
$$\approx [1+P_1+(L_s-n)P_y](1+P_1+P_y) \cdot \frac{T_s}{2}$$
$$\approx [1+2P_1+(L_s-n)P_y] \cdot \frac{T_s}{2}$$
$$t_a \approx [1+2P_1+(L_s-n)P_y] \cdot \frac{T_s}{2}$$

The expression (2-7) is a general one for the average acquisition time. Its tenable condition is that no protection action has been taken in the course of acquisition, but the effects of code type of the frame alignment signal are considered and the optimum code type is selected. That is to say, when out-of-frame occurs but still in the overlapped section, it is guaranteed that the virtual alarm phenomena for the code type of the frame alignment will not occur.

In which, P_1 is the virtual missing probability on the synchronization bit positions; P_y is the virtual alarm probability on non-synchronization bit positions; L_s is the frame length, n is the frame alignment code length; T_s is frame circle.

For instance, in CCITT Rec. G.742, L_s=848, n=10; at P_e=1×10^{-3}:

$$P_1 \approx 0.01$$
$$(L_s - n)P_y \approx 0.8 \qquad (2\text{-}8)$$

$$t_a \approx T_s = \frac{L_s}{f_h} \qquad (2\text{-}9)$$

in which, f_h is the bit rate of combined digital signal and it is a known value. So, the average acquisition time essentially depends on the taken value of the frame length. Thus, after the average acquisition time t_a as a design requirement is defined the quantity level of the frame length value to be taken will be defined. But the transmission efficiency easy generation of frame timing sequence and other existing factors should be considered when the value of frame length is taken. The value of frame length being taken and the relevant approximate calculated values for the average acquisition time in CCITT Recommendations are as follows:

Rec	Rate(kbit/s)	Frame length	Average acquisition time(μs)
G.732	2048	256	184.1
G.742	8448	848	92.3
G.751	34 368	1536	56.1
G.751	139 264	2928	36.4
G.922	564 992	2688	7.96

2.7 OPTIMUM LENGTH OF FRAME ALIGNMENT SIGNAL

The formula to calculate the average acquisition time, in which the protection action is not taken in the course of acquisition, is derived in the previous chapter:

Chapter 2 Synchronous Multiplexing

$$t_a \approx \frac{T_s}{2}[1+2P_1+(L_s-n)P_y]$$

$$\approx \frac{T_s}{2}\left[1+2nP_e+(L_s-n)\cdot\frac{1}{2^n}\right] \quad (2\text{-}10)$$

$$\approx \frac{T_s}{2}\left(1+2nP_e+\frac{L_s}{2^n}\right)$$

In which, T_s is frame circle, L_s frame length, P_e average bit error rate, n frame alignment signal length. After the bit rate, frame length and considered average bit error rate are defined for the combined digital signal, the frame alignment signal length will determine the average acquisition time. As the n value is getting bigger, the virtual alarm probability on the non-synchronization bit positions ($P_y=1/2^n$) will become smaller but the virtual missing probability on the synchronization bit positions bigger. Thus, there is an extreme value among the average acquisition time along with the different value being taken as the frame alignment signal length.

In accordance with:

$$\frac{dt_a}{dn} \approx \frac{T_s}{2}\left(2P_e - \frac{L_s \ln 2}{2^n}\right) \begin{cases} >0, n>n_0 \\ =0, n=n_0 \\ <0, n<n_0 \end{cases} \quad (2\text{-}11)$$

to obtain:

$$n_0 = \frac{\ln\left(\frac{L_s \ln 2}{2P_e}\right)}{\ln 2} \quad (2\text{-}12)$$

So, there is a minimum value among ta along with the different value being taken as n:

$$t_{a\min} \approx \frac{T_s}{2}\left(1+2n_0 P_e + \frac{L_s}{2^{n_0}}\right) \quad (2\text{-}13)$$

After frame circle (T_s) /frame length (L_s) and the considered average BER P_e is defined, the optimum length (n_0) and the minimum value of average acquisition time ($t_{a\min}$) can be obtained in accordance with the formulae (2-12) and (2-13).

The formula to calculate the optimum length of the frame alignment signal is able to be derived with another approach. It is known from the discussion in previous chapters that there is one synchronization bit position and (L_s−1) non-synchronization bit positions in each frame; and the virtual alarm phenomena may occur on (L_s−n) non-synchronization bit positions in the non-overlapped section; and the synchronization acquisition procedure will last for half circle in average. Thus, the one-time successful probability for synchronization probability is:

$$P = (1-P_y)^{\frac{1}{2}(L_s - n)} \cdot (1 - P_1)$$

$$\approx (1-P_y)^{\frac{L_s}{2}} \cdot (1 - P_1)$$

$$\approx \left(1 - \frac{L_s}{2} P_y\right) \cdot (1 - P_1) \quad (2\text{-}14)$$

$$\approx 1 - \frac{1}{2} \cdot \frac{L_s}{2^n} - nP_e$$

$$\frac{dP}{dn} \approx \frac{1}{2} \cdot \frac{L_s \ln 2}{2^n} - P_e \begin{cases} < 0, n > n_0 \\ = 0, n = n_0 \\ > 0, n < n_0 \end{cases} \quad (2\text{-}15)$$

to obtain:

$$n_0 = \frac{\ln\left(\frac{L_s \ln 2}{2P_e}\right)}{\ln 2} \quad (2\text{-}16)$$

As a result, a maximum value exists among the one time successful probability (P) of the synchronization acquisition along the with the variation of the frame alignment signal length(n):

$$P_{max} \approx 1 - \frac{L_s}{2 \cdot 2^{n_0}} - n_0 P_e \quad (2\text{-}17)$$

After the frame length (L_s) and considered average BER (P_e) are defined, the optimum length of the frame alignment signal (n_0) and the maximum value of one-time acquisition successful probability (P_{max}) can be calculated in accordance with formulae (2-16) and (2-17).

Checking the formula (2-12) against formula (2-16), we can see that both of them are same, i.e. the approximate calculation formulas derived from two approach are same, but those two approaches have different definitions on optimum value of n_0. For the former, n_0 is defined as n value which makes the synchronization acquisition have the minimum average acquisition time when the protection action is not employed, and for the latter, n_0 is defined as n value which let the asynchronization acquisition have maximum one-time successful probability. Obviously, the latter definition has nothing to do with the protection actions being done in the course of acquisition. In this way, the definition n_0 is more popular.

The calculating result of the optimum length formula of the frame alignment signal is shown in Figure 2-6; the result calculated in cases of relatively serious BER ($P_e=5\times10^{-2}$) is compared with the values stipulated in CCITT Recommendations as follows:

CCITT Rec	Rate (kbit/s)	Frame length	Frame alignment signal	
			Value in CCITT Rec	Calculated optimum Value ($P_e=5\times10^{-2}$)
G.732	2048	256	7	$10.5 \cong 11$
G.742	8448	848	10	$12.3 \cong 12$
G.751	34 368	1536	10	$13.1 \cong 13$
G.751	139 264	2928	12	$14.1 \cong 14$
G.922	564 992	2688	12	$14.0=14$

It is clear that, the value in CCITT Rec. is quite close to the optimum value calculated in cases of relatively serious BER; when BER is relative low, the calculated optimum value of the frame alignment code length become bigger, under such circumstance, the value in the CCITT Rec. seems to be lower (refer to Figure 2-6). In consideration of engineering, the optimum value of frame alignment code length designed in accordance with relatively serious BER seems to be more reasonable because the relatively light BER has light effect on the acquisition performance, in addition, it is proper that the value of the frame alignment code length is shorter than the optimum value in consideration of the available channel efficiency.

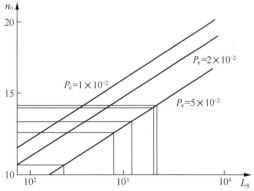

Figure 2-6 The optimum length of frame alignment signal

2.8 SYNCHROMIZATION STATUS PROTECTION

(1) The necessity of the synchronization status protection

The system can provide the normal service only after it get into the synchronization status, but such synchronization status may be destroyed and get into the acquisition status again due to the bit errors in the received bit stream. If the out-of-frame is too frequently or we say that the time interval (t_f) between two out-of-synchronizations owing to the bit errors is too short, the system can not guarantee the quality of service. In case the protection action is not taken as discussed in the previous chapters, in the n bit code of the frame alignment signal, so long as the bit error occurs at one bit. it will cause the out-of-frame immediately. The expression for the out-of-frame probability $P_1 \approx nP_e$ has been given in the previous chapters; its product with frame frequency (F_s) is the times of the out-of-frame occur per second in average; obviously, the reversed number of the product is equal to the average time interval (t_f) between the two out-of-frames:

$$t_f = \frac{1}{F_s \cdot P_1}$$
$$\therefore t_f = T_s / nP_e$$

(2-18)

It is the formula to calculate the average interval between the two out-of-frames without the protection action of synchronization status taken. For example, In CCITT Rec. G.742:T_s=100μs, n=10, when BER in the bit stream P_e=1×10^{-6}, $t_f \approx$ 10s is obtained. so we know that it is not acceptable in terms of engineering that one out-of-frame occurs every ten seconds even if the BER is rather low. Therefore, it is necessary to take the synchronization status protection action.

(2) The protection method for the synchronization status

The protection method normally used for the synchronization status is as follows: having entered into the synchronization status, it is checked if me phenomena of the frame alignment signal losses happen or not at the stipulated time (i.e. at every one complete frame after the establishment of the synchronization status). If the frame alignment signal is found to be lost for β times continuously, it is confirmed to enter into the out-of-frame status. After such protection action having been taken, the expression for the probability of out-of-frame and the time interval between the two out-of-frames is:

$$P_1^\beta \approx (nP_e)^\beta$$
$$t_f \approx T_s/(nP_e)^\beta \quad (2\text{-}19)$$

For example, in CCITT Rec. G. 742, $T_s=100\mu s$, $n=10$, $\beta=4$, $t_f=3$hour is obtained at $P_e=1\times10^{-3}$, i.e. in cases of rather high BER, out-of-frame occurs only every three hours. It is clear that such protection action of the synchronization status is rather effective.

(3) Average acknowledgment time of out-of-frame

It is obvious that implementation of the synchronization protection is favorable to the system. The accidental losses of the frame alignment signal less than β times will not destroy the synchronization status, but there is an unfavorable side, i.e. in case the judgment can not be made at the time when the system already have the out-of-frame, which needs to be observed for B times at least, then the out-of-frame will be acknowledged to start the acquisition. The time duration from the out-of-frame occurring really to the acknowledgment of the out-of-frame is called the average acknowledgment time of out-of-frame.

As shown in Figure 2-7, after the out-of-frame really occurs, the false code type of the frame synchronization many be formed from the message code at the original stipulated moment, which will cause the virtual alarm phenomena.

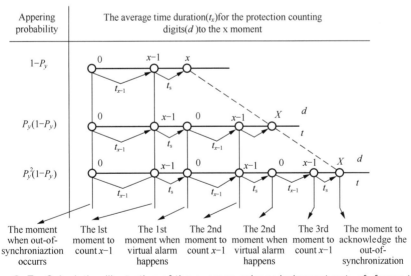

Figure 2-7 Calculating illustration of the average acknowledgment out-of-frame time

In the course of recognition, the times of virtual alarm appearing may be unequal, and virtual alarm events may occur at the different moment. Referring to Figure 2-7, the normal expression for the average time duration (t_x) from the moment when the out-of-frame appears until the moment when the counted digits (β) at the protection counter reaches to X:

$$t_x = (1-P_y)(t_{x-1}+T_s)$$
$$+ P_y(1-P_y)2(t_{x-1}+T_s)$$
$$+ P_y^2(1-P_y)3(t_{x-1}+T_s)+\cdots$$
$$= (1-P_y)(t_{x-1}+T_s)\sum_{i=1}^{\infty} iP_y^{i-1}$$
$$= \frac{t_{x-1}+T_s}{1-P_y}$$

When $x=1$, considering that the moment of out-of-synchronization may occur random at any instance in one frame, so the average time duration at protection countered digits $x=1$ is:

$$t_1 = (1-P_y)\left(\frac{T_s}{2}\right) + P_y(1-P_y)\left(\frac{T_s}{2}+T_s\right)$$
$$+ P_y^2(1-P_y)\left(\frac{T_s}{2}+2T_s\right)+\cdots$$
$$= (1+P_y)\frac{T_s}{2}\sum_{i=0}^{\infty} P_y^i + P_y(1-P_y)T_s\sum_{j=0}^{\infty}(j+1)P_y^j$$
$$= \frac{1+P_y}{1-P_y}\cdot\frac{T_s}{2}+\frac{T_s}{1-P_y}-\frac{T_s}{2}$$

When $x = \beta$, i.e. protection counter counting to β, thus, the average time duration(t_d) undergone to acknowledge the out-of-synchronization is the obtained average acknowledgment time of the out-of-frame:

$$t_d = \frac{t_{\beta-1}+T_s}{1-P_y}$$
$$= \frac{T_s}{1-P_y}+\frac{t_{\beta-2}+T_s}{(1-P_y)^2}$$

$$= \frac{T_s}{1-P_y} + \frac{T_s}{(1-P_y)^2} + L + \frac{t_1+T_s}{(1-P_y)^{\beta-1}}$$

$$= \frac{T_s}{1-P_y} + \frac{T_s}{(1-P_y)^2} + L + \frac{T_s}{(1-P_y)^{\beta-1}}$$

$$+ \frac{T_s}{(1-P_y)^\beta} - \frac{T_s}{2(1-P_y)^{\beta-1}} \quad (2\text{-}20)$$

$$\therefore t_d = \left[\sum_{j=1}^{\beta} \frac{1}{(1-P_y)^j} - \frac{1}{2(1-P_y)^{\beta-1}}\right] \cdot T_s$$

The formula (2-20) is an common expression for the average acknowledgment time of the out-of-frame, when $P_y=0$,

$$t_d = \left(\beta - \frac{1}{2}\right)T_s \quad (2\text{-}21)$$

The formula (2-21) is an approximate expression for average acknowledgment time of the out-of-frame, which normally ignores the effect of virtual alarm probability (P_y). When $P_y \neq 0$, (2-20) can be written as:

$$t_d = \left\{\frac{(1-P_y)^{-(\beta+1)}-1}{(1-P_y)^{-1}-1} - \left[1+\frac{1}{2}(1-P_y)^{-(\beta-1)}\right]\right\}T_s$$

2.9 ACQUISITION PROCEDURE CHECKING

(1) The necessity to check the acquisition procedure

If the protection action was not adopted for the synchronization status, the check on the acquisition would not be employed. It produces unfavorable effect on the synchronization acquisition performance to adopt the protection action for the synchronization status. when the checking measure is not taken for the acquisition procedure, once the false synchronization code block is found over the non-synchronization bits position, the synchronization checking circuit decides immediately to have entered into the synchronization status. When the synchronization status protection is not adopted, so long as no false synchronization code block is discovered at the next stipulated moment, it can have a searching toward the next bit position at once; after the synchronization status protection has been employed, only if no false synchronization code block occurs for continuous β times at least, the acquisition can move toward the next bit position. It is thus clear that once the virtual alarm event takes place on the non-synchronization bit position, the acquisition can move toward next bit position at least β frame circles time passed. Obviously the synchronization acquisition time has been extended. Therefore, it is necessary to adopt the relative checking measure in the acquisition procedure.

(2) The checking method of the acquisition procedure

The checking method of acquisition procedure in normal use is introduced as follows: In the course of acquisition, after the first frame synchronization code type is found at the specified moments (i.e. at the moments in alternate frames and from the moments when the first code type of frame synchronization is discovered), we check the code type for frame synchronization appear or not. Only if the continuous discovery is defined for α times in total, which is verified to have entered into the synchronization status. At this moment if we consider that entering synchronization status only once caused by virtual alarm probability P_y, the formula to calculate the average retention time on the non-synchronization bit position is as follows:

$$\Delta t_a^1 = (1-P_y)T_h + P_y(1-P_y)(T_h+T_s) + P_y^2(1-P_y)(T_h+2T_s)$$
$$+ \cdots + P_y^{\alpha-1}(1-P_y)[T_h+(\alpha-1)T_s] + P_y^\alpha[T_h+(\alpha-1)T_s+t_d]$$

In which, the average acknowledgment time for the out-of-frame is:

$$t_d \approx \left(\beta - \frac{1}{2}\right)T_s$$

$$\Delta t_\alpha^1 = T_h[(1-P_y) + P_y(1-P_y) + P_y^2(1-P_y) + \cdots$$
$$+ P_y^{\alpha-1}(1-P_y) + P_y^\alpha] + T_s[P_y(1-P_y) + 2P_y^2(1-P_y)$$
$$+ \cdots + (\alpha-1)P_y^{\alpha-1}(1-P_y) + (\alpha-1)P_y^\alpha] + t_d P_y^\alpha \qquad (2\text{-}22)$$
$$= T_h + T_s(P_y + P_y^2 + \cdots + P_y^{\alpha-1}) + t_d P_y^\alpha$$
$$= T_h + \frac{P_y + P_y^\alpha}{1-P_y} \cdot T_s + t_d P_y^\alpha$$

$$\therefore \Delta t_\alpha^1 = T_h + \frac{P_y + P_y^\alpha}{1-P_y} \cdot T_s + P_y^\alpha t_d$$

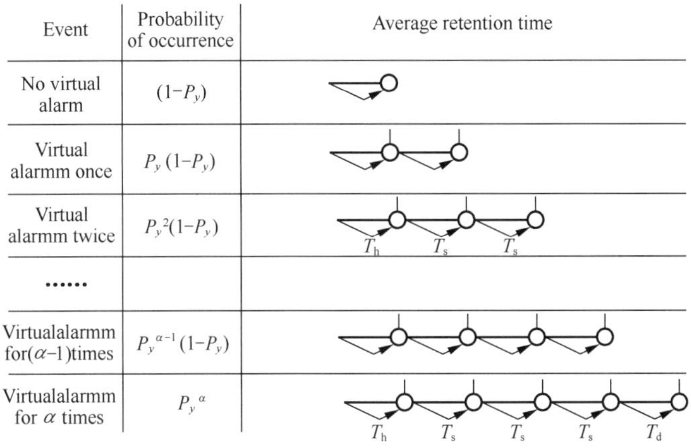

Event	Probability of occurrence	Average retention time
No virtual alarm	$(1-P_y)$	
Virtual alarmm once	$P_y(1-P_y)$	
Virtual alarmm twice	$P_y^2(1-P_y)$	$T_h \quad T_s \quad T_s$
......		
Virtualalarmm for $(\alpha-1)$times	$P_y^{\alpha-1}(1-P_y)$	
Virtualalarmm for α times	P_y^α	$T_h \quad T_s \quad T_s \quad T_s \quad T_d$

Figure 2-8 The calculating illustration of the average retention time on the non-synchronization bits after adopting the protection/checking actions

Considering it is a common practice to guarantee $P_y \ll 1$, $P_y^{\alpha+1} \ll P_y$, at this time:

$$\Delta t'_\alpha \approx T_h + \frac{P_y}{1-P_y} \cdot T_s \quad (2\text{-}23)$$

This is the average retention time after the acquisition checking and synchronization protection actions are adopted. Obviously, it is same as corresponding formula (2-1) when the relevant actions is not adopted, Certainly, it is an approximate result. This shows that no obvious effect is produced on the average retention time on the non-synchronization bit position after having adopted the synchronization protection and acquisition checking measure. Therefore, no matter if checking measure is adopted or not, the same calculating formula (2-7) can be used to calculate the average acquisition time.

(3) Average acknowledgment synchronization time

It is obvious favorable for the system to adopt checking measure in the acquisition procedure, which will make the system not to such an extent as to have obvious longer acquisition procedure due to the adoption of synchronization protection measure. But it also has an unfavorable aspect for the system to take such measure, i.e. if the system already enters into the synchronization status practically, but which can not be acknowledged immediately and it is necessary to make a judgment after a period time, then the synchronization protection will be employed after the acknowledgment of synchronization. Such time duration is average acknowledgment synchronization time (t_w). It is considered that when the counting

(w) of the checking counter has reached to $X < \alpha$ in the course of acquisition and checking followed by a virtual missing phenomena, the checking counter will be reset to zero immediately and start the acquisition from beginning. Thus, after the adoption of checking measure, the average acknowledgment synchronization time (t_w) is the whole time duration beginning from acquisition to the synchronization.

Refer to Figure 2-9, the normal expression of the average time duration (t_y) from beginning acquisition until the checking counter to y can be written follows:

$$t_y = (1-P_1)(t_{y-1} + t'_{s\max}) + P_1(1-P_1)2(t_{y-1} + t'_{s\max})$$
$$+ P_1^2(1-P_1)3(t_{y-1} + t'_{s\max}) + L$$
$$= (1-P_1)(t_{y-1} + t'_{s\max})\sum_{j=1}^{\infty} jP_1^{j-1}$$
$$= \frac{t_{y-1} + t'_{s\max}}{1-P_1}$$

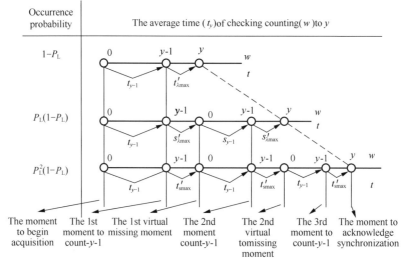

Figure 2-9 The calculating illustration of the average acknowledgment synchronization time (t_w)

In which, t_{y-1} is the average time duration when the checking counter reaching to $(y-1)$; $t'_{s\max}$ is the maximum time duration transited from non-synchronization bit position to synchronization; P_1 is virtual alarm probability. Considering that the moment to begin the acquisition may occur at any place relative to the input signal frame and in addition, the similar relation of formula (2-23), it can be obtained that the average time duration as checking counter counting to 1 with checking measure used is approximately equal to the average time (t_α) without the checking measure used:

$$t_1 = \frac{1+P_1}{1-P_1} \cdot \frac{t'_{s\max}}{2}$$

When $y = \alpha$, the average time to acknowledge synchronization is the required average acknowledgment synchronization time:

$$\begin{aligned}
t_W &= \frac{t_{\alpha-1} + t'_{s\max}}{1-P_1} \\
&= \frac{t'_{s\max}}{1-P_1} + \frac{t_{\alpha-2} + t'_{s\max}}{(1-P_1)^2} \\
&= \frac{t'_{s\max}}{1-P_1} + \frac{t'_{s\max}}{(1-P_1)^2} + L + \frac{t_1 + t'_{s\max}}{(1-P_1)^{\alpha-1}} \\
&= \frac{t'_{s\max}}{1-P_1} + \frac{t'_{s\max}}{(1-P_1)^2} + L + \frac{t'_{s\max}}{(1-P_1)^{\alpha-1}} + \frac{t'_{s\max}}{(1-P_1)^\alpha} - \frac{t'_{s\max}}{2(1-P_1)^{\alpha-1}}
\end{aligned} \quad (2\text{-}24)$$

The formula (2-24) is a normal expression for average acknowledgment synchronization time. When $P_1 = 0$,

$$\begin{aligned}
t_W &= \left(\alpha - \frac{1}{2}\right) t'_{s\max} \\
&\approx \left(\alpha - \frac{1}{2}\right) T_s
\end{aligned} \quad (2\text{-}25)$$

Formula (2-25) is a approximate expression for average acknowledgment time when the effects of virtual alarm probability (P_y) and virtual missing probability (P_1) are ignored usually. When $P_1 \neq 0$, he formula (2-24) can be rewritten as:

$$t_W = \left\{\frac{(1-P_1)^{-(\alpha+1)} - 1}{(1-P_1)^{-1} - 1} - \left[1 + \frac{1}{2}(1-P_1)^{-(\beta-1)}\right]\right\} t'_{s\max}$$

2.10 FRAME ALIGNMENT PROTECTION PARAMETERS

According to the conclusion discussed in previous two chapters, it is able to induce a complete logical drawing for the frame alignment protection (See Figure 2-10).

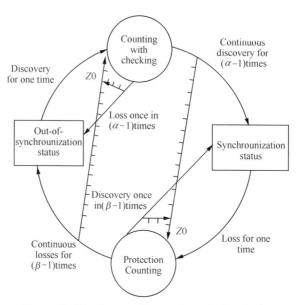

Figure 2-10　Frame alignment protection logic.

When it is in the out-of-synchronization, once the code type of frame synchronization is discovered by the detecting circuit of the frame synchronization cook (abbreviated as discovery once), the checking counter begins counting. Then, if it is discovered continuously at the stipulated moments for $(\alpha-1)$ times, i.e. the synchronization will be acknowledged as checking counter counts to α, while α status is maintained for the checking counter and the protection counter is returned to zero; if the code type of frame alignment is not discovered just for one time at the stipulated continuous $(\alpha-1)$ moments soon after wards (abbreviated as loss once), the checking counter is returned to zero to start the acquisition again.

In case of synchronization status, as long as the detecting circuit of frame synchronization code does not discover the frame alignment code for one time, the protection counter counts for 1. Then, in case it is not discovered continuously at the stipulated moments for $(\beta-1)$ times, i.e. the protection counter counts to β, it is acknowledged to have entered the out-of-synchronization, and at same time, β status counted by protection counter is maintained and the checking counter is returned to zero; but is the frame synchronization code is discovered at continuous $(\beta-1)$ stipulated moments, the protection counter will be returned to zero and the synchronization monitoring will be implemented once again.

At this time, the occurrence probability of the virtual missing events (P_1^β) at synchronization bit position and the occurrence probability of the virtual alarm events (P_y^α) will be respectively:

$$P_1^\beta \approx (nP_e)^\beta \quad (2\text{-}26)$$

$$P_y^\alpha = \left(\frac{1}{2}\right)^{n\alpha} \quad (2\text{-}27)$$

In the engineering design, in order to make it not so complicated for the synchronization acquisition/maintenance circuit, it is enough to limit the virtual missing probability and virtual alarm probability to 1×10^{-8} level. If the inapplicable threshold ($P_e=1\times10^{-3}$) of the channel BER recommended by CCITT is considered, the following formula can be derived:

$$P_1^\beta = P_y^\alpha = 1\times10^{-8}$$

$$\alpha = \frac{8}{0.3n} \quad (2\text{-}28)$$

$$\beta \approx \frac{8}{3-\lg n} \quad (2\text{-}29)$$

Thus, after the code length of frame alignment signal is decided, the relevant frame alignment protection parameters—the values of protection times β and the checking times α can be obtained according to the two formulas mentioned above. The corresponding calculation curves is shown in Figure 2-11. The values in CCITT Rec. are given in the following table and the relative calculating result is also listed.

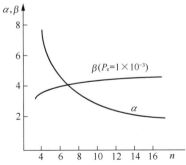

Figure 2-11 α,β calculating curves ($P_y^\alpha = P_1^\beta = 1\times10^{-8}$)

CCITT Rec	Rate (kbit/s)	n(bit)	Recommended Value		Calculated Value	
			α	β	α	β
G.742	8448	10	3	4	$2.7 \cong 3$	$4.0 = 4$
G.751	34 368	10	3	4	$2.7 \cong 3$	$4.0 = 4$
G.751	139 264	12	3	4	$2.2 \cong 2$	$4.2 \cong 4$
G.922	564 992	12	3	4	$2.2 \cong 2$	$4.2 \cong 4$

2.11 AVERAGE OUT-OF-FRAME TIME AND AVERAGE SYNCHRONIZATION TIME

(1) Average out-of-frame time (t_r)

The out-or-frame phenomenon taking place in the digital multiplexing system can be divided into two types. One kind of out-of-frame is due to the structure damages of multiplexing frame status or demultiplexing frame status. As a result, the out-of-frame is caused. For instance, either the out-of-frame happening in the high-ray group multiplexer or the generator of multiplexing frame being interfered may destroy the frame structure of the multiplexing frame; either the interruption of receiving clock or the generator of local demultiplexing frame being interfered may destroy the frame structure of local demultiplexing frame. Such kind of out-of-frame is really one, abbreviated as structural out-of-frame. Other kind of out-of-frame is abbreviated as synchronization out-of-frame, which means that the out-of-frame does not occur in the system itself, the code type of frame alignment is destroyed only due to bit errors of the combined digital signal to cause the recognition being missed which makes the system enter into the acquisition and acknowledgment procedure.

It is clear from above-said statement that the average out-of-frame times introduced by the two out-of-frames mentioned above are different. The average out-of-frame time (t'_r) of the structural out-of-frame includes both average acknowledgment out-of-frame time (t_d) and average acknowledgment synchronization time (t_w):

$$t'_r = t_d + t_w \qquad (2\text{-}30)$$

in which,

$$t_d \approx \left(\beta - \frac{1}{2}\right)T_s$$

$$t_w \approx \left(\alpha - \frac{1}{2}\right)T_s \qquad (2\text{-}31)$$

$$t'_r \approx (\alpha + \beta - 1)T_s$$

For the average out-of-frame time of synchronization out-of-frame (t''_r), it is really to enter the out-of-frame status at the time when acquisition has begun, so it only includes the time item of average acknowledgment synchronization:

$$t''_r \approx \left(\alpha - \frac{1}{2}\right)T_s \qquad (2\text{-}32)$$

(2) Average synchronization time (t_1)

The average time interval between two adjacent-occurred out-of-frames is obtained in the chapter of synchronization status protection:

$$t_f \approx T_s / (nP_e)^\beta \qquad (2\text{-}33)$$

From the average intervals between two adjacent-occurred out-of-frames, the average retention time of the out-of-frame is deducted and the remaining is average synchronization retention time:

$$t_1 = t_f - t_r \qquad (2\text{-}34)$$

When structural out-of-frame occurs, the average synchronization time is:

$$t'_1 = t_f - t'_r$$
$$\approx \left[\frac{1}{(nP_e)^\beta} - \left(\alpha + \beta - \frac{1}{2}\right)\right]T_s \qquad (2\text{-}35)$$

when synchronization out-of-frame happens, the average synchronization time is:

$$t''_1 = t_f - t''_r$$
$$\approx \left[\frac{1}{(nP_e)^\beta} - \left(\alpha - \frac{1}{2}\right)\right]T_s \qquad (2\text{-}36)$$

2.12 THE BIT ERRORS OF SYNCHRONIZATION OUT-OF-FRAME

The additional bit errors introduced by synchronization out-of-frame to transmitted bit stream are called bit errors of synchronization out-of-frame, P_{ef} is used to express average bit error rate of the synchronization out-of-frame. If there was no bit error in the transmitted bit stream, the synchronization out-of-frame would not exist according to the definition in the previous chapter, of course, the bit errors of synchronization out-of-frame would not exist too. Practically, the bit errors exist in the transmitted bit stream. When bit errors cause the occurrence of synchronization out-of-frame, the additional bit errors of synchronization out-of-frame will be produced. Therefore, the bit errors of synchronization out-of-frame always exist together with the bit errors of the channel, i.e. in the basis of bit errors of the channel, the extra quantity of bit errors are added.

According to the definition of average bit error code: the number of bit errors in the course of out-of-frame plus that in the course of synchronization, then, divided by the total numbers of code elements is to obtain the total average BER in the output code stream:

$$P_E = \frac{\frac{1}{2} f_1 t_r'' + P_e f_1 t_1}{f_1 \cdot t_f}$$

$$= \frac{\frac{1}{2} t_r'' + P_e (t_f - t_r'')}{t_f} \quad (2\text{-}37)$$

$$= \left(\frac{1}{2} - P_e\right) \frac{t_r''}{t_f} + P_e$$

$$P_{ef} = P_E - P_e$$

$$= \left(\frac{1}{2} - P_e\right) \frac{t_r''}{t_f} \quad (2\text{-}38)$$

$$\approx \frac{t_r''}{2 t_f}$$

$$\therefore t_r'' \approx \left(\alpha - \frac{1}{2}\right) T_s \quad (2\text{-}39)$$

$$t_f \approx T_s / (n P_e)^\beta$$

$$\therefore P_{ef} \approx \frac{1}{2}\left(\alpha - \frac{1}{2}\right)(nP_e)^\beta$$

$$\frac{P_{ef}}{P_e} \approx \frac{1}{2}\left(\alpha - \frac{1}{2}\right)n^\beta \cdot P_e^{\beta-1}$$

(2-40)

The formulas (2-39) and (2-40) are approximate ones to calculate the bit error rate of synchronization out-of-frame in the demultiplexing tributary. In those formulas, α is the acquisition and checking parameter of the demultiplexer, β the synchronization protection parameter, n the code length of frame alignment signal for multiplexing code stream, P_e the average bit error rate of multiplexing code stream. For example, $\alpha = 3$, $n = 10$, $\beta = 4$; When $P_e \leqslant 1 \times 10^{-2}$, we obtain $\frac{P_{ef}}{P_e} \leqslant 1.25 \times 10^{-2}$. Thus, even if the channel bit error rate is rather serious, the out-of-frame bit error rate introduced by it is rather low, only equal to 1% level of the original bit error rate of the channel.

2.13 FRAME ALIGNMENT ACQUISITION/ MAINTENANCE LOGIC

The principle drawing of prototype frame alignment acquisition/maintenance logic is shown in Figure 2-12[18]. It consists of synchronization detecting/checking counter, out-of-synchronization detecting/protecting counter, and status initiator. The recognizer of frame alignment signal provides the logic circuit of frame alignment acquisition/maintenance with the status signal (B_F) of multiplexing frame alignment signal, while the time-sequence generator of local demultiplexing frame supplies the status signal (b_F) of demultiplexing frame alignment signal to the logic circuit of frame alignment acquisition/maintenance. The logic circuit of frame alignment acquisition/maintenance controls the time-sequence generator of local demultiplexing frame according to if the B_F and b_F being overlapped or not. If those two not being overlapped, the time-sequence of demultiplexing frame will be set to b_F (at high level) status, i.e. waiting status; If those two being overlapped, the status initiator will be commanded to let clock go through, to drive time-sequence generator of demultiplexing frame operating synchronously. The procedure in detail is as follows.

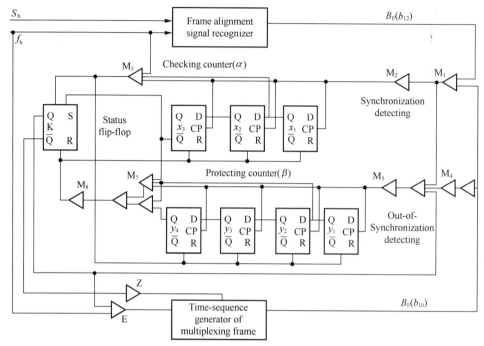

Figure 2-12 Logic diagram of flame alignment acquisition/protection

(1) Initial acquisition procedure

In the status of out-of-synchronization, the time-sequence generator of demultiplexing frame is set to b_F being at high level status; as long as the frame alignment recognizer send out B_F signal, the synchronization detector (M_2) will feed out positive pulse to make checking counter counts for $\alpha=1$; and also make M_3 output low level and set the status initiator at $Q=1$ status, thus, gate E is opened and through gate E, clock f_h will drive time-sequence generator of demultiplexing frame operate. Its time waveform is shown in Figure 2-13.

(2) Acknowledgment of the synchronization procedure

When it is guaranteed that B_F and b_F are overlapped for continuous three frames, the checking counter counts for $\alpha=3$ while the checking counter X3 outputs $Q=0$ status. The status initiator (K) is set again at 1 and M_7 is closed, it is guaranteed that non-overlapping by chance within 3 times will not cause the status change of status initiator (K). At this time, the system enters into the acknowledgment synchronization status with its time waveform shown in Figure 2-14.

Chapter 2 Synchronous Multiplexing

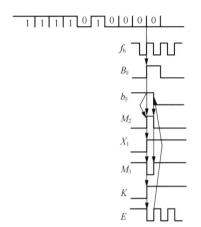

Figure 2-13 Initial acquisition procedure

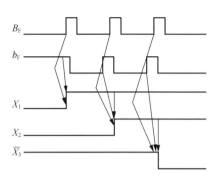

Figure 2-14 Acknowledgment procedure diagram of synchronization

(3) Acknowledgment procedure of out-of-synchronization

In the status of synchronization, M_5 of out-of-synchronization detector is open, if b_F is not overlapped with B_F, b_F pulse will pass through M_5 and counts as $\beta=1$ in the protection counting, If they are not overlapped for continuous four times, the protecting counter will counts fully for $\beta=4$, so as to make M_6 output low level, reset checking counter and set the status initiator to $Q=0$ status, then, the status initiator shuts gate E, i.e. stop transmitting clock to the time-sequence of local demultiplexing frame and set it to b_F status. At this moment, the system enters out-of-synchronization status. Its time waveform is shown in Figure 2-15.

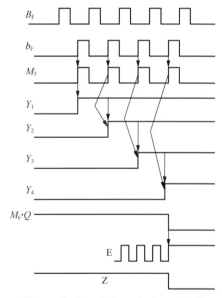

Figure 2-15 Acknowledgment of out-of-synchronization

From prototype acquisition/maintenance logical working procedure, it is thus clear that the checking counter and protecting counter is not activated simultaneously, while one is counting, the other memorizes some certain statues: synchronization status or out-of-synchronization status. Therefore, it is possible to use only one counter, in addition, one memory initiator is equipped to memorize the synchronization status and out-of-synchronization status. it can be see also from the original circuit that Pin D of first D flip-flop of the counter is idie. Pin D and pin P of the D flip-flop can be used as input pins of B_F and b_F, in this way, the first D flip-flop can detect if the B_F is overlapped with b_F, i.e. playing a role of synchronization/out-of-synchronization status detection. By simplifying the two mentioned above, the simplified acquisition/maintenance logical circuit is obtained (Figure 2-16)[37]. The connection relations between such simplified circuit and the recognizer of frame alignment signal, and between such simplified circuit and time-sequence generator of the demultiplexing frame are exactly same as the original ones and the two circuits are totally same in function, only the equipment is simplified.

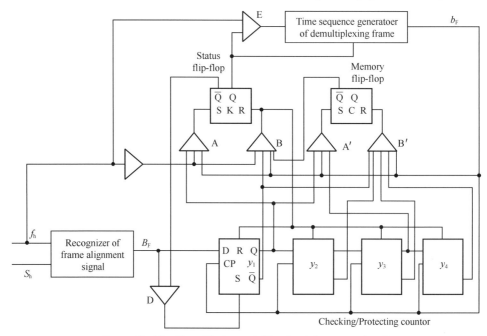

Figure 2-16 Simplified acquisition/maintenance circuit

Figure 2-17 shows the time waveform drawing of initial acquisition of the simplified acquisition/maintenance circuit. In case of out-of-synchronization status b_F set to high level, when high level occurs at the recognizer output B_F of frame alignment signal, B_F sets the first D flip-flop (y_1) of the counter to $Q = 1$ through gate D, thus gate A will be opened; then the clock sets the status initiator (k) to $Q = 1$ through gate A, as a result, gate E will be opened; at last, clock (f_h) drives the time-sequence generator of demultiplexing work through gate E.

Chapter 2 Synchronous Multiplexing

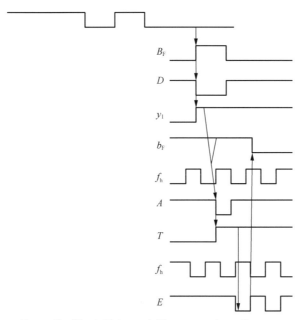

Figure 2-17 Initial acquisition procedure diagram

The working procedures of acknowledgment of synchronization and acknowledgment of out-of-synchronization are similar too. Like the prototype circuit, in case B_F is overlapped with b_F for continuous $\alpha=3$ times and the first three D flip-flop (y_1, y_2 and y_3) of the counter are all at status $Q=1$, gate A′ outputs low level to set memory initiator (C) to status $Q=0$ and a low level is output from Q to close gate B, that means to have entered the synchronization status already; then, if B_F is not overlapped with b_F for continuous $\beta=4$ times and all the four d flip-flop (y_1, y_2, y_3 and y_4) of the counter are at status $Q=1$, gate B′ outputs low potential to set memory initiator (C) to status $Q=1$ and the low level output is from gate B to set the status initiator (K) to status $Q=0$, close gate E to stop transmitting clock and the time-sequence generator of the demultiplexing frame is set to status B_F, that means to have entered out-of-synchronization status.

2.14 GUARANTEEING THE SYNCHRONIZATION CIRCUMSTANCE

It is prerequisite to realize the synchronization multiplexing that tributary digital signals to be multiplexed must be guaranteed to synchronize with multiplexing clock, which is the main issue in the multiplexing technology. It is well known that the favor of the synchronization multiplexing is obviously, e.g. relative higher multiplexing efficiency and smaller multiplexing impair. Only if the synchronization circumstance is guaranteed, the synchronization multiplexing equipment can be used. Under some concrete conditions, it is relative easy to provide synchronization circumstance, in which it is suitable to adopt the synchronization multiplexing; Under other concrete conditions it is not easy to supply synchronization circumstance unless some technical or economic cost is paid, in such case the application of the synchronization multiplexing will be restricted somewhat. A few kinds of typical application and the concrete method to guarantee the synchronization are described as follows:

(1) Delta modulation encoding synchronization multiplexing

As shown in Figure 2-18, by means of the time relation in the time sequence of encoding and sampling, it is very easy to arrange the code elements from tributaries in multiplexing order to realize synchronization multiplexing. In such case, if synchronization multiplexing technology is used, the equipment is quite simple and the channel availability relative higher.

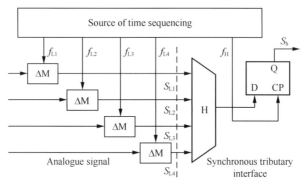

Figure 2-18 ΔM encoding synchronous multiplexing

(2) Synchronization multiplexing used for digital switch

Refer to Figure 2-19, by means of frame regulator all the clocks of input code stream is converted to local clocks and tributary frames are aligned with local frame structure through delay regulation. Thus, it is very convenient to realize synchronization multiplexing. Although, relevant complex frame regulator needs to be equipped at this place, it is an essential equipment inside sensible for the switching network, which is not specially installed for synchronous multiplexing. Therefore, in this case, it is suitable to employ synchronous multiplexing in respect of technology and economy.

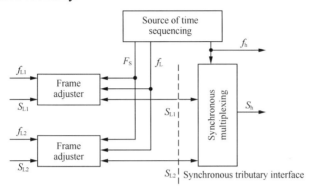

Figure 2-19 Synchronous multiplexing used for switch

(3) Synchronous multiplexing for short range transmission

See Figure 2-20, under such condition, the clock loop can be used to provide synchronous circumstance for the synchronous multiplexing. That is to say, the tributary clock is generated in the multiplexer and will be sent to opposite end through the clock loop so as to guarantee the tributary clock frequencies under multiplexing equal. When range is short, the code stream wander introduced by transmission lines is smaller, In addition, Usually, there is no frame structure for the voice channel code stream in the short transmission. Normally, there is no special requirement to multiplex such voice channel code stream, so it will not

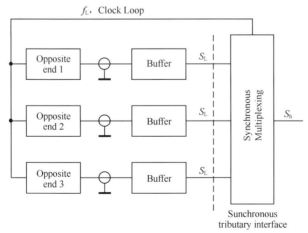

Figure 2-20 Synchronization multiplexing used in short range transmission

bring specialized difficulty to the synchronous multiplexing. In conclusion, to provide synchronous multiplexing for such short range transmission code stream, it is only required to install synchronous clock loop and simplified buffer. now, the synchronous multiplexing is satisfied for the engineering project.

(4) Group multiplexing in the long range transmission network

For the group multiplexing in the long range transmission network, it will meet some practical difficulties if the synchronous multiplexing is still used. Firstly, it is not yet a simple thing to provide clock to all nodes in the long range transmission network; Secondly, cause the code stream wander introduced by long distance line is relative bigger, e.g.in CCITT Rec. G.822 it is stipulated that the maximum, wander is 18μs for domestic network or international link (2048kbit/s and 8448kbit/s), in order to absorb such big wander, the buffer with capacity of 35 bit to 145 bit needs to be installed. Obviously, it is not recommendable in consideration of economy and it is not allowed to have so big delay impair of the network. Thirdly, it is generally required for the synchronous group multiplexing to recognize the frame alignment signals from tributary code streams before the implementation of multiplexing, then the signals will be put on the dedicated locations in the multiplexing frame. For this purpose, it is necessary to have specialized frame regulator. To sum up, in the long range transmission network, at lest so far the synchronous multiplexing is not adopted normally and plesiochronous multiplexing technology is common used internationally. As for the digital multiplexing used in connection with point-to-point high capacity optic fiber, that is another topic to be discussed separately.

2.15 CCITT RECOMMENDATIONS

Up to July of 1988, concerning the series of 2048kbit/s, CCITT recommended seven kinds of synchronization multiplexing equipment in total, which are all channel ones; in addition, one kind of second group synchronization multiplexing equipment are written in the annex.

Among the seven kinds of recommended channel multiplexing equipment, the five (G.732, G.73A, G.73B, G.73C and G.734) are primary group synchronization multiplexing equipment. and their frame structures are unified (see Figure 2-21). The frame frequency is 8kHz, frame length 32 time slots with 256 bit. the alignment is used in alternate frame with frame alignment signals 0011011.

Chapter 2　Synchronous Multiplexing

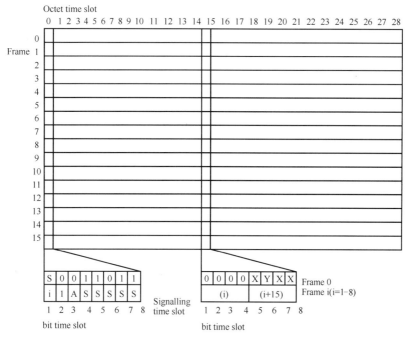

Figure 2-21　(1) Primary digital group (2048kbit/s) multiframe (G.704)

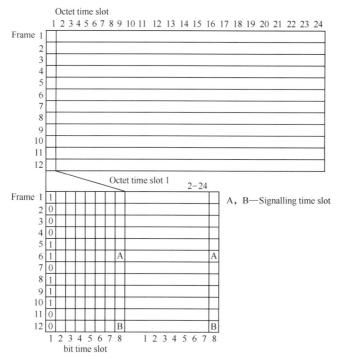

Figure 2-21　(2) Primary digital group (1544kbit/s) multiframe (12frame) (G.704)

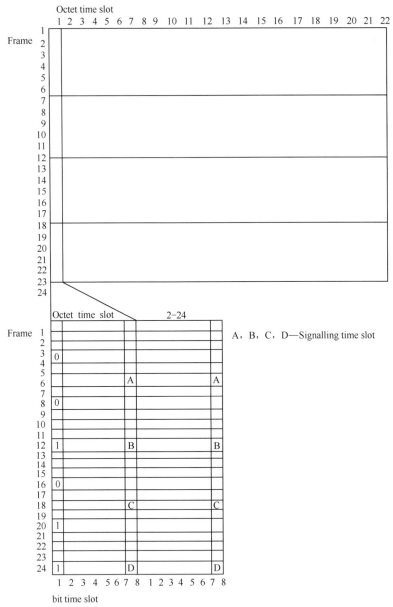

Figure 2-21 (3) Primary digital group (1544kbit/s) multiframe (24frames) (G.704)

G.732 is an analogue input PCM multiplexing equipment to multiplex 30 channels in all and transmit signaling in 16th time slot. When channel associated signaling is used, 16 basic frames will form a multiframe. and the 16th time slot in the zero frame is used to transmit the multiframe alignment (0000) and service bit; the other 16th time

Chapter 2 Synchronous Multiplexing

slots in the remaining 15 frames are allocated respectively to 30 voice channels to transmit the signaling. i.e. each voice channel occupies 2kbit/s signaling path; G.737 is a PCM multiplexing equipment with integrated both analogue and digital input, its time slot allocation is exactly same as that of G.732; G.738 is 64kbit/s digital multiplexing, in which 31 channels are multiplexed in total. i.e. 16th time slot is also used to transmit signaling and 0 time slot allocation is exactly same as that of G.732; G.739 is a outside access equipment, i.e. in the 2048kbit/s transmission channel. If required, several 64kbit/s digital bit streams can be inserted or taken out and its time arrangement is as same as that of G.732; G.734 is a digital multiplexing equipment used for switching and its time slot arrangement is as same as that of G.732.

In the seven types of recommended channel multiplexing equipment, two of them (G.744 and G.746) are second group digital multiplexer. Their frame structures are also unified (see Figure 2-22). The frame frequency is 8kHz, and frame length 132 time slots counting for 1056 bit, all of them are used to multiplex 120 voice channel bit stream (with 64kbit/s), the first 6 bit in time slot zero and 66th time slot are used to transmit frame alignment signal 11100110100000, the last two bits in 33rd, 99th and 66th time slot are standby or used to transmit service digits. time slot 1-4 are idle and 67th-70th time slot are used to transmit the signaling.

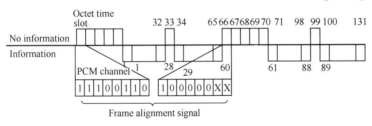

Figure 2-22　(1) PCM 2-ary digital group(8448kbit/s)(G.704)

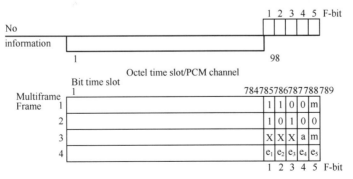

Figure 2-22　(2) PCM 2-ary digital group(6312kbit/s)(G.704)

The frame structure of the second group synchronous multiplexer written in the Annex 1 of Rec. G.741 is totally same as that shown in Figure 2-22. It is worthy to say that, concerning the multiplexing method, the frame alignment signal in all the 2048kbit/s bit streams need to be identified specially before the multiplexing and which will be multiplexed into the preassigned 1st-4th time slots of 8448kbit/s frames. The advantages to do so is very obviously, once the 2048kbit/s tributary code stream is multiplexed into 8448kbit/s code stream. immediately it is "Solubilized" into second group synchronous multiplexing frames. then it is able to be compatible with G.744 and G.746 thoroughly, for example, it is not necessary to go through 2048kbit/s but demultiplexed into 64kbit/s voice channel code stream directly. At this time, it is very convenient for second group branch and demultiplexing. But to realize such multiplexing—before the multiplexing, the frame alignment signals are identified and then, they will be put on the assigned location, of the multiplexing frame—is not a easy thing. CCITT is still studying the group synchronization multiplexing actively.

The above introduction shows that the synchronization multiplexing currently recommended by CCITT are all based on PCM word multiplexing. For sure, they depend on the frame structure of Rec. G.732; the other kinds of synchronization multiplexing frame structure are essentially unified and thoroughly compatible will them. There is still broad leeway for research and prospects for development.

Chapter 3   Positive Justification

3.1   PLESIOCHRONOUS MULTIPLEX

Plesiochronous multiplex is that the tributary binary digit clocks and multiplex binary digit clocks are nominally equal within a certain definite tolerance. Strictly speaking, if the corresponding effective instant of two signals is at one nominal bit rate, and any changes of rate are limited within the specified range, the two signals are plesiochronous with each other. For example, two signals generated by different clock sources and having the same nominal bit rate are usually plesiochronous. Where, the nominal bit rate and its tolerance are specified unifiedly beforehand.

Since plesiochronous multiplex permits the clock frequency change arbitrarily within the specified tolerance range, there will be no limitation naturally on the phase relationship of tributary clocks multiplexed. Hence, it need not to supply special condition for the plesiochronous multiplex. As long as the nominal clock and its tolerance accord with the specification, the plesiochronous multiplex will be realized. Since plesiochronous multiplex has such special characteristics, in some particular applications, for example, in telecommunication transmission network (especially in high order group multiplex), it is simple and economical to use plesiochronous multiplex technique to realize multiplex. Up to May 1980, all group multiplex in transmission recommended by CCITT used the plesiochronous multiplex technique.

To realize plesiochronous multiplex various suitable particular technical methods may be used. For example, low speed plesiochronous multiplex usually uses simple high speed sample technique, i.e. it uses many time slots to transmit a tributary bit. The error of binary element width will not exceed a multiplex time slot width. In order to reduce the error of binary element width, we should suitably increase quantity of time slot. Obviously, the characteristics of this method are simple equipment and low multiplex efficiency. Hence, It is suitable for low speed data transmission but not for high speed digits multiplex. For example, in moderate rate data or digits multiplex, it usually used pulse skip coding several years ago. That is, only every skip edge of the tributary binary digits are transmitted to the opposite side, then according to the corresponding changes of skip edges the binary digits are recovered. Consider the requirement of precision and quantity of equipment, we use usually more than 3 bits to describe the detail of skip edge; i.e.whether a skip happens, such a skip from high level to low level or vice versa;the time when skip happens is either in the first half of multiplex time slot or the second half. Hence, using pulse skip coding to realize plesiochronous multiplex, it need three times capacity of tributary rate, which depends on precision. Unfortunately, quantity of equipment of this method is not low enough to be satisfied. Today, high speed digits usually use various code justification techniques. The quantity of equipment is equivalent to that of pulse skip coding equipment, but the channel capacity occupied is only several percent more than the rate of binary digits transmitted. Therefore, such a technique is suitable for high speed binary digits plesiochronous multiplex.

The justification technique actually is to regulate all the tributary binary digits of plesiochronous multiplex into synchronous binary digits, then synchronously multiplex these synchronous binary digits. Therefore, plesiochronous multiplex technique is only a particular method which provides a synchronous environment to realize synchronous multiplex. The plesiochronous multiplex is a total concept of the justification and the synchronous multiplex. The justification technique can be classified into three kinds of positive justification, positive/negative justification and positive/zero/negative justification. The most popular technique is positive justification, hence this book will emphasize on discussion of the positive justification technique.

3.2 PRINCIPLE OF POSITIVE JUSTIFICATION

The plesiochronous multiplexer using the positive justification is shown in Figure 3-1. Every binary digits multiplexed firstly pass an independent justification device to convert the plesiochronous binary digits into synchronous binary digits, then enter the synchronous multiplex device to be multiplexed. In the receiving

Chapter 3 Positive Justification

side, firstly execute synchronous demultiplex, then through the recovery device the original plesiochronous multiplex tributary binary digits is recovered.

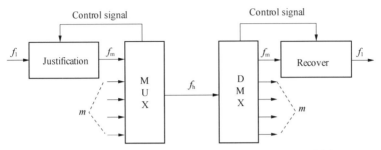

Figure 3-1 Diagram of justification plesiochronous multiplexer

Main part of justification is a buffer. And some necessary control circuits. The input clock frequency is the plesiochronous tributary clock frequency f_1, and the output clock frequency is the synchronous multiplex tributary frequency f_m. In the positive justification technique, the output frequency f_m is greater than the input frequency f_1, it is called positive justification because $f_m > f_1$.

Assuming the buffer is originally half filled. As time goes on, because $f_m > f_1$ the information in the buffer become less and less. Finally, if a special method is not used, the buffer will be empty so that the false information will be read out. This process is very similar with water flowing in and out in a reservoir. If the flowing out rate is greater than that of flowing in, as time goes on, the water level will drop, then finally the water is drain. If a suitable control system is used, see Figure 3-2, when water level drops down to a definite level, the control signal will be sent out to close the output pipe in a unit time, then the water level will increase. When the output pipe is open. The above process will be repeatedly. In such a way of control, all water will pass through the reservoir, neither increase nor decrease. The justification process is also in such a way, when information bits in the buffer decrease to the minimum poing, then send out a control signal to let the read out clock stop one beat, then the information in the buffer will increase one bit. Repeated in such a way, the binary digits will be passed out through the buffer, neither add nor lose any information. The process of positive justification may be described as sequential beat, see Figure 3-3. In a justification device, whenever binary digits decrease to the specified threshold, f_m will stop a beat. Corresponding to this beat time, no information will be read out from buffer. This is a positive justification. In the recovery device, at the corresponding beat time (because no information is read out), no information is written into the buffer either, this is the corresponding recovery. With such justification and recovery, it can be sure to get error free transmission.

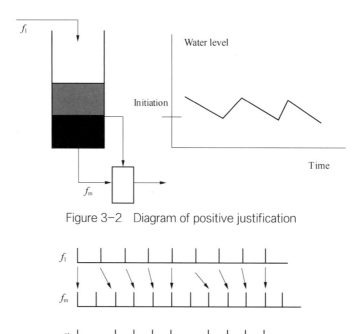

Figure 3-2 Diagram of positive justification

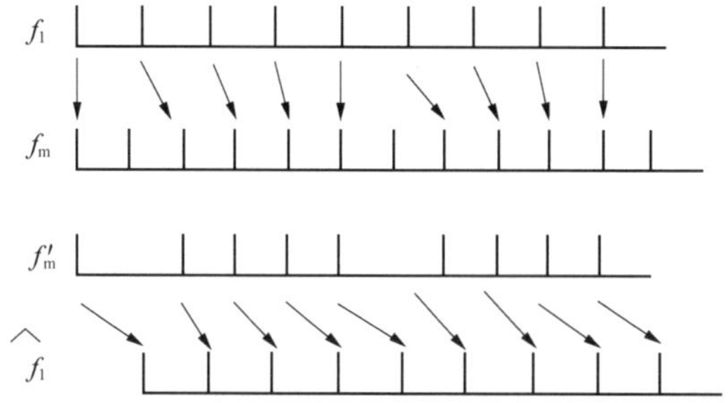

Figure 3-3 Principle diagram of positive justification

3.3 BASIC FORMULA OF POSITIVE JUSTIFICATION

To realize positive justification, various particular methods may be used. To simplify the device, usually in every multiplex frame a specific time slot is specified to provide a chance for particular tributary to do positive justification. If the tributary does not need justification, the time slot transmits ordinary information as usual. Otherwise, the time slot will be idle once. This specific time slot is called bits from the tributaries for the positive justification. Sometimes it is

Chapter 3 Positive Justification

called justification digit times slot, abbreviated to SV.

It is necessary to tell the recovery device whether the specific time slot (SV) is idle (justification) or carries information (no justification). Hence, in a multiplex frame, a specific time slot is reserved to transmit indication signal of justification. Obviously, such an indication signal is very important, in case it has error, the tributary binary digits will lose a bit or insert a wrong bit, resulting a structural impair, i.e. slip. Therefore, it is usual to set indication code more than three bits for debugging. For example, 111 denotes for justification, 000 denotes for no justification and using majority decision. Such an indication signal is called justification service digits or stuffing service digits, abbreviated to SZ.

In the multiplex frame the arrangement of SV and SZ is shown in Figure 3-4. The notions in Figure 3-1 and Figure 3-4 are described as following:

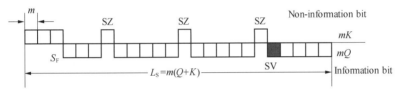

Figure 3-4 Structure of multiplex frame

f_1—Nominal tributary bit rate;

f_h—Nominal multiplex bit rate;

Q—Bits per tributary per frame;

K—Non-information bits corresponding to every tributary per frame, i.e. total bits of non information in every frame are mK bit, in which;

m—number of tributaries multiplexed;

L_s—Frame length, i.e. the total of information bits and non-information bits in every frame, hence

$$L_s = m(Q+K) \qquad (3\text{-}1)$$

from Figure 3-1 and Figure 3-4, define the following:

Frame rate-frames per unit time:

$$F_s = \frac{f_h}{L_s} \qquad (3\text{-}2)$$

Synchronous multiplex Tributary rate—maximum multiplex rate of each synchronous multiplex tributary provided by synchronous multiplexer

$$f_m = \frac{Q}{L_s} \cdot f_h \qquad (3\text{-}3)$$

Nominal justification rate—when tributary rate and multiplex rate are both equal to their nominal, the rate of deleting or inserting justification digit, is also called nominal stuffing rate:

$$f_s = f_m - f_1 \qquad (3\text{-}4)$$

Maximum justification rate—maximum rate of possibly inserting or deletion justification digit. Usually in each multiplex frame only one justification bit is set. Hence

$$f_{s\,max} = F_s \qquad (3\text{-}5)$$

Justification ratio—the ratio of practical justification rate to maximum justification rate, also called stuffing ratio:

$$S = \frac{f_s}{f_{s\,max}} \qquad (3\text{-}6)$$

From the above definition and formula, it is not difficult to derive:

$$\left(\frac{1}{m} - \frac{f_1}{f_h}\right) L_s = K + S \qquad (3\text{-}7)$$

Formula (3-7) is the basic formula of positive justification. In the formula, tributary rate f_1, multiplex rate f_h and number of multiplex m are known value. Frame length L_s, non-information bit K and justification ratio S, are all basic design value, where usually frame length L_s is an argument. Whenever L_s takes a value, we can calculate the right side value by basic formula (3-7), whose integer part is K and decimal part is S. The range of L_s and the available scheme of K and permissive S, we will discuss in other chapter.

From the above definition, it is easy to understand that the units of $f_1, f_h, L_s, f_m, f_s, f_{s\,max}$ is Hz, and Q, K, m, L_s, S are dimensionless.

3.4 JUSTIFICATION DESIGN

The diagram of the justification device is shown in Figure 3-5. It consists of buffer and justification control circuits.

Chapter 3 Positive Justification

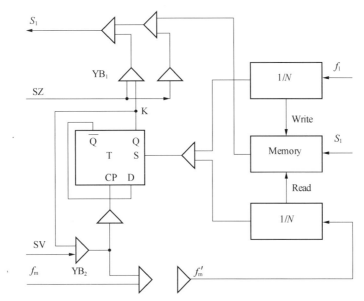

Figure 3-5 Diagram of positive justification

The tributary clock (f_1) writes the tributary signal code (S_1) into the buffer bit by bit; the synchronous multiplex clock (f_m) reads the signal code bit by bit from the buffer, and sends it to the synchronous multiplex unit. Here the tributary clock (f_1) is uniforml; the synchronous multiplex clock (f_m) is not uniform, and is n all the beats to function as control. Particularly speaking, the beat corresponding to SV of the synchronous multiplex clock (f_m) will not function when a positive justification is executed; whereas it will function normally when no justification is executed. This depends completely on whether the frame need a justification. The need of a justification depends on the leading time of the write in pulse to the read out pulse of the buffer, i.e. depends on the number of the signal codes in the buffer. If the signal code bits stored decreases to the lowest number allowed, i.e. decrease to a threshold, then the read/write time difference recognizer outputs a control pulse; see Figure 3-6.

The negative pulse P sets the trigger T to let the Q end of T output the high level (K) which opens the NAND gates YB_1 and YB_2; the control signal SV deduct a beat, corresponding to SV, of the synchronous multiplex clock (f_m) through YB_2; i.e. no signal code is output at this beat, this completes a positive justification; at the same time, the control signal SZ functions through YB_1 to forcedly insert symbol

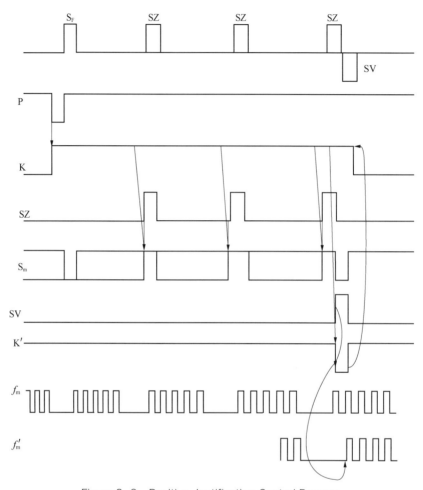

Figure 3-6 Positive Justification Control Process

111 into the three time slots corresponding to SZ in the output tributary signal code (S_m), which is used to express no transmission of signal code at SV of this frame. Here, the control signals f_m, SV and SZ are all generated by the synchronous multiplex timing unit, and are sent to the justification devices of all the tributaries in parallel. After the above control process completes, the trigger T is recovered to its original state using the trailing edge of the pulse SV; if the stored signal code bits have not decreased to the threshold, the recognizer will not output pulse P, hence the control trigger T will not act. Then the control signal SV will not function, i.e. the beat corresponding to SV of the synchronous multiplex clock f_m functions as usual. This is non justification work state. In this case, the control

signal SZ forcedly inserts 000 state into several slots corresponding to SZ of the output tributary binary digits S_m, which is used to express no justification in this frame. We can see from the above process, in the case that a justification is correctly done, consider the average of the tributary buffer in a proper time interval, the read out in formation bits should be just equal to the write in information bits. That is, the read out clock frequency (f_m) is equal to the tributary (write in) clock frequency (f_1), then the tributary binary digits pass through the justification mechanism, neither losing signal code nor producing false signal code.

3.5 JUSTIFICATION TRANSITION PROCESS

As it is known to all, the prerequisite that the buffer works normally is to write in the information bits at first, then to read out the information bits. The lead of the write in time relative to the read out time is called the read/write time difference. When the buffer is working normally, the read write time difference changes always as the time changes. The justification transition process just describe the change process of the read/write time difference as the time, definitely speaking, it studies the developing law that in the cases of various initial states, the read/write time difference changes and tends to the final state.

Normally, a two dimensional plane is used to describe the justification transition process, the abscissa takes the time when reading out signal code as the measurement dial, at the time to read out the xth signal code, the abscissa is written as x; the ordinate expresses the read/write time difference, written as Δt_x, it expresses the read/write time difference when reading out the xth signal code. When $x=0$, written as t_o, it is called the initial read/write time difference.

Known from the explanation about the physical process of the positive justification in the previous sections, if do not consider the non-information bits and the time when the read out clock beat stops caused by the justification bit, i.e. in the case of uniformly reading out, the write in rate of the tributary buffer is f_1, the read out rate is $\dfrac{f_h}{m}$. In the other word, the tributary buffer writes in a information bit every T_1 time interval; reads out an information bit every mT_h time interval. Hence, between the two consecutive read out times, the change of the read write time difference is $(mT_h - T_1)$; from the 0th read out time to the xth read out time, the change of the read write time difference is $(mT_h-T_1)x$; in an actual justification process, the write in always at the rate f_1, i.e. if do not consider the jitter impairment, it always uniformly writes in an information bit every T_1 time; whereas the read out beats are not uniform, at the non-information bits(e.g. frame alignment signal, justification service digits, and the other service bits)and the justification bits, the read out clock stops. Relative to the above uniform read out process, whenever the read out clock stops a beat, the read/write time difference increases an mT_h time. Assume between the 0th read out time and the xth read out time, the read out clock stops g beats altogether, then corresponding to the xth read out time, the read write time difference is:

$$\Delta t_x = \Delta t_0 + (mT_h - T_1)x + mTg$$

$$\Delta t_x = \Delta t_0 + \left[\frac{mT_h}{T_1}(x+g) - x\right] \cdot T_1 \quad (3\text{-}8)$$

The formula(3-8)is the general expression of the positive justification transition process. Where, the number of the multiplex tributaries(m), the tributary signal code period(T_1)and the total signal code period(T_h)are given; the stopped beats(g)of the read out clock in the transition process considered is decided by the frame structure; the initial read/write time difference Δt_0 is decided by the history before the transition process, it may be believed as a random within a limited range, according to the different values of Δt_0, it may form various different justification transition processes, this is just the contents that will be discussed in the following. Among the above variables, T_h, T_1, Δt_x and Δt_0 have dimension of time; m, x and g are dimensionless; for use conveniently later, take T_1 as the measurement unit of the read/write time difference Δt_x and Δt_0. After T_h, T_1, m, g and Δt_0 are determined, based on the variable x according to the formula(3-8), may get the corresponding the read/write time difference.

For example. CCITT Recommendation G.742: f_h=8448kHz, f_1=2048kHz, m=4. the flame structure is shown in Figure 3-7.

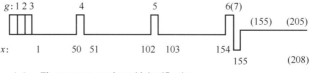

( )—The sequence number with justification

Figure 3-7 G.742 Frame Structure Diagram

Assume $\Delta t_0=0$, then:

$$\frac{\Delta t_x}{T_1} = \frac{mf_1}{f_h}(x+g) - x \qquad (3\text{-}9)$$

$$\frac{\Delta t_x}{T_1} = 0.97(x+g) - x$$

The calculation result is shown in the following table, the calculation curve is shown in Figure 3-8.

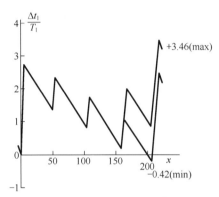

Figure 3-8 Justification process(G.742, $\Delta t_0=0$)

The following is to get the net change of the read write time difference after a frame of the justifieation process. Still take G.742 as an example, according to the formulas(3-8)as well as(3-1) to(3-5), may get the difference between the read/write time difference(Δt_f) at the end of a frame end and the read/write time difference(Δt_0)at the beginning of a frame.

(1) No justification within a frame

$$\Delta t_f - \Delta t_0 = \left[\frac{nf_1}{f_h}(x+g) - x\right] T_1$$

$$x = Q, g = K$$

$$\frac{\Delta t_f - \Delta t_0}{T_1} = \frac{mf_1}{f_h}(Q+K) - Q = \frac{f_1}{f_h}m(Q+K) - Q$$

$$= f_1 \cdot \frac{L_s}{f_h} - Q = \frac{f_1}{F_s} - \frac{f_m}{F_s}$$

$$= -\frac{f_m - f_1}{f_{s\max}} = -\frac{f_s}{f_{s\max}} = -S$$

$$\Delta t_f - \Delta t_0 = -ST_1 \qquad (3\text{-}10)$$

The formula(3-10)describes that the read write time difference will decrease a net ST_i when non-justification is within a frame.

x	g		$\dfrac{\Delta t_x}{T_1}$
	justification	non-justification	
0	0		0
1	3		2.88
50	3		1.39
51	4		2.33
102	4		0.79
103	5		1.73
154	5		0.18
155		6	1.12
155	7		2.09
205	7		0.58
206		6	−0.42(min)
206	10		3.46(max)
207		9	2.46

(2) A justification within a frame

$$\Delta t_f - \Delta t_0 = \left[\dfrac{mf_1}{f_h}(x+g) - x\right]T_1$$

$$x = Q-1, g = K+1$$

$$\dfrac{\Delta t_f - \Delta t_0}{T_1} = \dfrac{mf_1}{f_h}[(Q-1)+(K+1)] - (Q-1)$$

$$= \left[\dfrac{mf_1}{f_h}(Q+K) - Q\right] + 1$$

$$= 1 - S$$

$$\Delta t_f - \Delta t_0 = (1-S)T_1 \qquad (3\text{-}11)$$

The formula (3-11) describes that the read/write time difference will increase a net $(1-S)T_i$ when a justification is within a frame.

Chapter 3 Positive Justification

3.6 THE CHANGE RANGE OF THE READ/WRITE TIME DIFFERENCE IN A STABLE JUSTIFICATION PROCESS

From Figure 3-8 we can see that if at the beginning($x=0$) of a frame the read/write time difference(Δt_0) is equal to 0, and no positive justification is within the frame, then near. the end of the frame, the read/write time difference is less than 0, i.e. the write out time is behind the read in time. That is, it is read out when nothing has been written in, then a false signal is read out. This phenomenon is called "catch". To avoid the catch phenomenon, a justification threshold(Δt_s) is set usually, see Figure 3-9. When the read write time difference(Δt_x) reduces to the threshold($\Delta t_x \leqslant \Delta t_s$), a positive justification will be done in the next frame, as long as the threshold(Δt_s) is properly set, the catch phenomena may be avoided.

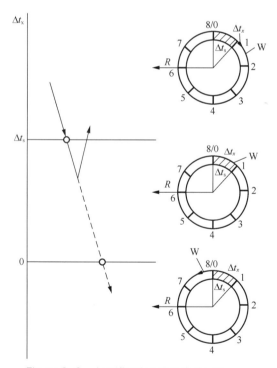

Figure 3-9　Justification threshold diagram

From Figure 3-8 we can see that if no positive justification is done, the lowest point of the read/write time difference(Δt_x) is always at the end of a frame. Hence, it is usual to decide if a justification should be done in the next frame according to the value of Δt_x at the end of a frame. It is specified that within a limited area(d) at the end of a frame, if $\Delta t_x > \Delta t_s$, no justification will be done; whereas if $\Delta t_x \leqslant t_s$, a justification will be done. Such a limited area(d) is called stable justification request area.

The following still takes CCITT Recommendation G.742 as an example to discuss the possible minimum($\Delta t_x > \Delta t_s$,) and the possible maximum(Δt_{max}) of the read/write time difference in a stable justification process, in order to get the change range(Δt_{pp}) of the read/write time difference in a stable justification area.

See Figure 3-10, assume that the read/write time difference(Δt_0) at the beginning($x=0$) of a frame is equal to the justification threshold(Δt_s). According to the above specification, the justification may be done within this frame. The beginning($x=0$) of the frame, i.e. the end of the previous frame, is the final time of the justification request area of the previous frame. If the previous frame has requested unsuccessfully, no justification will be done in the current frame, then the current frame must be the earliest one which requests a justification, then get the earliest cross point of the read/write time difference and the justification threshold. Obviously, this point is just the beginning of the justification request area in a stable justification process; it is self-evident that the end of the justification request area of the stable justification process is just the end of the current frame. The value of the read/write time difference line where is the earliest request of justification and which reaches at the end of the current frame, is the minimum possible read/write time difference(Δt_{min}); the value of the read/write time difference line where the latest request of justification is and which reaches at the end of the current frame, is the maximum possible read/write time difference which is provided to the next frame, hence at $x=1$ of the next frame, it certainly gets the transient value of the maximum read/write time difference(Δt_{max}).

(1) Δt_{min}

According to the formula (3-10),

$$\Delta t_f - \Delta t_0 = -ST, \Delta t_0 = \Delta t_s$$
$$\Delta t_f = \Delta t_s - ST_1, \Delta t_{min} = \Delta t_s - ST_1 \qquad (3\text{-}12)$$

The formula(3-12) expresses the minimum possible transient value of the read/write time difference in a stable justification process.

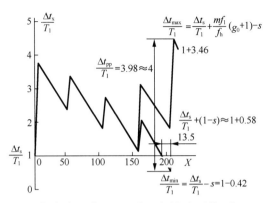

Figure 3-10 Parameter calculation diagram of a stable justification process(g_0=3, S= 0.42)

(2) Δt_{max}

According to the formula(3-8),

$$\frac{\Delta t_f - \Delta t_0}{T_1} = \frac{mf_1}{f_h}(x+g) - x$$

where

$$\frac{\Delta t_0}{T_1} = \frac{\Delta t_s}{T_1} + (1-S)$$

$$x = 1, g = g_0$$

$$\frac{\Delta t_{max}}{T_1} = \frac{\Delta t_s}{T_1} + (1-S) + \frac{mf_1}{f_h}(1+g_0) - 1$$

$$= \frac{\Delta t_s}{T_1} + \frac{mf_1}{f_h}(g_0+1) - S$$

$$\Delta t_{max} = \Delta t_s + \left[\frac{mf_1}{f_h}(g_0+1) - S\right]T_1 \quad (3\text{-}13)$$

The formula(3-13)expresses the maximum possible transient value of the read/write time difference in a stable justification process.

(3) Δt_{pp}

$$\Delta t_{pp} = \Delta t_{max} - \Delta t_{min}$$

$$= \left[\Delta t_s + \frac{mf_1}{f_h}(g_0+1)T_1 - ST_1\right] - \left[\Delta t_s - ST_1\right] \quad (3\text{-}14)$$

$$= \frac{mf_1}{f_h}(g_0+1)T_l$$

Based on the formula(3-14), may get the maximum change range of the read/write time difference transient value in a stable justification process. Where, g_0 is the number of stopped beats of the tributary binary digits of the beginning($x=1$) of the frame. For example, in G.742, $g_0=3$, in this case, $\Delta t_{pp} \approx 4T_i$. In the buffer design, this value should be considered.

(4) d

See Figure 3-10:

$$d = -ST_1 / \left(\frac{mf_1}{f_h} - 1\right) T_1$$

$$d = S / \left(1 - \frac{mf_1}{f_h}\right) \quad (3\text{-}15)$$

The formula(3-15) is the calculation formula of the justification request area(d) in a stable justification process.

3.7 CLASSIFICATION OF JUSTIFICATION TRANSITION PROCESS

Known from the discussion in the above sections, when the tributary rate f_l, the multiplex rate f_h, the number of tributaries m, the justification threshold Δt_s, the buffer size N and the frame structure(the distribution of S, g_0 and g) are determined, the justification transition process Δt_x is decided by the initial read/write time difference Δt_0. Corresponding to different initial read/write time difference, different justification transition processes may appear.

See Figure 3-11, given the initial read/write time difference Δt_0, according to the formula(3-8), may make the various typical read/write time difference curve Δt_x. Firstly, let us discuss the case that the initial of the read/write time differenc, obeys:

$$N > \frac{\Delta t_0}{T_1} > N > \frac{mf_1}{f_h}(g_0 + 1) - 1 \quad (3\text{-}16)$$

See the curve(1) of Figure 3-11: in the transition process the catch phehomenon appears continuously. Here, the transition process also obeys the justification principle: within the justification request area at the end of a frame, if $\Delta t_x > \Delta t_s$, no justification is requested; if $\Delta t_x \leqslant \Delta t_s$, a justification is requested. In such a process, if no justification is done, then the read/write time difference will decrease a net ST_i

after every frame;if a justification is done, then the read/write time difference will increase a net$(1-S)T_i$ after every frame. As long as the read/write time difference $\Delta t_x > \Delta t_s$, no justification will appear;only when the read/write time difference $\Delta t_x \leq \Delta t_s$, a justification will be done. Hence, in the case of $\Delta t_x > \Delta t_s$, no justification is done, the read/write time difference Δt_x will decrease at the rate of ST_i per frame;whereas in the case of $\Delta t_x \leq \Delta t_s$, a justification is done, the read/write time difference Δt_x will increase at the rate of$(1-S)T_i$ per frame. Obviously, in the case of satisfying the formula(3-16), although a catch phenomenon appears, the transition process always transits to the stable direction. The area satisfying the formula(3-16)is called catch transition area.

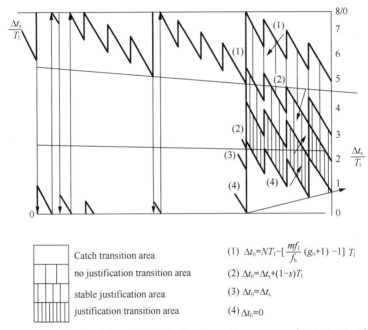

Figure 3-11 Partition of justification transition process(G.742, $N = 8$)

When the initial of the read write/time difference obeys:

$$N - \frac{mf_1}{f_h}(g_0 + 1) - 1 > \frac{\Delta t_0}{T_1} > \frac{\Delta t_s}{T_1} + (1 - S) \qquad (3\text{-}17)$$

no catch phenomenon happens;because $\Delta t_x > \Delta t_s$, no justification will be done, then the read/write time difference Δt_x will decrease at the rate of ST_i per frame. The area satisfying the formula(3-17)is called non justification transition area.

When the initial of the read/write time difference obeys:

$$\frac{\Delta t_s}{T_1} > \frac{\Delta t_0}{T_1} > 0 \tag{3-18}$$

no catch phenomenon happens;because $\Delta t_x < \Delta t_s$, hence justification will be done in every frame, thus the read write time difference Δt_x will increase at the rate of $(1-S)T_i$ per frame. The area satisfying the formula(3-18) is called justification transition area.

When the initial of the read/write time difference obeys:

$$\frac{\Delta t_s}{T_1} + (1-S) > \frac{\Delta t_0}{T_1} > \frac{\Delta t_s}{T_1} - S \tag{3-19}$$

no catch phenomenon happens;because Δt_x may be either greater than Δt_s or less than Δt_s, hence a justification may be done in some frame but not in the others; thus the initial value of the read/write time difference Δt_x will change between $\Delta t_x - ST_i$ and $\Delta t_x + (1-S)T_i$. The area satisfying the formula(3-19) is called stable justification area.

In a word, no matter which area the initial read/write time difference is, the justification transition process always transits to the stable justification area finally. It transits from the catch transition area to the non-justification transition area or to the justification transition area;then transits to the stable justification area from the non-justification transition area or the justification transition area. Finally, the system will be keeping normal justification process in the stable justification area.

3.8 BIT RATE RECOVERY DESIGN

The recovery device consists of control circuit, buffer and phase-lock loop. Its structure is shown in Figure 3-12 and the time waveform is shown in Figure 3-13.

If the number of 1s in time slot SZ of S_m is more than that of 0s, the output of the SZ detection trigger is high level of K′ ;a beat of SV in f_m is deducted. Hence, the content of time slot SV in S_m will not be written into the buffer, that is, a positive recovery control is completed. Then at the trailing edge, the SZ detector reset;if the 0s in time slot SZ of S_m is more than 1s, SZ detector does nothing, hence the SV will not function, the beat of SV in S_m will work as usual, that is, the content of SV in S_m is written into the buffer. After the tributary information is written into the buffer, by the phase-lock loop it gets the average $\hat{f}_1$ of the write clock f'_m, and reads out the tributary information from the buffer using $\hat{f}_1$ as the tributary output clock. The multiplex process is completed finally.

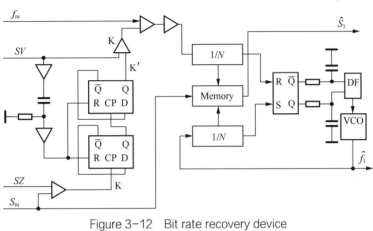

Figure 3-12 Bit rate recovery device

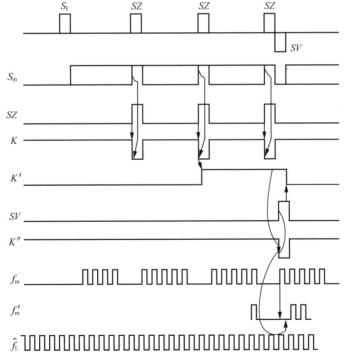

Figure 3-13 Bit rate recovery control process

Figure 3-14 is the diagram of phase-lock loop of the bit rate recovery. It is a popular phase-lock loop, the only difference is that there is a modulo N circuit at the input of the phase discriminator. Where N stands for the buffer size, it plays a function of division N. This modulo N circuit is necessary when writing into and reading from the buffer. However in the phase-lock loop it plays an important function incidentally when the change of read write time difference Δt_z exceeds over one beat of the input clock f'_m, the phase discriminator will still works normally. In Figure 3-14, the frequency tolerance of the input clock is $\Delta f'_m$;the output clock frequency is $\hat{f}_1$ and its tolerance is $\Delta \hat{f}_1$. The transfer coefficient of the phase discriminator which contains two modulo N circuits is K_{PD} with dimension of volt/rad;the transfer coefficients of the RC integrator circuit and the DC amplifier are K_{PLF} dimensionless;the transfer coefficient of the voltage-controlled oscillator is K_{VCO} with dimension of rad/sec. volt. The gain of phase-lock loop is K with dimension of 1/sec, i.e. Hz;the loop time constant is T with dimension of sec, the relationship of the dimension parameters and them is as the following:

$$K = K_{PD} \cdot K_{LPF} \cdot K_{VCO}$$

$$T = RC$$

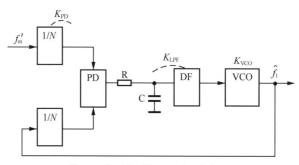

Figure 3-14 Phase-Lock loop

The recovery phase-lock loop is a normal second-order phase-lock loop, recovery hence it can be calculated by a classic formula. But in the case of bit rate recovery application, only the suppress and trace characteristics are required. Precisely speaking, stricter requirements are made to loop suppress and trace characteristics, the others are ignored;meanwhile, consider that the CCITT recommendations have made specific requirement to suppression, it is necessary to derive the specific application formula based on the general formula, in order to make engineering design conveniently.

3.9 PHASE-LOCK PARAMETERS DESIGN

(1) Loop Phase Jitters Suppression Characteristics

The diagram of phase jitter suppression characteristics analysis loop is shown in Figure 3-15, where the input phase function is the read/write time difference mentioned previously(Δt_x);the output jitter, i.e. phase residual wave after loop filtered will be discussed in detail followed. The phase jitter in the output binary digits is called stuffing jitter(A_j).

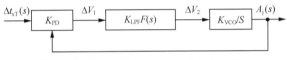

Figure 3-15 Jitter suppression function

Definition: $\dfrac{A_j(S)}{\Delta t_x(S)} = H(S)$;write the suppression characteristics expression from Figure 3-15 directly:

$$H(S) = \frac{KF(S)}{S + KF(S)}$$

$$K = K_{PD} \cdot K_{LPF} \cdot K_{VCO} \quad (3\text{-}20)$$

$$F(S) = \frac{1}{ST+1}$$

$$T = RC$$

Rewrite formula (3-20) into the general form:

$$H(S) = \frac{1}{\left(\dfrac{S}{f_n}\right)^2 + 2\zeta\left(\dfrac{S}{f_n}\right) + 1} \quad (3\text{-}21)$$

$$f_n = \sqrt{\frac{K}{T}}, \zeta = \frac{1}{2\sqrt{KT}}$$

Formula(3-21) is a general expression of second-order phase-lock loop suppression characteristics using simple RC integrator circuit. Where f_n is the loop natural harmonic frequency with dimension of 1/sec; ζ is the loop damping coefficient dimensionless. Consider the CCITT specific requirement to the phase jitter suppress of recovery phase-lock loop, it is necessary to derive the relationship among the jitter gain(A_{dB}), the initial suppress frequency(f_0) and the suppers slope, and the loop elementary parameters(K, ζ, ω_n).

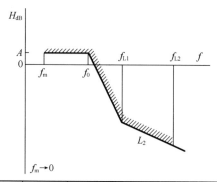

CCITT Recommendation MUX	f_i(kbit/s)	A(dB)	f_0(Hz)	L_1(dB/oct)	f_{L1}(kHz)	L_2(dB/oct)	f_{L2}(kHz)
G.742 8448/2048	2048	0.5	40	20	0.4		
G.743 6312/1544	1544	0.5	350	40	2.5	20	15
G.751 34 368/8448	8448	0.5	100	20	1.0		
G.751 139 264/34 368	34 368	0.5	300	20	3.0		
G.751 139 264/8448	8448	0.5	100	20	1.0		
G.752 32 064/6312	6312	0.5	65	20			
G.752 44 736/6312	6312	0.1	500	40	2.5	20	15
G.752 97 728/32 064	32 064	0.5	320	20			

Figure 3-16　Jitter suppression sample

From(3-21)derive:

$$H(f) = 1 / \sqrt{\left[1 - \left(\frac{f}{f_n}\right)^2\right]^2 + 4\zeta^2 \left(\frac{f}{f_n}\right)^2} \quad (3\text{-}22)$$

$$H(f)_{max} = 1/2\zeta\sqrt{1-\zeta^2}, \zeta \leq 0.707 \quad (3\text{-}23)$$

$$\lim_{f \to \infty} H(f) = f_n^2 / f^2 \quad (3\text{-}24)$$

Then get the relationship among the values recommended by CCITT and loop parameters:

$$A_{dB} = 20\lg H(f)_{max} \qquad (3\text{-}25)$$
$$= 20\lg(1/2\zeta\sqrt{1-\zeta^2})$$

$$L_{dB \atop \text{oct.}} = 20\lg\left[\lim_{f\to\infty} H(f) / \lim_{f\to\infty} H(2f)\right]$$
$$= 20\lg\left[\frac{f_n^2}{f^2} / \frac{f_n^2}{(2f)^2}\right] \qquad (3\text{-}26)$$
$$= 12$$

According to Figure 3-16: $H(f_0)_{max} = \lim_{f\to\infty} H(f)|_{f=f_0}$ then get:

$$f_0 = \sqrt{2\zeta\sqrt{1-\zeta^2}} \square f_n \qquad (3\text{-}27)$$

When $\zeta = 0.707, f_0 = f_n$; when ζ is near 0.707,
$$f_0 \approx f_n$$

Known from(3-26)that such a simple RC integrator phase-lock loop can satisfy CCITT recommended suppress slope;when A_{dB} and f_0 are given, loop elementary parameters and f_n can be gotten from(3-25)and(3-27);then the specific loop parameters K and T can be gotten from ζ and f_n, hence the circuit can be designed to satisfy me loop suppress characteristics.

(2) Loop Trace Phase Error

Because of the transition caused by the error between the loop input frequency f_m and the center frequency of the loop voltage controlled oscillator, the error of the residual output phase to the input phase when stable is the loop trace phase error:

$$\varphi_\varepsilon = \lim_{t\to\infty} \varphi_e(t)$$
$$\varphi_e(t) = \Delta t_x - A_j$$

According to LaPlace Theorem, from LaPlace transform expression get the solution directly:

$$\varphi_\varepsilon = \lim_{t\to\infty} \varphi_e(t)$$
$$= \lim_{s\to 0} S \square \Phi_e(S)$$
$$= \lim_{s\to 0} \frac{\Delta\omega}{S + KF(S)}$$
$$= \frac{\Delta\omega}{KF(O)}$$

When using a simple RC filter, $F(0)=1$, hence

$$\varphi_\varepsilon = \frac{\Delta\omega}{K}$$

$$\varphi_\varepsilon = \frac{2\pi(\Delta f'_m + \Delta f_{VCO})}{K}$$

$$\Delta f'_m = \Delta f_i$$

$$\varphi_\varepsilon = \frac{2\pi(\Delta f_i + \Delta f_{VCO})}{K} \qquad (3\text{-}28)$$

formula (3-28) is the loop phase trace error expression of bit rate phase-lock loop using RC integrator. Where Δf_i is the frequency tolerance of the tributary bit stream, Δf_{VCO} is the frequency tolerance of the voltage controlled oscillator, K is the loop gain.the dimensions of Δf_i, Δf_{VCO} and K are 1/sec, therefore, the dimersion of the trace phase error is radian.

(3) Loop Parameters Design

When A_{max}, f_{0max} and $\varphi_{\varepsilon max}$ are given, may get the elementary parameters ζ_{min}, $f_{n\ max}$, K_{min} from the formulas(3-25), (3-27)and(3-28);meanwhile get K_{max} from equation based on ζ_{min} and $f_{n\ max}$

$$\begin{cases} f_n = \sqrt{\frac{K}{T}} \\ \zeta = \frac{1}{2\sqrt{KT}} \end{cases}$$

$$K_{max} = \frac{f_{n\ max}}{2\zeta_{min}} \qquad (3\text{-}29)$$

As the result of the above calculation, if $K_{min}<K_{max}$, there is a solution, that is, the loop can satisfy the requirements of both suppression and trace characteristics;if $K_{min}>K_{max}$, there is no solution, that is, the loop can't satisfy the suppression and trace characteristics simultaneously. There are two schemes to solve the problem in this case:either modify the given conditions or improve the loop structure, until the condition of $K_{min}<K_{max}$ is satisfied to solve the problem.

As shown in Figure 3-17, the finite region surrounded by K_{min}, ζ_{min} and $\omega_{n\ max}$ on the K-ζ plan is the range of loop elementary parameters. Where, each point represents a group of parameters, i.e. a design scheme. within the range, any value can satisfy both the loop suppression and trace characteristics. If reduce ζ_{min}

further, K_{max} increases then, the jitter gain peak $H(f)_{max}$ increases correspondingly;if f_{max} is increased, K_{max} increases then, however, the width f_0 will increase usually;if reduce K_{min}, the trace error will increases. If the condition of $K_{min}<K_{max}$ still can't be satisfied by modifying the suppression and trace characteristics, improve the loop structure to satisfy $K_{min}<K_{max}$. For example, reduce the K_{min} to satisfy the above condition;reference to Figure 3-18, when $\varphi_\varepsilon \geqslant \varphi_{\varepsilon\max}$, an extra voltage is added to the voltage controlled oscillator, to make the center frequency of the oscillator shift a fixed value, as the result, the loop trace suppression characteristics is partitioned within the range of tributary frequency tolerance. In this way, the loop gain may be reduced obviously, while the phase trace error is still not over the given maximum $\varphi_{\varepsilon\max}$.

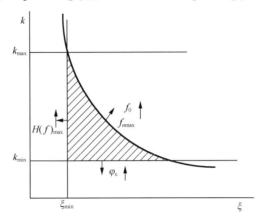

Figure 3-17　Loop elementary parameter range

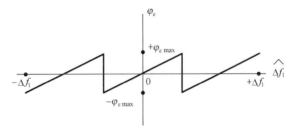

Figure 3-18　Example of reduction of φ_ε

3.10　VOLTAGE CONTROLLED OSCILLATOR DESIGN

(1) Using positive justification technique, the only analog circuits are the DC amplifier and the voltage controlled oscillator in the phase-lock loop, the design of the voltage controlled oscillator will be discussed. Consider the tributary rate of the standard multiplexer is between 2048-13 926kbit/s;the rate tolerance is at the level of 50ppm-10ppm, hence it is suitable to choose crystal voltage controlled oscillator

(XVCO) which is good in both technique and economy. There are usually two kinds of typical XVCO driven by linear amplifier package and driven by switch circuit package. The main advantage of the former is frequency temperature stability, but sometimes its frequency control range is not easy to satisfy the requirement;the later is just contrary,its frequency control range is easy to be satisfied, but the frequency temperature stability is poorer. Hence in engineering design, different techniques should be used for different schemes, to get performance as good as possible.

(2) Frequency control range extension of linear amplifier driven XVCO

The principle of a typical linear amplifier driven XVCO is shown in Figure 3-19. To extend its frequency control range the following main aspects should be considered.

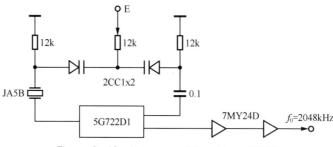

Figure 3-19　Linear amplifier driven XVCO

1. Feedback branch design:the capacity of the feed through capacitor should be bigger than that of the volticap in order to fully use voltage control sensitivity of the volticap;two of them connected back to back are used to conceal the negative feedback by AC detection;

2. Justification of oscillation intensity:on the premise of oscillating in the full frequency band, the oscillation state should be adjusted weak such that the input impedance of the linear amplifier does not change too much, so that it will not affects control sensitivity;if the oscillation is too strong, the impedance will change at different voltage control work state, which partially conceals control function of the volticap;

3. Fully use of the volticap:the work point usually is set at a higher bias voltage, in this case the slope of the volticap can not be fully used;when the work point is set at a positive bias voltage and the bias resistor is smaller(or use color coded inductor), a bigger slope of varying capacitance may be gotten. But consider

the temperature stability, the bias work point should be usually set at a negative bias voltage near zero;

4. Choice of crystal harmonic oscillator:the voltage control of a voltage control oscillator using standard master harmony crystal is usually too small, even using the above methods it may silll not be satisfied. In this case the voltage control oscillator is the best to use voltage control crystal harmonic oscillator. The voltage control crystal has bigger value of C/L, whereas the harmonic crystal has bigger value of $Q = \sqrt{\dfrac{L}{CR}}$;

5. Additional parallel inductor:if it is still not satisfied using the above methods, an additional parallel inductor may be used that the relative frequency offset could be in the level of $n \times 10^{-3}$. But it should be mentioned that the additional parallel inductor is usually not suitable;if the additional parallel inductor has to be used, the criterion of adding parallel inductor must follow to satisfy the control frequency extension on the promise of less sacrifice in stability:

(3) The frequency temperature stability of switch circuit driven XVCO

The principle of a typical switch circuit driven XVCO is shown in Figure 3-20. To improve the frequency temperature stability of the XVCO, the following aspects should be considered.

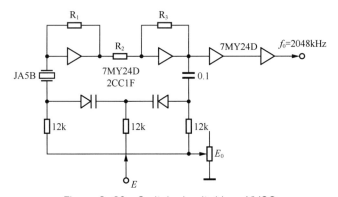

Figure 3-20 Switch circuit driven XVCO

1. Choice of the additional resistors of the switch circuit:to insure the XVCO works normally, each of three resistors has a permitted range. Where, R_1 affects the frequency most greatly;R_3 affects the frequency in the same direction of R_1;R_2 affects the frequency just in contrary to R_1, that is why R_2 is a resistor but a

capacitor. Hence, R_1 should be a high stable resistor;R_2 is a normal resistor but the temperature coefficient is of the same polarity as R_1:R_3 is also a normal resistor but its temperature coefficient is in contrary to the polarity of R_1. Thus, within the XVCO partial temperature comprehensive compensation is realized;

2. Choice of work point of the volticap:the work point of the volticap is set at a negative bias voltage, on the promise of satisfying frequency range reduce the bias current as much as possible to increase temperature stability;

3. Temperature compensation with DC amplifier: as the environment temperature changes the center frequency of the XVCO changes(Δf_i), and the center value of output control voltage of the DC amplifier E also changes, as the result, it makes the center frequency of the XVCO additional change(Δf_2). Justifying the direction of the volticap, to partially conceal Δf_1 using Δf_2, and finally to make the DC amplifier—XVCO a good temperature stability;

4. Frequency temperature compensation:by using the above three methods, generally speaking, the XVCO can satisfy the requirement of the temperature stability. Otherwise, a special temperature stable method may be used. Usually a simple network of thermistor or even a single resistor may satisfy the requirement. The choice of temperature compensation point should be considered, usually it is not suitable to choose the point that is sensitive to frequency. The usual compensation point is at R_3 or E_0, for example. But if it has to, usually it is not suitable to use temperature compensation independently.

3.11 BUFFER SIZE

If the buffer size is too large, the transmission delay time will increase and it is not economic:on the other hand, the buffer size may not be too small, at least it should contain the changes of the read the write time difference caused by the following factors.

(1) The read/write time difference peak-peak changes caused by steady-state bit rate justification:from the justification formula(3-14), get

$$\Delta t_{pp} = \frac{mf_1}{f_h}(g_0 + 1)T_1$$

In the CCITT recommendation, usually $g_0=3$, $\frac{mf_1}{f_h} \approx 1$, hence

Chapter 3 Positive Justification

$$\Delta t_{pp} = 4T_1$$

(2) Phase jitter of the input binary digits: according to CCITT recommendation G.703, it should be as the sample specification as Figure 3-21, i.e. peak-peak jitters is:

$$\Delta t_{app} = 1.5T_1$$

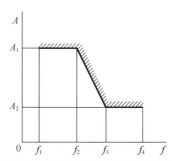

CCITT Recommendation MUX	f_1(kbit/s)	f_1(Hz)	f_2(Hz)	f_3(Hz)	f_4(Hz)	A_1(UI)	A_2(UI)
G.823	64	20	600	3	20	0.25	0.05
G.823	2048	20	2400	18	100	1.5	0.20
G.823	8448	20	400	3	400	1.5	0.20
G.823	34 368	100	1000	10	800	1.5	0.15
G.823	139 264	200	500	10	3500	1.5	0.075
G.743 6312/1544	1544	10	200	8	40	2.0	0.05
G.752 32 064/6312	6312	60	1600	32	160	1.0	0.05
G.752 44 736/6312	6312	10	600	24	120	2.0	0.05
G.752 97 728/32 064	32064	10	800	8	400	2.0	0.20

Figure 3-21 Input jitter template

(3) In the recovery phase-lock loop, because of the trace error peak-peak changes caused by the input tributary frequency offset and the center frequency offset of the voltage control oscillator, the maximum phase offset $\varphi_{max} < 0.3T_i$ is usually designed, hence the trace error peak-peak is:

$$\Delta t_{\varphi pp} = 0.6T_1$$

Because of the superposition relationship of the phase or the read/write time difference caused by the above three factors is shown in Figure 3-22. The W/R time difference peak-peak change is $6.1T_1$, where positive change is $4.05T_1$;the negative change is $1.55T_1$. Hence the initial read write time difference of the buffer is best $3T_1$:thus if there is an extra near $\pm 1T_1$, the buffer size N is 8bits.

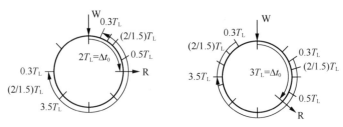

Figure 3-22 Arrangement of buffer size

3.12 CCITT Recommendations

Up to May, 1980, all the group multiplexers, except the Russian series, recommended by CCITT used the positive justification plesiochronous multiplex techniques. The particular recommendations are shown in Figure 3-23.

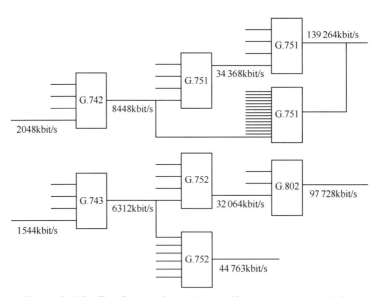

Figure 3-23 The Series of positive justification group multiplex

The group multiplex frame structure of 2048kbit/s rate series is shown in the

Figure 3-24 and Table 3-1. The bit rate justification indicator takes 3 bit or 5 bit; the nominal justification rate is between 0.419-0.439.

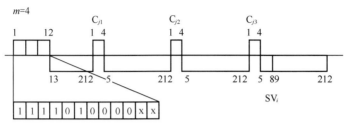
Figure 3-24　(1)MUX(8448/2048×4)(G.742)

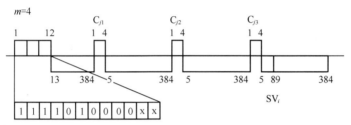
Figure 3-24　(2)MUX(34 368/8448×4)(G.751)

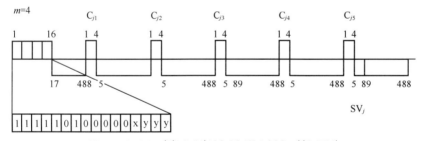
Figure 3-24　(3)MUX(139 264/34 368×4)(G.751)

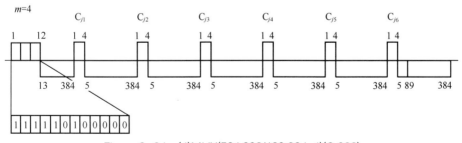
Figure 3-24　(4)MUX(564 992/139 264×4)(G.922)

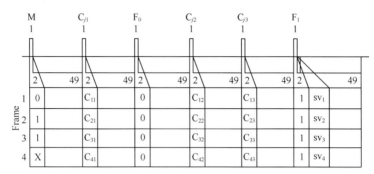

Figure 3-24 (5)MUX(6312/1544×4)(G.743)

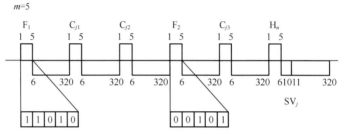

Figure 3-24 (6)MUX(32064/6312×5)(G.752)

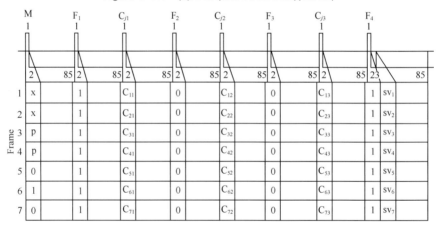

Figure 3-24 (7)MUX(44736/6312×7)(G.752)

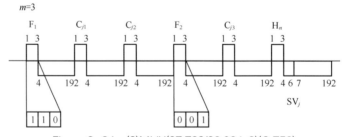

Figure 3-24 (8)MUX(97 728/32 064×3)(G.752)

Table 3-1　　　　The group multiplex frame structure

CCITT Recommendation	G.742	G.751	G.751	G.922	G.743	G.752	G.752	G.752
f_h(kbit/s)	8448	34 368	139 264	564 992	6312	32 064	44 736	97 728
f_L(kbit/s)	2048	8448	34 368	139 264	1544	6312	6312	32 064
$m(-)$	4	4	4	4	4	5	7	3
L_s(bits)	848	1536	2928	2688	1176	1920	4760	1152
Q(bits)	206	378	723	663	288	378	672	378
K(bits)	6	6	9	9	6	6	8	6
f_m(kbit/s)	2052.226	8457.750	34 387.934	139 356.285	1545.796	6312.600	6315.671	32 067.000
f_s(kbit/s)	4.226	9.750	19.934	92.285	1.796	0.600	3.671	3.000
$f_{s\,max}$(kbit/s)	9.962	22.375	47.563	210.190	5.367	16.7	9.398	84.833
$S_0(-)$	0.424	0.436	0.419	0.439	0.334	0.036	0.391	0.035
	(1)	(2)	(3)	(4)	(5)	(6)	(7)	(8)

Chapter 4　Impairment of Positive Justification

In the positive justification of plesiochronous multiplex, by the process of multiplex/demultiplex, two kinds of additional impairment will be introduced into tributary code, i.e. stuffing jitter and stuffing error.

4.1　STUFFING JITTER

4.1.1　Physical Concept of Stuffing Jitter

Jitter is a short time deviation of digital signal from the ideal time position in the effective instants. Normally, jitter peak to peak amplitude A_{jpp} is used to represent the magnitude of a jitter.

The multiplex/demultiplex process introduces jitter into the tributary binary code. From code justification transient curve we know that the read/write time difference in code justification can be decomposed into three independent processes, the read/write time difference changing in period of subframe and caused by insert/abstract process of service & control digit, the read/write time difference changing in period of frame and caused by frame alignment signal, and the read/write time difference caused by signal justification, see Figure 4-2 in detail. A jitter introduced into the tributary binary code by the former two processes is called multiplex jitter. It has definite frequency which is in the same order of frame frequency. For example, the frame frequency is 10Hz, and the passband of phase-locked loop of code recover is in the order of $n \times 10$Hz. So the periodical variation of read write time difference is suppressed at least 40dB, that is, multiplex jitter is actually neglectable.

In Chapter 3, we have discussed the three independent processes, i.e. insert/abstract process of service & control bit, insert/abstract process of frame alignment signal and process of signal justification. As described above, only the process of signal justification is actually effective. Except specially mentioned, the code justification discussed later only denotes signal justification.

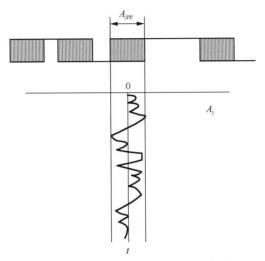

Figure 4-1   Jitter and its representation

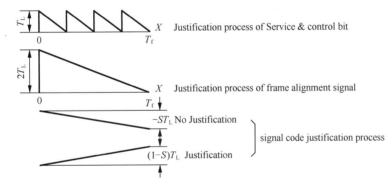

Figure 4-2   Decompose diagram of justification process

As mentioned in the section of parameter design of phase-locked loop in previous chapter, stuffing jitter is the phase ripple of residual part in the output tributary binary code, which is caused by the read/write time difference(Δt_x) and filtered by phase-locked loop in the process of signal justification. Because the residual phase ripple basically is caused by insert/abstract process, we call it stuffing jitter. Since signal justification process is considerably slowly, i.e. its justification period is quite long, code recover phase-locked loop is difficult to suppress the stuffing jitter. Hence the signal justification process directly decides the characteristics of stuffing jitter. To analyze the stuffing jitter, we must analyze signal justification process at first.

4.1.2 Signal justification process justifying only q times in p frames

In Chapter 3, we specified a code justification rule: when the read/write time difference is equal to or lower than justification threshold, the following code justification bit will be justified; and specified that: there is only one possible code justification bit in a frame, only one idle pulse is inserted when a justification happens. As mentioned above, when justification has happened, the influence of insert/abstract bit is not considered, i.e. except the code justification bit all binary codes are uniform.

When we discuss about the process of code justification, y is the sequential number of code justification times, Δt_y is the read/write time difference at the y time. In this section we only discuss about the basic case of signal justification process justifying only q times in p frames, that is, the p and q are relative prime, and ratio of code justification is equal to:

$$S=q/p, (q,p)=1 \qquad (4\text{-}1)$$

According to the formula (3-10) and (3-11) in chapter 3, in the process of code justification after every frame, if justification has not happened, read write time difference will be abstracted ST_1; otherwise, read write time difference will be added $(1-S)\,T_1$. According to this justification rule, when the read/write time difference is just equal to the justification threshold(Δt_s) in the justificalion time and it makes a justification immediately, at the end of the justification the read write time difference will be the possible maximum value, hence

$$\Delta t_{y\max} = \Delta t_s + T_1 \quad \Delta t_{y\max} = \Delta t_s + T_1 T_1$$
$$\left.\frac{\Delta t_y}{T_1}\right|_{\max} = \frac{\Delta t_s}{T_1} + 1 \qquad (4\text{-}2)$$

If at the start time ($y=0$), the read/write time difference start value is just equal to the its maximum possible value, then at the y time the formula of the read/write time difference will be

$$\frac{\Delta t_y}{T_1} = \left(\frac{\Delta t_s}{T_1} + 1\right) - (Sy - [Sy]) \qquad (4\text{-}3)$$

where $[Sy]$ is the integer part of Sy, i.e. every time when Sy increases to an integer a justification will happen, in the other word, whenever Δt_y decreases to or lower than Δt_s, a justification will happen, so $Sy-[Sy]$ is the decimal part of Sy. Then the formula (4-3) can be rewritten as

$$\left(\frac{\Delta t_s}{T_1}+1\right)-\frac{\Delta t_y}{T_1}=(Sy-[Sy]) \qquad (4\text{-}4)$$

$$\left(\frac{\Delta t_s}{T_1}+1\right)-\frac{\Delta t_y}{T_1}=\frac{qy}{p}-\left[\frac{qy}{p}\right]$$

From Figure 4-3, when $y=0$, both sides of formula (4-4) are equal to zero, the read/write time difference will be the maximum possible value. When next time the read/write time difference gets the maximum value, i.e. the second time both sides of formula is equal to zero.

$$\frac{qy}{p}-\left[\frac{qy}{p}\right]=0, \frac{qy}{p}=\left[\frac{qy}{p}\right] \qquad (4\text{-}5)$$

In formula (4-5), since $q \neq p$, then

$$y = p \qquad (4\text{-}6)$$

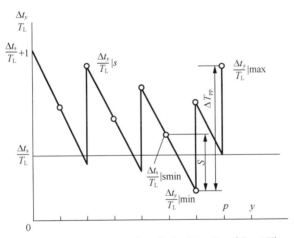

Figure 4-3 Process of code justification ($S = 4/7$)

The formula (4-6) describes that after p frames, process of code justification goes back to its initial state, i.e. the read/write time difference is equal to its maximum uadue again. It shows that period of process of code justification (T_p) is equal to p times frame period, or the frequency of process of code justification (F_p) is equal to frame frequency T_s divided by p,

$$T_p = pT_s, \ F_p = F_s/P \quad (4\text{-}7)$$

After justified at all possible code justification bits, the discrete expression of the read/write time difference is

$$\left(\frac{\Delta t_s}{T_1} + 1\right) - \frac{\Delta t_y}{T_1}\bigg|_s = \frac{qy}{p} - \left[\frac{qy}{p}\right] \quad (4\text{-}8)$$

Because in formula (4-8) y is a positive integer which is less than or equal to p; p and q is relative prime each other:

$$y=1,2,3,...p, \ (p,q)=1$$

So $\left(\dfrac{qy}{p} - \left[\dfrac{qy}{p}\right]\right)$ is integer multiple of $1/p$, i.e. the following relationship exists:

$$\frac{qy}{p} - \left[\frac{qy}{p}\right] = \frac{m}{p}, \ m=0,1,2...(p-1)$$

$$\therefore \frac{\Delta t_y}{T_1}\bigg|_s = \left(\frac{\Delta t_s}{T_1} + 1\right) - \frac{m}{p} \quad (4\text{-}9)$$

The formula (4-9) describes that after code justification, the read/write time difference may has m discrete values. Obviously, when $m=(p-1)$, it will get the minimum value:

$$\frac{\Delta t_y}{T_1}\bigg|_{s\min} = \left(\frac{\Delta t_s}{T_1} + 1\right) - \frac{p-1}{p}$$

$$\frac{\Delta t_y}{T_1}\bigg|_{s\min} = \frac{\Delta t_s}{T_1} + \frac{1}{p} \quad (4\text{-}10)$$

The formula (4-10) describes that after justification (if it has already done justification once), the minimum value of the read/write time difference in every possible code justification bit is at the level of $\dfrac{1}{p}T_1$ over the justification threshold.

Obviously, at this level after one more frame, i.e. the read/write time difference continuously decreases $\dfrac{q}{p}T_1$, it certainly will intersect with threshold and then a justification will happen. Hence, before the following justification, the read/write time difference certainly gets its minimum value:

$$\frac{\Delta t_y}{T_1}\bigg|_{s\,min} = \frac{\Delta t_y}{T_1}\bigg|_{s\,min} - S$$

$$= \frac{\Delta t_s}{T_1} - \frac{q-1}{p} \tag{4-11}$$

$$\Delta t_{y\,min} = \Delta t_s - \frac{q-1}{p} \cdot T_1$$

That is, in the process of code justification, the minimum read/write time difference is $\frac{q-1}{p}T_1$ smaller than the threshold.

The maximum range of the read/write time difference caused by the code justification process, i.e. its peak to peak value, is:

$$\Delta T_{pp} = \Delta t_{y\,max} - \Delta t_{y\,min}$$

$$= (\Delta t_s + T_1) - \left(\Delta t_s - \frac{q-1}{p} \cdot T_1\right)$$

$$= \left(1 + \frac{q-1}{p}\right)T_1 \tag{4-12}$$

$$\Delta T_{pp} = \left(1 + \frac{q-1}{p}\right)T_1$$

i.e. in the process of code justification the peak to peak value is $\left(1+\frac{q-1}{p}\right)T_1$, the frequency is F_s/p. A typical process of code justification is shown in the Figure 4-4.

In a physical equipment, at each code justification bit, the resolution of the read/write time difference is limited, i.e. the real quantization level(m) of the read write difference (Δt_y) is limited, hence the maximum period of the code justification is also limited. When the maximum value of m is 21, then P_{max} is 20. For example, the frame frequency F_s=10kHz; the code recover parameters of phase-locked loop are: f_0=30Hz, L=12dB/oct. When $p=P_{max}$=20 and s=3/7, from formula we can find out code justification envelop frequency(F_p)and amplitude envelop(A_p):

$$F_p = \frac{F_s}{p} = 500\text{Hz}$$

$$A_p = \frac{q-1}{p} \cdot T_1 = \frac{2}{7}T_1$$

$$L(F_p) \approx 48\text{dB}$$

$$20\lg\frac{A_p}{A_j} = L(F_p)$$

$$A_j \approx \frac{A_p}{250} \approx \frac{T_1}{1000}$$

It is obviously that when p is smaller than $P_{max}=20$ and $S=q/p$, the actual stuffing jitter A_j of code justification is smaller than the frictional percentage points of T_1. In such case, the influence of stuffing jitter of code justification is neglectable from point view of engineering.

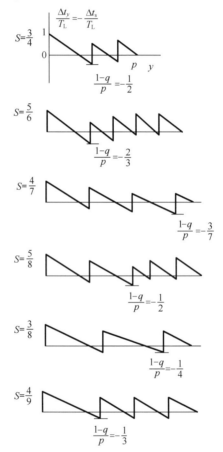

Figure 4-4 An example of code justification

4.1.3 Justifying *q* times in *p* frames with a residue

The residue here means after q times of justification in p frames there is a small (positive or negative) residue of read write time difference. We can describe such a case as follows: the ratio of code justification (S) is equal to the (q/p) plus a residue (Δs). Where Δs can be expressed by two positive relative prime integers and satisfy the following conditions:

$$s = \frac{q}{p} + \Delta s \qquad (4\text{-}13)$$

$$\Delta s = \frac{b}{a}; (q,p) = 1; (b,a) = 1; (p,a) = 1;$$

$$\Delta s \ll \frac{q}{p}; a \gg p$$

We call divide the process of code justification into two independent processes, the process decided by $S=q/p$ and the process decided by $\Delta s=b/a$. Then describe them as a unified process. In the above section, the first process has already been discussed, it has almost no influence on the stuffing jitter. In the following we discuss about the influence of the second process.

Consider the influence of Δs, according to the formula (4-9), at the yth justification point the general formula of the read write time difference call be rewritten into the following:

$$\frac{\Delta t_y}{T_1} = \left(\frac{\Delta t_s}{T_1} + 1\right) - \frac{m}{p} - \Delta S y; \qquad (4\text{-}14)$$

$$m = 0,1,2...(p-1)$$

That is, consider the influence of Δs, the read/write time difference of the yth justification point will decrease $\Delta S y \cdot T_1$ more than that in the case of $\Delta s=0$. The absolute value of this additional item will increase to:

$$|\Delta S y| = \frac{1}{p}$$

$$\frac{\Delta t_y}{T_1} = \left(\frac{\Delta t_s}{T_1} + 1\right) - \frac{m+1}{p}; m=0,1,2...(p-1) \qquad (4\text{-}15)$$

At that time, the process of code justification will add an additional process. Hence, the average time between the additional justifications, i.e. the period of additional justifications (T_p'), can be calculated according to the formula (4-15):

$$|\Delta S| y' = \frac{1}{p}$$

$$y' = \frac{1}{|\Delta s| \cdot p} \quad (4\text{-}16)$$

$$T'_p = y' T_s = \frac{T_s}{|\Delta S| \cdot p}$$

$$F'_p = p \cdot |\Delta S| \cdot F_s \quad (4\text{-}17)$$

Formula (4-17) shows the frequency of additional justification caused by deviation of the justification ratio (Δs). It is equal to the product of p in the code justification ratio, deviation (Δs) and frame frequency(F_s).

Hence as long as the additional read write time difference $\Delta s \cdot y$ increases to $\frac{1}{p} T_1$ an additional justification then happens, hence the envelop amplitude (A'_p) caused by the additional justiflcation is equal to $\frac{1}{p} T_1$, that is

$$A'_p = \frac{T_1}{p} \quad (4\text{-}18)$$

Formula (4-18) shows that the envelop amplitude of additional justification is equal to the width of tributary binary element T_1 divided by p. From the result of the last section, $p < p_{max} = 20$.

Figure 4-5 is the justification process curve drawn strictly according to the code justification specification. In the figure, it shows two kinds of procsses: the $S=1/3$ and the $S = \frac{1}{3} + \frac{1}{31}$. In the process of $S=1/3$, after $p=3$ frames a justification happens and strictly recovers to the initial state of justification process i.e. the period of frames is regular. In the process of $S = \frac{1}{3} + \frac{1}{31}$, at the beginning a justification happens in every $p=3$ frames and will not recover to the initial state, then the residual time difference regularly increases. when it is accumulated till the 11th frame, a justification happens in advance, i.e. an irregular point appears. When $\Delta S > 0$, a justification happens in advance, whereas when $\Delta S < 0$, it will be delayed. Figure 4-5 also shows that calculating results of the formula (4-17) and (4-18) are different from the strict results of regular justification. About the period of the Irregular points, the calculating result is $T'_p = \frac{T_s}{\Delta S \cdot P} \approx 10 T_s$, and the strict result is $T_p'' = 11T_s$; about the envelop amplitude, the calculating result is $A_p' = \frac{1}{3}T_1$, and the strict result is $A_p'' = \frac{9}{31}T_1$. Hence we see that formula (4-17) and (4-18) are approximate.

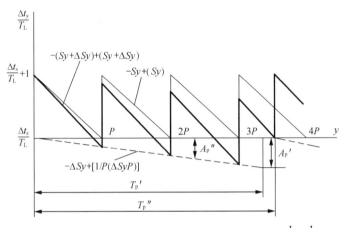

Figure 4-5 The process of code justiffcation ($S = \frac{1}{3} + \frac{1}{31}$)

Because the relationship between q/p and Δs is indefinite, the process caused by $S=q/p$ and the process caused by $\Delta s=b/a$ are independent each other. When Δs takes difference values, the additional justification possibly appears in different justification points in the range of pT_1. Considering that p and q are relative prime, hence y' is not integral times of p, the additional justification points can not coincide with the regular justification points. In such a case, the calculation error from formula (4-17) is less than one regular justification period, i.e. the absolute calculation error is,

$$\Delta T_p' < pT_s \qquad (4\text{-}19)$$

and the relative calculation error is

$$\frac{\Delta T_p'}{T_p'} < \frac{PT_s}{\frac{T_s}{|\Delta S| \cdot P}} = |\Delta S| \cdot p^2 \qquad (4\text{-}20)$$

In the same way, from the code justification envelop calculation formula (4-18) the boundary of absolute calculation error and the boundary of relative calculation error are as the following respectively:

$$\Delta A'_p < |\Delta S| \cdot pT_1 \quad (4\text{-}21)$$

$$\frac{\Delta A'_p}{A'_p} < \frac{|\Delta S| \cdot pT_1}{\dfrac{T_1}{p}} = |\Delta S| \cdot p^2 \quad (4\text{-}22)$$

When $p<P_{max}=20$ and $\Delta s=1\times10^{-4}$, from formula (4-17) and formula (4-18) the relative calculating error are 4%. Hence, when Δs is rather small, the results from formula (4-17) and (4-18) are reasonable approximate values for engineering; but when Δs increases, the calculating error will increase proportionally. In the example of figure (4-5), the relative calculating error is 29%. Obviously, it is too big, but for the purpose of illustration clearly in the figure, we take such value.

To sum up, the envelop amplitude and its frequency of the read write time difference caused by irregular justification are,

$$A_p = \frac{T_1}{p}$$

$$F_p = p \cdot |\Delta S| \cdot F_s$$

where, T_1 is the width of tributary binary element(UI), F_s is the frame frequency, P_{max} is a positive integer not exceeding 20, $|\Delta s|$ is a small value near zero. Hence, A_p is smaller than T_1, F_p is possibly near zero. Since F_p is possibly far smaller than the passband (f_0) of phase-locked loop of code recover, the main frequency spectral lines of read write time difference envelop can pass the passband of the loop. That is, the envelop changes will be introduced into the tributary binary code completely, then stuffing jitter happens. In this case, the peak to peak amplitude (A_{jpp}) and jitter frequency introduced into the tributary binary code are respectively:

$$A_{jpp} = \frac{1}{p} \cdot UI \quad (4\text{-}23)$$

$$F_j = p|\Delta S|F_s \quad (4\text{-}24)$$

When $S = \dfrac{p}{q} + \Delta S$ and Δs approaches zero, formula (4-23) and (4-24) show stuffing jitter introduced by the justification of tributary binary code. Since Δs may be either positive or negative, Δs approaches zero no matter from positive or negative direction, the formula (4-23) and (4-24) are both valid. Obviously, when $\Delta s=0$ (i.e. S=q/p), stuffing jitter A_{jpp} is also zero.

Chapter 4 Impairment of Positive Justification

To sum up the above discussion, we get the formula of maximum peak to peak envelop amplitude of positive justification stuffing jitter:

$$A_{jpp} = \frac{1}{p} \cdot UI$$

$$S = \frac{q}{p}$$

$$(q,p)=1$$

These formulas mean that: the peak to peak amplitude (A_{jpp}) of positive justification stuffing jitter is a function of code justification ratio (S) only; parameters p and q (relative prime each other) specify the possible approximate position along the axis S, and the possible discrete value of jitter peak along the A_{jpp} axis. Because directly using these formulas is not convenient, it is necessary to find a direct showing method.

Now we introduce a simple illustration method: from the basic formula it is easy to find a set of equation to make the illustration:

$$A_{jpp} = \frac{1}{p} \cdot S$$

$$A_{jpp} = \frac{1}{p-q} \cdot (1-S)$$

Using these equations and two dimension space illustration made from effective p and q, we can get the maximum peak value A_{jpp} according to a specific S. Connecting these peak points we can get peak envelop curve $A_{jpp}=F(s)$. Calculating values of envelop curve are shown in Table 4-1, and its curve in Figure 4-6.

Table 4-1 $A_{jpp}=F(S)$ calculating table

p	Q	S	$A_{jpp} = \frac{1}{q} S$	$A_{jpp} = \frac{1}{p-q}(1-S)$	A_{jpp}(UI)
1	/	1			1
2	1	1/2	$A = S$	$A = 1-S$	0.500
3	1	1/3	$A = S$	$A = \frac{1}{2}(1-S)$	0.333
3	2	2/3	$A = \frac{1}{2}S$	$A = 1-S$	0.333
4	1	1/4	$A = S$	$A = \frac{1}{3}(1-S)$	0.250
4	3	3/4	$A = \frac{1}{3}S$	$A = 1-S$	0.250

continued

p	Q	s	$A_{jpp} = \dfrac{1}{q}S$	$A_{jpp} = \dfrac{1}{p-q}(1-S)$	$A_{jpp}(UI)$
5	1	1/5	$A = S$	$A = \dfrac{1}{4}(1-S)$	0.200
	2	2/5	$A = \dfrac{1}{2}S$	$A = \dfrac{1}{3}(1-S)$	
	3	3/5	$A = \dfrac{1}{3}S$	$A = \dfrac{1}{2}(1-S)$	
	4	4/5	$A = \dfrac{1}{4}S$	$A = 1-S$	
6	1	1/6	$A = S$	$A = \dfrac{1}{5}(1-S)$	0.167
	5	5/6	$A = \dfrac{1}{5}S$	$A = 1-S$	
7	1	1/7	$A = S$	$A = \dfrac{1}{6}(1-S)$	0.143
	2	2/7	$A = \dfrac{1}{2}S$	$A = \dfrac{1}{5}(1-S)$	
	3	3/7	$A = \dfrac{1}{3}S$	$A = \dfrac{1}{4}(1-S)$	
	4	4/7	$A = \dfrac{1}{4}S$	$A = \dfrac{1}{3}(1-S)$	
	5	5/7	$A = \dfrac{1}{5}S$	$A = \dfrac{1}{2}(1-S)$	
	6	6/7	$A = \dfrac{1}{6}S$	$A = 1-S$	
8	1	1/8	$A = S$	$A = \dfrac{1}{7}(1-S)$	0.125
	3	3/8	$A = \dfrac{1}{3}S$	$A = \dfrac{1}{5}(1-S)$	
	5	5/8	$A = \dfrac{1}{5}S$	$A = \dfrac{1}{3}(1-S)$	
	7	7/8	$A = \dfrac{1}{7}S$	$A = 1-S$	
9	1	1/9	$A = S$	$A = \dfrac{1}{8}(1-S)$	0.111
	2	2/9	$A = \dfrac{1}{2}S$	$A = \dfrac{1}{7}(1-S)$	
	4	4/9	$A = \dfrac{1}{4}S$	$A = \dfrac{1}{5}(1-S)$	
	5	5/9	$A = \dfrac{1}{5}S$	$A = \dfrac{1}{4}(1-S)$	
	7	7/9	$A = \dfrac{1}{7}S$	$A = \dfrac{1}{2}(1-S)$	
	8	8/9	$A = \dfrac{1}{8}S$	$A = 1-S$	

Chapter 4 Impairment of Positive Justification

continued

p	Q	s	$A_{\text{jpp}} = \dfrac{1}{q}S$	$A_{\text{jpp}} = \dfrac{1}{p-q}(1-S)$	$A_{\text{jpp}}(\text{UI})$
10	1	1/10	$A = S$	$A = \dfrac{1}{9}(1-S)$	0.100
	3	3/10	$A = \dfrac{1}{3}S$	$A = \dfrac{1}{7}(1-S)$	
	7	7/10	$A = \dfrac{1}{7}S$	$A = \dfrac{1}{3}(1-S)$	
	9	9/10	$A = \dfrac{1}{9}S$	$A = 1-S$	

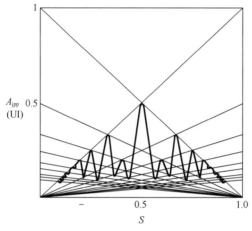

Figure 4-6 Stuffing calculating value $A_{jpp}=F(S)$

In the Table (4-1) it is only calculated to the $p \leqslant 10$, the corresponding curve in Figure (4-6) is also only to the $A_{\text{jpp}} \geqslant 10\% \text{UI}$. It is enough for the illustration, but still insufficient for engineering. Figure 4-6 also shows that curve $A_{\text{jpp}}=F(S)$ is symmetrical to axis S=0.5. The precise envelop curve in engineering applications use normally only the half side S=0-0.5(see Figure 9-1 and Table 9-1).

4.1.4 The relationship between the number of code justification detectors and stuffing jitter

In the discussion about stuffing jitter in the previous two sections, we have to emphasize one point in detail:the mechanism of code justification must continuously monitor the change of the read write time difference, i.e. the signal of justification once appears, then the signal is confirmed simultaneously, and justification is executed immediately in the next frame. The conclusion of stuffing jitter above discussed assumes that the detection of the read write time is without delay. To fulfill the assumption in the equipment, the number of detectors (k) have to be equal to the bits of buffer(N). In the following we will first discuss the relationship between number of detectors (k) and the detection delay of the read write time difference (τ); and then the relationship between the detection delay (τ); and stuffing jitter (A_{jpp}).

The principle diagram of the read write time difference detector is shown in the Figure 4-7. The tributary clock (f_1) drives the modulo N writing circuit to generate N writing pulses (W_i). The N writing pulses are pushed into the buffer and the detector simultaneously. In the same way, the multiplex clock (f'_m) drives the modulo N reading circuit to generate N reading pulses (R_i). The N reading pulses are pushed into the buffer and the detector simultaneously. There is at least one detector, or k detectors, but maximum N in the circuit. No matter how many detectors are used, the detection function can be fulfilled. Detection delay of different number of detectors (k) is shown in the Figure 4-8. For different k detectors, there is a delay from the request of justification, i.e. the pulse W_i coincide with the pulse R_i, to the time when the detectors find the signal to output a control pulse (P), it is the detection delay(τ_k). In the example of Figure 4-8,

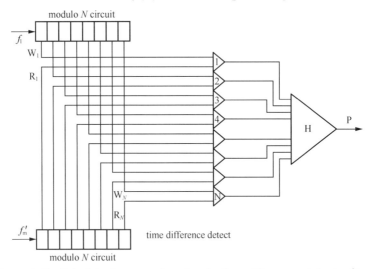

Figure 4-7 Principle diagram of read write time difference detector ($k=N$)

when one detector ($k=1$) is used, the maximum detection delay is ($N-1$)T_1, whereas N detectors ($K=N$) are used, the detection delay is zero. In general, k detectors (k is a factor of N) are used and uniform distributed, the general formula of detection delay is:

$$\tau_k = \left(\frac{N}{k} - 1\right)T_1 \quad (4\text{-}25)$$

In the following we discuss the relationship between the detection delay and stuffing jitter. From Figure 4-9 in a typical case, a justification request occurs already (the pulse W_i coincide with the pulse R_i) at the time $\frac{\tau_s}{T_1}$ before the end of a frame, because there is a detection delay $\frac{\tau_s}{T_1}$ big enough, hence me read write time difference curve will not intersect with the justification threshold until the end of the frame. Assume at this time the request is not confirmed, then in the next frame, the justification request will be started $\frac{\tau_s}{T_1}$ earlier. Hence when the detection delay occurs, justification request area will extend to:

$$D = d + \left(\frac{N}{k} - 1\right) \quad (4\text{-}26)$$

In this case, after the frame, the maximum of the read write difference below the justification threshold is:

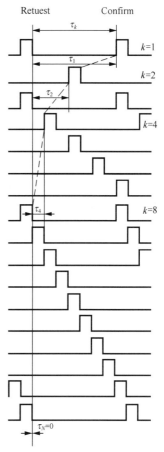

Figure 4-8　Sketch diagram of detecting delay [$\tau_k = F(K)$]

$$\Delta t_x - \Delta t_0 = -D\left(1 - \frac{mf_1}{f_h}\right)T_1$$

$$= -\left[d + \left(\frac{N}{k} - 1\right)\right]\left(1 - \frac{mf_1}{f_h}\right)T_1 \quad (4\text{-}27)$$

$$= -ST_1\left[1 + \frac{1}{S}\left(\frac{N}{k} - 1\right)\left(1 - \frac{mf_1}{f_h}\right)\right]$$

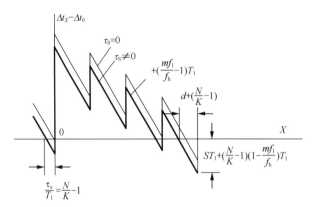

Figure 4-9 Relationship between detection delay and stuffing jitter

From the first two sections of the chapter, the amplitude of stuffing jitter is in direct proportion to the quantity of which the read write time difference is below the justification threshold. That is, when the envelop frequency when the read write time difference changes is far smaller than the cut-off frequency of code recover phase-locked loop, the amplitude of stuffing jitter is in direct proportion to the envelop amplitude of the read write time difference. If the peak to peak amplitude of stuffing jitter without detection delay ($\tau_s=0$) is A_{jpp}, the peak to peak amplitude of stuffing jitter with detection delay ($\tau_s \neq 0$) is:

$$\frac{A_{\text{jpp}}^k}{A_{\text{jpp}}} = \frac{ST_1\left[1+\frac{1}{S}\left(\frac{N}{k}-1\right)\left(1-\frac{mf_1}{f_h}\right)\right]}{ST_1} \quad (4\text{-}28)$$

$$\therefore A_{\text{jpp}}^k = \left[1+\frac{1}{S}\left(\frac{N}{k}-1\right)\left(1-\frac{mf_1}{f_h}\right)\right] \cdot \frac{1}{p} \cdot UI$$

Formula (4-28) is the general formula of stuffing jitter using K detectors of read write time difference. In the formula the ratio of code justification is $S=\frac{q}{p}$, N is stages of the buffer, m is the number of multiplexed tributary binary, f_1 is the tributary binary rate, f_h is the multiplex rate. Obviously, when $k=N$, $A_{\text{jpp}}^k = A_{\text{jpp}}$ is the result of above discussion.

For example, from recommendation G.742 of CCITT, f_1=2048kbit/s, f_h=8448kbit/s, m=4, when S=3/7, N=8, the calculation results are the following:

k	A_{jpp}^k (%UI)
1	21.5
2	17.4
4	15.3
8	14.3

4.1.5 Distribution of stuffing jitter

From the envelop curve $A_{jpp}=F(S)$ we can see that the stuffing jitter A_{jpp} is the function of code justification ratio (S) only. If S takes different values, the value of A_{jpp} will change rapidly. Code justification ratio(S)is the function of the tributary code rate(f_l), the multiplex rate(f_h), and the structure parameters (the frame length L_f and the number of the tributary bits Q). When the structure of frame is known. L_f and Q are definite values also. The value of S is dependent on f_l and f_h only. If f_l and f_h are known, then the S is known also. On the=F(S) curve we can find A_{jpp}.

Actually, the tributary code rate (f_l) and the multiplex rate (f_h) are changing within the tolerance ranges (Δf_l, Δf_h) about each own nominal values respectively:

$$f_1 = f_{lo} \pm \Delta f_1$$
$$f_h = f_{ho} \pm \Delta f_h$$
(4-29)

Because each pair of f_l and f_h has its corresponding S, hence in the two dimension range of tolerances of f_l and f_h, S has a set of values. It holds the range of S between S_{max} and S_{min}, hence correspondingly the stuffing jitter (A_{jpp}) also has a set of values. Since $A_{jpp}=F(S)$ changes rapidly following S, hence the difference between peaks is quite large, and it is not easy to design. In this case, distribution diagram of stuffing jitter is very useful.

The distribution of stuffing jitter is the distribution of A_{jpp} in the two dimension plan of $\Delta f_l/f_l$, $\Delta f_h/f_h$. The stuffing jitter is a three dimension space model. Actually, it is not necessary to do so, normally it uses the method of the contour on a map to describe such a stuffing jitter distribution. From the basic formula of positive justification:

$$S = Q - \frac{f_1}{f_h} \cdot L_s; f_1 = \frac{Q-S}{L_s} \cdot f_h; f_{lo} = \frac{Q-S}{L_s} \cdot f_{lo}$$

$$\therefore \frac{f_1 - f_{lo}}{f_{lo}} = \left(\frac{Q-S}{Q-S_0}\right)\left(\frac{f_h - f_{ho}}{f_{ho}}\right) - \frac{S-S_0}{Q-S_0}$$
(4-30)

where, the f_{lo}, f_{ho}, S_0 are the nominal of f_l, f_h, S. Usually $Q \gg S_0$, $S-S_0 \ll Q-S_0$, then the formula can be simplified:

$$\frac{f_l - f_{lo}}{f_{lo}} \approx \frac{f_h - f_{ho}}{f_{ho}} - \frac{S - S_o}{Q} \quad (4\text{-}31)$$

Let $f_l - f_{lo} = \Delta f_l$, $f_h - f_{ho} = \Delta f_h$, and substitute into the formula (4-31), we can get:

$$\frac{\Delta f_l}{f_l} \approx \frac{\Delta f_h}{f_h} + \frac{S_o - S}{Q} \quad (4\text{-}32)$$

Formula (4-31) is the general formula of stuffing jitter. According to this formula, we can get distributed S in the $\Delta f_l/f_l$, $\Delta f_h/f_l$ plan; and according to the $A_{jpp}=F(S)$ we can find the corresponding distributed A_{jpp}. Figure 4-10 is a typical distribution diagram of stuffing jitter.

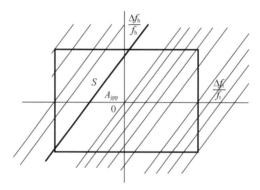

Figure 4-10 Stuffing jitter distribution

Generally, only after final regulating of distributed stuffing jitter, we may complete a overall design. If in the result of primary design we find there are large peaks of stuffing jitter in the two dimension of tolerance range, which do not permit to be taken; then we can appropriately regulate justification tolerance to avoid the maximum peak A_{jpp}. Regulated tolerance will not any essential influence to the any other characteristics of code justification. If the nominal of the tributary rate and the multiplex rate and their tolerances are specified already, and in the primary design we find stuffing jitter peak is not permitted in the tolerance domain, then we need modify parameters to redesign it until the criterion is satisfied. If the code rate and its tolerance are given, and parameters of equipment are selected already, then from the distributed diagram of stuffing jitter we can evaluate if the design is good or not. For example, from the recommendation of G.751, the

Chapter 4 Impairment of Positive Justification

34 368/139 264kbit/s code justification design is not a good design[19].

The table of code justification design parameters recommended by CCITT is shown in the Table 4-2, and distributed diagram of multiplex jitter which corresponds with Table 4-2 is in Figure 4-11.

Table 4-2   multiplex parameters recommended by CCITT

CCITT Rec	G.742	G.751	G.751	G.922	G.743	G.752	G.752	G.752
f_h (kbit/s)	8448	34 368	564 992	564 992	6312	32 064	44 736	97 728
$\dfrac{\Delta f_h}{f_h}$ (ppm)	30	20	15	15	30	10	20	10
f_l (kbit/s)	2048	8448	34 368	139 264	1544	6312	6312	32 064
$\dfrac{\Delta f_l}{f_l}$ (ppm)	50	30	20	15	50	30	30	10
L_s (bit/s)	848	1536	2928	2688	1176	1920	4760	1152
Q (bit/s)	206	378	723	663	288	378	672	378
S_0 (-)	0.424 24	0.435 75	0.419 12	0.439 06	0.334 60	0.035 92	0.390 56	0.035 36
ΔS_{max} (-)	0.016 48	0.018 90	0.025 31	0.019 89	0.023 04	0.015 12	0.033 60	0.007 56
S_{max} (-)	0.440 72	0.454 65	0.444 43	0.458 95	0.357 64	0.051 05	0.424 16	0.042 92
S_{min} (-)	0.407 76	0.416 85	0.393 81	0.419 17	0.311 56	0.020 81	0.356 96	0.027 80
A_{jppmax} (UI%)	14.3	14.3	20.0	14.3	33.3	5.0	20.0	4.0
S' (-)	0.4286	0.4286	0.4000	0.4286	0.3333	0.0500	0.4000	0.0400
	(1)	(2)	(3)	(4)	(5)	(6)	(7)	(8)

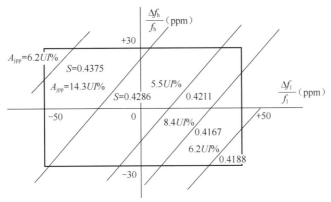

Figure 4-11 (1) MUX (564 992/139 264×4kbit/s)

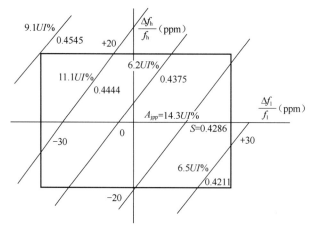

Figure 4-11　(2) MUX (34 368/8448×4kbit/s)

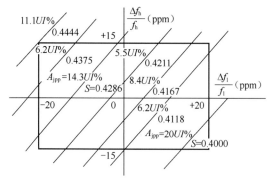

Figure 4-11　(3) MUX (139 264/34 368×4kbit/s)

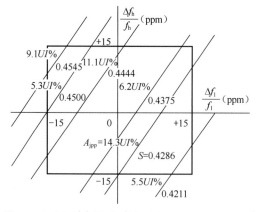

Figure 4-11　(4) MUX (564 992/139 264×4kbit/s)

Chapter 4 Impairment of Positive Justification

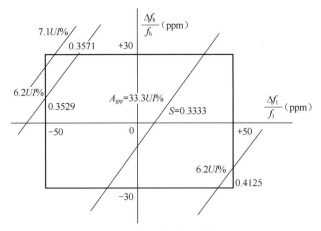

Figure 4-11　(5) MUX (6312/1544x4kbit/s)

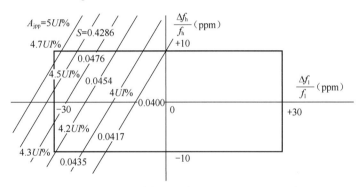

Figure 4-11　(6) MUX (32 064/6312x5kbit/s)

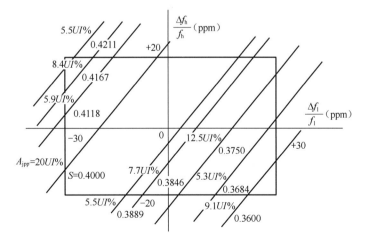

Figure 4-11　(7) MUX (44 763/6312x7kbit/s)

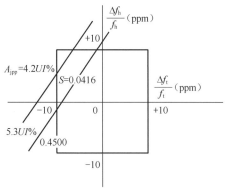

Figure 4-11 (8) MUX (97 728/32 064×3kbit/s)

4.2 STUFFING ERROR

4.2.1 Physical Concept of Stuffing Error

Stuffing error is an additional error, which is caused by mistake of code justification and introduced into the tributary binary code. See Figure 4-12, because there has channel error in the nth order transmission, it possibly produces an error in code justification indicating signal, and in turn causes a wrong operation of code justification recover, misunderstanding stuffing code as a signal code, or misunderstanding signal in SV as a stuffing bit. This causes a bit slip in the $(n-1)$th order binary code, then the $(n-1)$th demultiplexer gets a frame loss. During frame loss, the $(n-1)$th order demultiplex binary code gets an additional error code, i.e. the stuffing error.

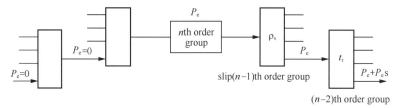

Figure 4-12 Relation diagram of stuffing error

4.2.2 Calculation of stuffing error

4.2.2.1 Average time between stuffing errors in the demultiplexer of *n*th order group

When stuffing indicating code is η' bits, tolerance is d' bits, the probability of stuffing errors in demultiplexer is:

$$P_s = \sum_{x=d'+1}^{\eta} C_{\eta'}^x (1-P_e)^{\eta'-x} \cdot P_e^x$$

The average stuffing errors per unit time in the nth multiplexer is:

$$\rho_s = F_s' \cdot P_s$$
$$= F_s' \sum_{x=d'+1}^{\eta'} C_{\eta'}^x (1-P_e)^{\eta'-x} \cdot P_e^x \quad (4\text{-}33)$$

When the error rate of the nth order is $P_e = 1 \times 10^{-3}$, formula (4-33) can be simplified as following:

$$\rho_s \approx F_s' C_{\eta'}^{d'+1} \cdot P_e^{d'+1} \quad (4\text{-}34)$$

$$t_s = \frac{1}{\rho_s} \approx T_s' / C_{\eta'}^{d'+1} \cdot P_e^{d'+1} \quad (4\text{-}35)$$

Formula (4-35) is the average time between stuffing errors in the nth order demultiplexer. Where T_s', η', and d' are the period of the nth order frame, the bit length of flame alignment signal and the number of the tolerance bits respectively.

For example: from the recommendation G.732 of CCITT, when $\eta'=3$, $d'=1$, $T_s' = 100\mu s$, the calculation result t_s is in the Figure 4-13.

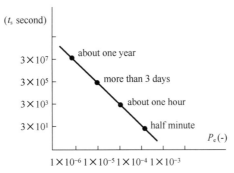

Figure 4-13　Average time between stuffing errors of Recommendation G.742 of CCITT

4.2.2.2　Average time of frame loss in the ($n-1$)th order demultiplexer

Since stuffing operating errors appear in the nth order binary code, the ($n-1$)th order binary code will produce a slip. The search process of ($n-1$)th order demultiplexer is to find firstly and then to confirm a stuffing frame loss which really has happened; then searches and finally confirms that synchronous state is really recover or not (See Figure 4-14). Average frame loss time is:

$$t_r = t_d + t_a + t_w \qquad (4\text{-}36)$$

where, t_d, t_a and t_w are the average frame loss deciding time, the average search time and the average synchronization deciding time respectively. Its approximate formula is:

$$t_d \approx \beta T_s ; t_a \approx \frac{1}{2} T_s ; t_w \approx (a-1) T_s$$
$$\therefore t_r \approx \left(\alpha + \beta - \frac{1}{2} \right) T_s \qquad (4\text{-}37)$$

Formula (4-37) is the average frame loss time of the (*n*−1)th order demultiplexer. Where T_s is the frame period of the binary code, α and β are the search check coefficient and the synchronization protection coefficient of the (*n*−1)th order demultiplexer respectively.

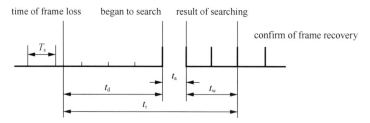

Figure 4-14 Average time of frame loss

4.2.2.3 Stuffing error code of the (*n*−2)order

From condition of the Figure 4-15, when frame loss happens, the error code of the (*n*−2)th order is:

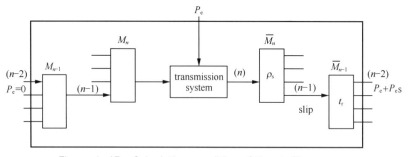

Figure 4-15 Calculation condition of the stuffing errors

$$P_e + P_{es} = \frac{f_1\left(\frac{1}{2}t_r\right) + f_1(P_e t_1)}{f_1 t_s}$$

$$= \frac{\frac{1}{2}t_r + P_e(t_s - t_r)}{t_s} \quad (4\text{-}38)$$

$$= \left(\frac{1}{2} - P_e\right)\frac{t_r}{t_s} + P_e$$

$$P_{es} = \left(\frac{1}{2} - P_e\right)\frac{t_r}{t_s}$$

$$= \left(\frac{1}{2} - P_e\right)\left(\alpha + \beta - \frac{1}{2}\right)T_s \Big/ \frac{T_s'}{C_{\eta'}^{d'+1} \cdot P_e^{d'+1}} \quad (4\text{-}39)$$

$$= \left(\frac{1}{2} - P_e\right)\frac{T_s}{T_s'}\left(\alpha + \beta - \frac{1}{2}\right)C_{\eta'}^{d'+1} \cdot P_e^{d'+1}$$

$$\frac{P_{es}}{P_e} \approx C_{\eta'}^{d'+1} \cdot \frac{T_s}{2T_s'}\left(\alpha + \beta - \frac{1}{2}\right)P_e^{d'} \quad (4\text{-}40)$$

Formula (4-39) and (4-40) are calculation formula of stuffing error rate of the $(n-2)$th order. Where α, β and T_s are the search check coefficient, the synchronization protection coefficient of the $(n-1)$th order demultiplexer, and the frame period of the $(n-1)$th binary code respectively; η', d', T_s' and P_e are the length of the stuffing indicating signal of the nth order binary code, the fault-tolerant bit of the stuffing indicating signal, and the frame period and error rate respectively.

For example, $n'=3, d'=1$, $T_s'=125\mu s$, $T_s=100\mu s$, $\alpha=3$, $\beta=4$, then $P_{es}/P_e \approx 12 P_e$; when $P_e \leqslant 1 \times 10^{-3}$, $P_{es}/P_e \leqslant 1.2\%$.

4.2.2.4 Total multiplex error rate of the $(n-2)$th order

From the multiplex system described in the Figure 4-16, the total error rate of the $(n-2)$th order should include the following items:

P_e—average error rate of the nth order sending to the $(n-2)$th order;

P_{ef}'—synchronization frame loss error rate of the $(n-1)$th order caused by synchronization frame loss of the nth order demultiplexer, all errors pass to the $(n-2)$th;

P_{es}—stuffing error rate of the $(n-2)$th caused by the error of stuffing operation of the nth order demultiplexer;

P_{ef}—frame loss of the $(n-1)$th order demultiplexer and error code of the synchronization frame loss of $(n-2)$th order caused by the $(n-1)$th order error code rate $(P_{\text{e}}+P'_{\text{ef}})$. The four kinds of errors which constitute the total errors$(P_\Sigma)$of the $(n-2)$th order are the channel error$(P_{\text{e}})$and errors caused by multiplex/demultiplex. The total multiplex error $(P_{\text{M}})$is:

$$P_\Sigma = P_{\text{e}} + P_{\text{M}} \tag{4-41}$$

$$P_{\text{M}} = P'_{\text{ef}} + P_{\text{es}} + P_{\text{ef}} \tag{4-42}$$

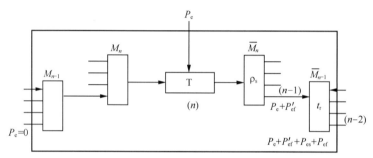

Figure 4-16 The total errors of $(n-2)$th order

From formula (2-40) and (4-40) we can get:

$$\begin{aligned}
P_{\text{M}} &= \frac{1}{2}\left(\alpha' - \frac{1}{2}\right)(n'P_{\text{e}})^{\beta'} + \frac{T_s}{2T'_s}\left(\alpha + \beta - \frac{1}{2}\right)^{P_{\text{e}}^{d'+1}} \cdot C_{\eta'}^{d'+1} + \frac{1}{2}\left(\alpha - \frac{1}{2}\right)[n(P_{\text{e}} + P'_{\text{ef}})]^{\beta} \\
&\approx \frac{1}{2}\left(\alpha' - \frac{1}{2}\right)(n'P_{\text{e}})^{\beta'} + \frac{T_s}{2T'_s}\left(\alpha + \beta - \frac{1}{2}\right)C_{\eta'}^{d'+1} \cdot P_{\text{e}}^{d'+1} + \frac{1}{2}\left(\alpha - \frac{1}{2}\right)(nP_{\text{e}})^{\beta} \\
&\quad + \frac{1}{2}\left(\alpha - \frac{1}{2}\right)\left[\frac{n}{2}\left(\alpha' - \frac{1}{2}\right)(n'P_{\text{e}})^{\beta'}\right]^{\beta} \\
&\approx \frac{1}{2}\left[\left(\alpha' - \frac{1}{2}\right)(n'P_{\text{e}})^{\beta'} + \frac{T_s}{T'_s}\left(\alpha + \beta - \frac{1}{2}\right)C_{\eta'}^{d'+1} \cdot P_{\text{e}}^{d'+1} + \left(\alpha - \frac{1}{2}\right)(nP_{\text{e}})^{\beta}\right]
\end{aligned} \tag{4-43}$$

when $\qquad d' \leqslant \beta - 1,\ \alpha'=\alpha,\ \beta'=\beta,\ F'_s = F_s$

$$P_{\text{M}} \approx \left(\alpha - \frac{1}{2}\right)(nP_{\text{e}})^{\beta} \tag{4-44}$$

The formula (4-44) shows that the total errors (P_{M}) mainly are decided by the bit length of frame alignment signal (n), the parameters of synchronous search (α, β) and the channel error rate (P_{e}).

4.2.3 Suppression of stuffing errors
4.2.3.1 Influence of stuffing errors

See Figure 4-16, all of the average error rates in the high order binary code pass to the lower order code; from the example of the calculation error rate we see that if the nth order average error rate is $P_e=1\times10^{-3}$, the stuffing error rate of the $(n-2)$th order is only 1.2% of the total error rate. Increasing such small percentage actually is neglectable; in the other way, from the recommendation of CCITT the average error rate $P_e=1\times10^{-3}$ is unacceptable threshold, and the standand average error rate threshold is $P_e=5\times10^{-6}$. Specifically speaking, in the 90% time $P_e \leqslant 5\times10^{-6}$ has to be satisfied, and the error rate in the range $5\times10^{-6}-1\times10^{-3}$ should not extend 10% of the work time. Generally speaking, for the tributary binary code stuffing error rate does not essentially influent on the error code.

4.2.3.2 Average time between stuffing errors

It is worth to consider that stuffing operation error of nth order code directly causes the slip of the $(n-1)$th order code, then damages the structure of binary code. CCITT recommends, assume in the case of 25 000km connection, the slip of 64kbit/s code not exceed once every five hours during 99% of the total time, and once every two minutes during 1% of the total time. And CCITT recommends that, the unreasonable connection is that the slip frequency of which exceed once every two minutes. The time of appearing of such phenomenon should not exceed 0.1% of the total work time.

From Figure 4-13, it is worth to consider that if $P_e=1\times10^{-3}$, the time interval between two stuffing errors is only 30 second; if $P_e=1\times10^{-4}$, the time interval between two stuffing errors is near an hour. Therefore, if the average error rate is rather high, it is necessary to use some proper method to increase time interval between two stuffing errors, and the result is very clear and obviously. There are at least two methods are available in various useful methods, first, strictly control the average error rate in the high order channel, second, suitably increase the number of the stuffing indicating code bits(η) and fault-tolerant bits(d). For example, the recommendation of CCITT is in multiplexer of 4th order and 5th order the stuffing indicating bits is 5 and the fault-tolerant bits is 2. From the recommendation G.745, calculation is:

$$\eta'=5,\ d'=2,\ T_s'=21\mu s,$$

$$t_s = \frac{21\times10^{-6}}{C_5^3 \cdot P_e^3} = \frac{21\times10^{-6}}{10 \cdot P_e^3} = \frac{2.1}{P_e^3}\mu s$$

If $P_e = 1\times10^{-3}$, $t_s = 2100$s, it is about 0.6 hours. Therefore, increasing stuffing indicating bits is a poweml method. When channel error rate is as high as 1×10^{-3}, slip index still keep in a suitable value. If we increase stuffing indicating bits and strictly control the average error rate in the high order channel, the frequency of appearing stuffing error will be limited within a reasonable range.

Chapter 5 Positive/Negative Justification

5.1 THE PRINCIPLE OF POSITIVE/NEGATIVE JUSTIFICATION

The principle of the positive/negative justification is the same as that of the positive justification. The only difference is the values of synchronous multiplex clock (f_m). In the positive/negative justification, the synchronous multiplex clock is equal to the nominal of the tributary clock, i.e. $f_m = f_1$. Consider f_m and f_1 changing within their own tolerances, there are three possible cases of their transient values: $f_m > f_1, f_m = f_1$, and $f_m < f_1$. In the first case, since the read rate is bigger than the write rate, it need stop f_m some bits for normal transmission. This is just the positive justification introduced in the previous chapters. In the second case, the read rate is equal to the write rate, obviously it does not need justification to keep normal transmission, which is called no justification state or synchronous state. In the third case, the read rate is smaller than the write rate, it need additional channel to transmit the superfluous part to keep normal transmission, which is called negative justification state.

Since it has three justification states, three kinds of corresponding justification instructions will be needed to send out control operation (positive justification, non justification, negative justification) of the sending side to the receiving side, and the receiving side does recovery operation according to the instructions. But doing three kinds of controls and transmitting three kinds of instructions in the sending side and using three kinds of strategies in the receiving side, will make the equipment complicated. Except for getting some special characteristic (e.g. extremely small stuffing jitter) or for some special case, it is not good to use such scheme.

In the positive/negative justification, positive justification and negative justification are two necessary justification states. Using positive and negative justification alternatively can replace non justification state. In such way, the system has only two states. Correspondingly, only two kinds of control instructions are needed. For example, using "111" stands for positive justification and "000" stands for negative justification, obviously it is very similar with the positive justification. In such system, the justification is quite simple: if the read/write time difference $\Delta t_x < 0$, do positive justification; if the read/write time difference $\Delta t_x > 0$, do negative justification; if the read write time difference $\Delta t_x = 0$, do positive justification and negative justification alternatively. This physical process is very similar with water in a reservoir flowing in and out, see Figure 5-1. If the water level is lower than a definite level, the output pipe will be closed for a unit time; If the water level is higher than a definite level, an additional water pipe will be open for a unit time. In such a way of regulation, the water in reservoir will float between the definite levels. The water in reservoir will regularly flow out.

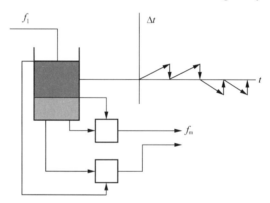

Figure 5-1 Schematic diagram of positive/negative justification

The positive/negative justification multiplexer is very similar with the positive justification. These two kinds of justifications both use frame as a unit to realize justifications. The frame structure arrangement is shown in Figure 5-2. Three time slots are used to transmit justification control instruction (SZ). A negative justification slot (−SV) is used, which is idle when no negative justification, otherwise transmits information code. A positive justification slot (+SV) is used, which transmit information code when no positive justification, otherwise is idle. Actually, there is either a positive or negative justification in a frame. So, two slots −SV and +SV either both are idle (positive justification) or both transmit information code (negative justification).

Chapter 5 Positive/Negative Justification

Figure 5-2 Frame structure of positive/negative justification

5.2 JUSTIFICATION BY FIX ED DECISION CONTROL[22]

According to the principle of the justification mentioned above, we can directly draw out the scheme of fixed decision control justification. The detail scheme is shown in the Figure 5-3. The main part of the scheme is a buffer, write clock f_l and read clock f_m; the corresponding write time is t_l and the corresponding read time is t_m. The read/write time difference is $\Delta t_x = t_1 - t_m$. The fixed threshold Δt_k has been set beforehand, if the read/write time difference $\Delta t_x > \Delta t_k$, when frame frequency pulses F_f come, the trigger T will output voltage level 0 which causes the read clock f_m to increase one beat, i.e. the read time will be advanced an UI (the tributary code element interval), on the other word, when the time difference Δt_x decreases an UI, a negative justification will be done; whereas if the read/write time difference $\Delta t_x < \Delta t_k$, when frame frequency pulses F_f come, the trigger T will output voltage level 1 which causes the read clock f_m to decrease one beat, i.e. the read time will be delayed an UI, on the other word, when Δt_x increases an UI, a positive justification will be done. At the same time of the positive justification, the corresponding control instruction (k) will be set into the SZ slot to transmit to the receiving side.

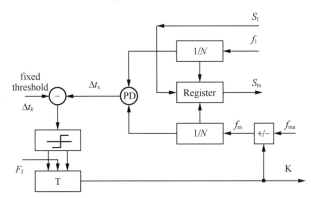

Figure 5-3 Diagram of fixed decision control justification

We specify the synchronous clock source f_{m0} and its effective time t_{m0} as the frequency reference and time reference of the justification system. We define $\Delta t_m = t_m - t_{mo}$; $\Delta t_1 = t_1 - t_{mo}$, hence, the read/write time difference $\Delta t_x = \Delta t_1 - \Delta t_m$. In the actual justification process, in the justification side, Δt_m traces Δt_1, Δt_x is the tracing error, then Δt_m is transmitted from the justification side to the recovery side. In the recovery side, $\Delta \hat{t}_1$ traces Δt_m, and the time difference $\Delta \hat{t}_x = \Delta t_m - \Delta \hat{t}_1$, whereas $\Delta \hat{t}_1 = \hat{t}_1 - t_{mo}$, where $\Delta \hat{t}_x$ is the tracing error at recovery time, $\hat{t}_1$ is the effective time of output clock. Obviously, $\Delta t_x + \Delta \hat{t}_x$ is the total tracing time difference.

When $f_1 = f_{lo} = f_{mo}$ and $t_1 = t_{lo} = t_{mo}$, $\Delta t_1 = 0$, Δt_1 is traced by Δt_m using square wave of amplitude UI and frequency $F_s/2$, which is a half of the frame frequency. The square wave of amplitude UI and frequency $F_s/2$ of the time difference Δt_x is shown in Figure 5-4. Hence, the frequency spectrum of Δt_x will be fully suppressed by the recovery phase-locked loop. Obviously, such an ineffective justification state, which is alternative of the positive justification and the negative justification, will not introduce stuffing jitter into the binary digits.

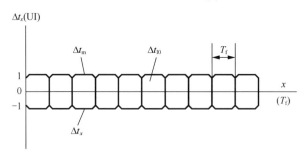

Figure 5-4 $\Delta t_x = F(x)$, ($f_1 = f_m$)

In general, if $f_1 \neq f_{mo}$; $t_1 \neq t_{mo}$, i.e. $\Delta t_1 \neq 0$, Δt_1 will be traced by Δt_m, in a relative complex process, hence the variation process of Δt_x is no longel a simple high speed square wave, see Figure 5-5. Generally speaking, the process of $\Delta t_x - x$ call be decomposed into a square wave whose amplitude is UI and frequency is $F_s/2$ and a triangle wave whose amplitude is UI and frequency is $\Delta f = f_m - f_1$. The former is an ineffective justification process, the latter is an effective justification process. As described above, an ineffective justification process actually will not introduce stuffing jitter, hence it is worth to discuss the latter only. The frequency spectrum of the read/write time difference $\Delta t'_x$ of an effective justification process is:

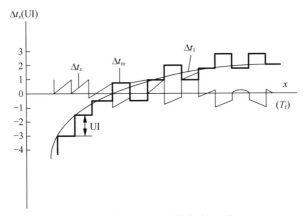

Figure 5-5 $\Delta t_x = F(x)$, $(f_i \neq f_m)$

$$\Delta t'_x = \frac{1}{\pi}\left[\sin 2\pi(f_m - f_1)t - \frac{1}{2}\sin 4\pi(f_m - f_1)t + \frac{1}{3}\sin 6\pi(f_m - f_1)t - \ldots\right]UI$$

(5-1)

This frequency spectrum is filtered by the recovery phase-locked loop, where the low passed part is the stuffing jitter. See Figure 5-6, if the frequency difference

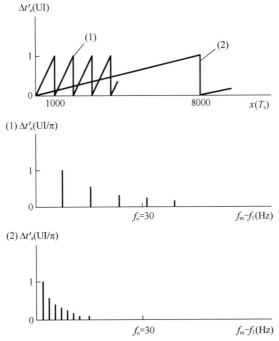

Figure 5-6 Frequency spectrum of effective justification process(F_s=8000Hz)

$f_1 - f_m$ is bigger, most of spectrum is suppressed, only small part passes the pass-band of (f_o=30Hz) phase-locked loop. In this case, stuffing jitter is smaller. If the frequency difference $f_1 - f_m$ is smaller, most spectra pass the phase-locked loop, only part of high spectrum is suppressed, stuffing jitter is bigger. In the extreme case, when the time difference $f_1 - f_m$ approaches zero, spectrum is basically pass the phase-locked loop without suppression, hence the stuffing jitter approaches the maximum, i.e. A_{jpp}=1·UI. The relationship between the stuffing jitter corresponding to the Figure 5-6 and the frequency difference $f_m - f_1$ is shown in Figure 5-7.

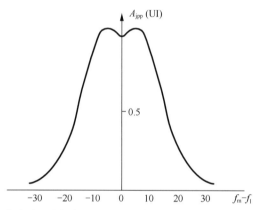

Figure 5-7 Relationship between stuffing jitter and frequency difference

As mentioned above, in this kind of fixed decision control justification scheme, if tributary frequency approaches the multiplexing frequency, maximum peak to peak value of stuffing jitter approaches a code element width (UI). Obviously, the stuffing jitter of the scheme is too high, it is unacceptable But the idea of this scheme enlighten to us that in at least two ways we can suppress stuffing jitter. Firstly, improve justification method, for example, using self-adaptive justification method. Secondly, improve control environment of the justification, for example, all kinds of designs are adopted to ensure that $|f_m - f_1|$ is bigger than a definite value.

5.3 ADAPTIVE JUSTIFICATION CONTROL

It has two adaptive justification methods: simple delta control justification and sigma delta control justification. The following will discuss these two kinds of schemes respectively.

The simple delta control justification is shown in Figure 5-8. In the diagram the integration of the control instruction is used as the justification decision reference Δt_k; the integration of the read/write time difference Δt_x is equivalent to the analog input signal of delta modulator; Δt_k is recovered signal after decoding; the justification control instruction K is equivalent to the digital output signal of delta modulator. The justification control signal k control the effective time of read clock, then the read time difference Δt_m traces the write time difference Δt_l, that is, the read clock f_m traces the write clock f_l.

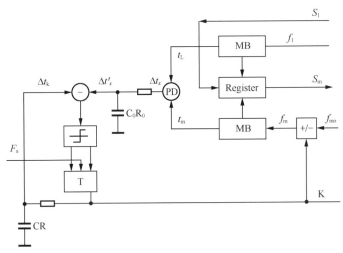

Figure 5-8 Block diagram of simple delta control justification

From the calculation formula of overload characteristics and quantified signal noise ratio of simple delta modulation we can directly derive out the jitter approximate formula of justification of the simple delta control justification. As shown in Figure 5-9, we can get the following result according to the corresponding delta modulation formula.

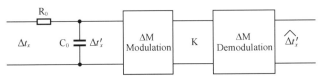

Figure 5-9 Simple delta control justification

If the non-overload condition is satisfied,

$$\Delta t'_x \leqslant \frac{F_s}{f_1 - f_m} \cdot UI \tag{5-2}$$

Stuffing jitter caused by justification tracing error,

$$\frac{\Delta t'_x}{A_j} = 0.2 \frac{F_s^{3/2}}{f_a^{1/2} \cdot (f_1 - f_m)} \tag{5-3}$$

$$f_a = \frac{1}{2\pi RC} \tag{5-4}$$

Where

F_s—frame frequency,
f_a—cut-off frequency of demodulator filter,
f_1—write clock, i.e. tributary clock,
f_m—read clock, i.e. synchronous multiplex clock.

Consider that,

$$\frac{\Delta t'_x}{\Delta t_x} = H(f_1 - f_m) = \frac{f_o}{\sqrt{f_o^2 + (f_1 - f_m)^2}} \tag{5-5}$$

$$f_o = \frac{1}{2\pi R_o C_o} \tag{5-6}$$

We can get non-overload condition and approximate jitter formula.

$$\Delta t_x \leqslant \frac{\sqrt{f_o^2 + (f_1 - f_m)^2}}{f_1 - f_m} \cdot \frac{F_s}{f_o} \cdot UI \tag{5-7}$$

$$\frac{\Delta t_x}{A_j} = \frac{\sqrt{f_o^2 + (f_1 - f_m)^2}}{f_1 - f_m} \cdot \frac{0.2 F_s^{3/2}}{f_a^{1/2} \cdot f_o} \tag{5-8}$$

If $\Delta t_x = 1 UI$, then

$$A_j = \frac{f_1 - f_m}{\sqrt{f_o^2 + (f_1 - f_m)^2}} \cdot \frac{5 f_a^{1/2} \cdot f_o}{F_s^{3/2}} \cdot UI \tag{5-9}$$

If $f_o \ll f_1 - f_m$, then

$$\Delta t_x \leqslant \frac{F_s}{f_o} \cdot UI \tag{5-10}$$

$$A_j = \frac{5 f_a^{1/2} \cdot f_o}{F_s^{3/2}} \cdot UI \tag{5-11}$$

Chapter 5 Positive/Negative Justification

If $f_1 - f_m \ll f_o$, then

$$\Delta t_x \leqslant \frac{F_s}{f_1 - f_m} \cdot UI \quad (5\text{-}12)$$

$$A_j = \frac{5 f_a^{1/2}(f_1 - f_m)}{F_s^{3/2}} \cdot UI \quad (5\text{-}13)$$

Obviously, for this simple delta control justification method, if the frequency difference of tributary is larger, the stuffing jitter and the overload range all have no relationship to the frequency difference of tributary; if the frequency difference of tributary is very small and as the frequency difference approaches zero, non overload range extends, hence the stuffing jitter decreases.

The sigma delta control justification scheme is shown in Figure 5-10. In the diagram, the integration of the read write time difference ($\Delta t'_x$) is equivalent to the analog input signal of the sigma delta modulator. Δt_k is the output control signal k which is equivalent to the output digital signal of the modulator. Signal k controls justification of read clock f_m, hence it causes the read clock f_m to trace the write clock f_1.

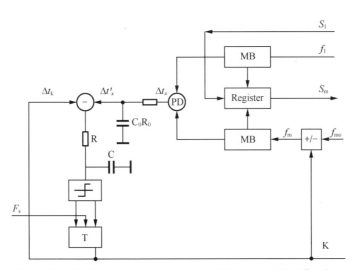

Figure 5-10 Block diagram of sigma delta control justification

From the formula of overload characteristics and quantified signal noise ratio of sigma delta modulation, we can derive out approximate formula of stuffing jitter of justification controlled by sigma delta modulation, as shown in Figure 5-11.

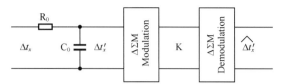

Figure 5-11 Sigma delta control justification

From the corresponding formula of sigma delta modulation, we can get: If non. overload condition is satisfied:

$$\Delta t_x \leqslant 1\text{UI} \tag{5-14}$$

Stuffing jitter A_j introduced by justification tracing error is

$$\frac{\Delta t_x}{A_j} = 0.35\left(\frac{F_s}{f_a}\right)^{3/2} \tag{5-15}$$

$$f_a = \frac{1}{2\pi RC} \tag{5-16}$$

Similar with the discussion of simple delta control, we get formula:

$$\frac{\Delta t_x'}{\Delta t_x} = \frac{f_o}{\sqrt{f_o^2 + (f_1 - f_m)^2}} \tag{5-17}$$

$$f_o = \frac{1}{2\pi R_o C_o} \tag{5-18}$$

Consider all above formulas, we get non-overload condition and jitter approximate formula:

$$\Delta t_x = \frac{\sqrt{f_o^2 + (f_1 - f_m)^2}}{f_o} \cdot UI \tag{5-19}$$

$$A_j \approx \frac{\sqrt{f_o^2 + (f_1 - f_m)^2}}{f_1 - f_m} \cdot 3\left(\frac{f_a}{F_s}\right)^{3/2} \cdot UI \tag{5-20}$$

If $f_o \ll f_1 - f_m$, then

$$\Delta t_x \leqslant \frac{f_1 - f_m}{f_o} \cdot UI \tag{5-21}$$

$$A_j \approx 3\left(\frac{f_a}{F_s}\right)^{3/2} \cdot UI \tag{5-22}$$

If $f_1 - f_m \ll f_o$, then

$$\Delta t_x \leqslant 1 \cdot \text{UI} \tag{5-23}$$

$$A_j \approx \frac{3f_o}{f_1 - f_m}\left(\frac{f_a}{F_s}\right)^{3/2} \cdot \text{UI} \tag{5-24}$$

Obviously, in this sigma delta control justification scheme, if the frequency difference of tributary is larger, non-overload range is in proportional to the frequency difference of tributary, and stuffing jitter has no relationship with the tributary frequency difference; if the frequency difference of tributary is smaller, non-overload range has no relationship with it, but stumng jitter is inverse proportional to the frequency difference of tributary.

From above basic formulas we can compare characteristics of two kinds of adaptive control justification, see Figure 5-12 and Figure 5-13. In the Figures, if the frequency difference of tributary approaches zero, non-overload range controlled by simple delta modulation approaches infinity, and stuffing jitter approaches zero; the range of non-overload controlled by sigma delta modulation approaches 1UI and stuffing jitter is infinity. Considering pass-band of phase-locked loop of the code recovery is only several decades Hz and suppression of every double frequency range, which is outside of pass-band, is at least 6dB. Therefore, if the frequency difference of tributary is larger, we are not interested in the suppression characteristics of the loop.

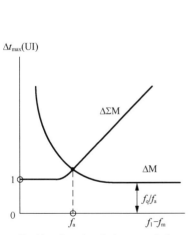
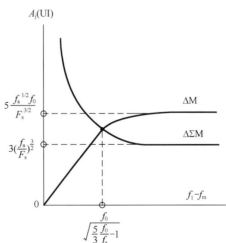
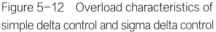

Figure 5-12 Overload characteristics of simple delta control and sigma delta control

Figure 5-13 Comparison between stuffing jitter of simple delta control and sigma delta control

From the idea mentioned above, we design the simple delta control circuit and sigma delta control circuit, shown in Figure 5-14 and Figure 5-15. Measuring result is shown in Figure 5-16. The Figure shows that the tendency of the oretical prediction is correct. If the tributary frequency difference (f_1-f_m) approaches zero, the stuffing jitter of the simple delta control is quite low; if the tributary frequency difference (f_1-f_m) increases gradually, the stuffing jitter of the sigma delta control justification scheme decreases rapidly. From Figure 5-12 and Figure 5-13 we can understand that as the tributary frequency difference increases, the stuffing jitter of the simple delta control scheme increases linearly, and the overload limit decreases to force the stuffing jitter to increase; but as the tributary frequency difference increases, the suppression of the recovery phase-locked loop will increase. Hence, as the tributary frequency difference increases, the total stuffing jitter of the simple delta control scheme decreases slowly. But it is not the case for the sigma delta control scheme, as the tributary frequency difference increases, the stuffing jitter will decrease inverse-proportionally, and the decrease of the overload limit also forces the stuffing jitter to decrease. Considering the suppression of the code recovery phase-lock loop also enhances as the tributary frequency difference increases. Hence, the total stuffing jitter of the sigma delta control scheme decreases rapidly, as the tributary frequency difference increases.

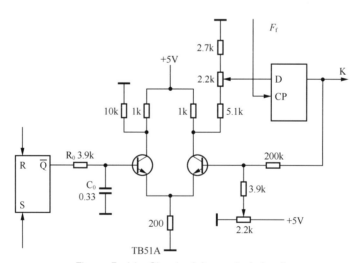

Figure 5-14   Simple delta control circuit

Chapter 5 Positive/Negative Justification

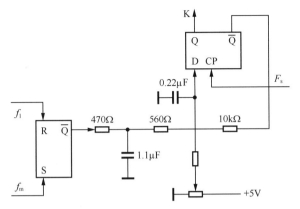

Figure 5-15 Sigma delta control circuit

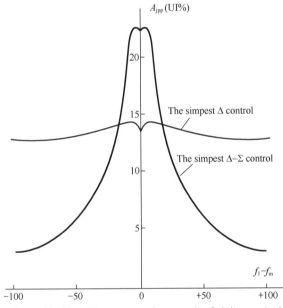

Figure 5-16 Jitter measuring result of delta control

In short, the stuffing jitter of the adaptive control justification is obviously lower than that of fixed control justification. The main reason is that the adaptive control has changed the frequency spectrum of the read/write time difference Δt_x of justification process. The spectrum structure of the read/write time difference Δt_x given in the appendix of the recommendation G.741 of CCITT (1973-1976), is shown in Figure 5-17. It shows that the maximum energy spectrum of the stuffing jitter concentrates in the range of (0.1-0.2) F_s; near the base frequency, as the frequency decreases the spectrum amplitude decreases exponentially. This is why

– 123 –

the adaptive control justification obviously suppresses the jitter. In the equipments of Soviet Union, the effective jitter is 3.0%UI, the peak to peak value of jitter is 15% UI in magnitude.

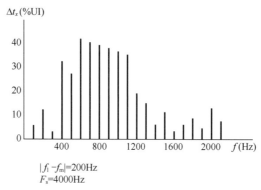

Figure 5-17   Jitter spectrum of adaptive justification control

5.4   POSITIVE/NEGATIVE JUSTIFICATION CONTROL CIRCUIT

The control circuit of the positive/negative justification is shown in Figure 5-18, and the wave form diagram of the positive justification is shown in Figure 5-19.

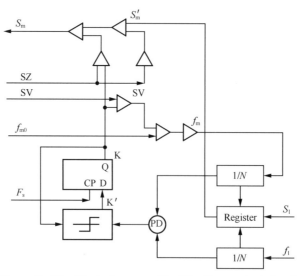

Figure 5-18   Block diagram of justification control circuit

Chapter 5 Positive/Negative Justification

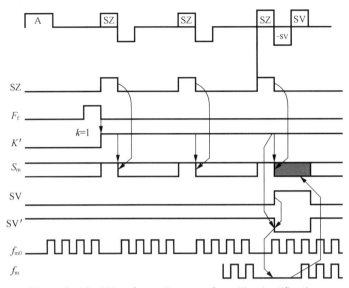

Figure 5-19 Waveform diagram of positive justification

If a positive justification is requested, after frame pulse T_s arrives, K turns to the high level, outputs SZ=111 to insert into the SZ position of binary digits (S_m), and sends out SV to form negative pulses SV′, deducts two beats of the multiplex clock, corresponding to +SV and −SV. At that time, the corresponding two information codes will not be read out. This completes a positive justification. If a negative justification (K′ =0) is requested, after frame pulse T_s arrives, K turns into the low level, which closes SZ pulse and input gate of SV. At that time, the time slot SZ of binary digits S_m is forcedly set to 000; the two multiplexing clock beats corresponding to +SV and −SV are effective as usual, i.e. information code can be read out from buffer. This completes a negative justification.

The recovery control is the same as that of the positive justification, and may use the common recovery plug-in units with the positive justification. The only difference is that the signal SV of the positive justification only corresponds to the time slot +SV; whereas the signal SV of the positive/negative justification corresponds to two time slots +SV and −SV. This kind of difference appears only in the synchronous demultiplex parts, therefore the recovery plug-in card may be used commonly.

The parameter design of the recovery phase-lock loop is the same completely. But for the positive justification further reducing pass-band may not get any benefit;

whereas for the positive/negative justification it may reduce the stuffing jitter. For example, the data given in the reference [39]: pass-band f_o=125Hz, stuffing jitter A_{jpp}=19.8%UI; f_o=95Hz, A_{jpp}=14.2%UI; f_o=65Hz, A_{jpp}=7.3%UI. Generally speaking, in the engineering we hope to use a universal card as many as possible. Actually, if we use a universal card for code recovery, the stuffing jitter of the positive/negative justification can be smaller than 15% UI. It is low enough. Therefore, the positive/negative justification and the positive justification usually use a common type of code recovery plug-in card.

5.5 TRANSITION PROCESS OF JUSTIFICATION

Assume the multiplex frame structure accord with the recommendation of G.745. In the Figure 5-20, f_h=8488kbit/s, f_I=2048kbit/s, m=4, L_s=1056bit, Q=256bit.

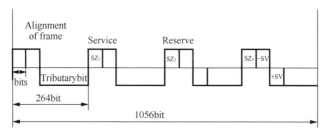

Figure 5-20   Frame structure of recommendation G.745

In the buffer of the justification, the read/write difference of reading out xth tributary code is written as Δt_x. After every m T_h read out one bit, and at every T_1 write in one bit. Whenever write in one bit and read out one bit, the read/write time difference will change with (mT_h-T_1), hence at the xth bit time the change of the read/write time difference is $(mT_h-T_1)x$. Whenever the read clock stops one beat, the read/write time difference will increase mT_h, and if it stops g beats the change of read/write difference is $mT_h \cdot g$. Assume at the beginning (x=0) the read/write difference is Δt_o, then the general formula of the read/write time difference is as the following:

$$\Delta t_x = \Delta t_o + (mT_h - T_1)x + mT_h g \qquad (5\text{-}25)$$

where, $T_h = \dfrac{1}{f_h}$; T_1 is usually written as UI, the formula can be rewritten as following:

$$\Delta t_x = \Delta t_o + \left[\frac{mf_1}{f_h}(x+g) - x\right] \cdot UI \qquad (5\text{-}26)$$

From the frame structure in Figure 5-20, if the transitional process gets to the end of the first subframe, i.e. $x = \frac{Q}{4}, g = \frac{K}{4}$,

$$\begin{aligned}
\Delta t_x &= \Delta t_o + \left[\frac{mf_1}{f_h}\left(\frac{Q}{4} + \frac{K}{4}\right) - \frac{Q}{4}\right] \cdot UI \\
&= \Delta t_o + \left[\frac{f_1}{4f_h} m(Q+K) - \frac{Q}{4}\right] \cdot UI \\
&= \Delta t_o + \left[\frac{f_1 \cdot L_s}{4 \cdot f_h} - \frac{Q}{4}\right] \cdot UI \qquad (5\text{-}27) \\
&= \Delta t_o + \left[\frac{Q}{4} - \frac{Q}{4}\right] \cdot UI \\
&= \Delta t_o
\end{aligned}$$

Obviously, after every subframe the net change of the read/write difference is zero. In this special example the transitional process may be discussed using subframe as a basic unit. At the beginning of each subframe, Δt_x is at the maximum point, and at the end of each subframe Δt_x is at the minimum point. Considering usual arrangement of jitter indication position (SZ) and jitter position (−SV and +SV), the decision time of positive/negative justification is suitable in the first subframe. Since in which position of first subframe does not really influence the justification process, it seems that the middle is more proper, in such a way, buffer capacity can be fully used. In this discussed scheme, the decision time of the positive/negative justification is set at the end of the first subframe. Figure 5-21 is transitional process curve of the positive/negative justification in such arrangement.

Plane of Δt_x−X may be partitioned into 5 regions: the region of tile initial read write difference Δt_o>+4UI and Δt_o<−2UI is called catching transitional region. The phenomenon that the write time catches the read time and the read time catches the write time appears within the region, but it take the state at the decision time as reference. Whenever a justification is done, it can get into the justification transitional region. The region of Δt_o in the +2UI—+4UI is called: negative justification transitional region and the region of Δt_o in the −2UI—+1UI is called positive justification transitional region. In the justification transitional region, after the corresponding justification it can automatically get into the steady justification region. The region of Δt_o in the +1UI—+2UI is called the steady justification region. Obviously, the positions of the above justification regions are closely related to selection of justification decision time. In Figure 5-21, the transitional process curve is based on the justification decision time which is at the end of first subframe. From the justification transitional process we can see that the positive/negative justification is stable, which can automatically transfer to the steady region. Its peak to peak value of steady region is 3UI, which is lower than that of the positive justification. If the justification time is selected at the end of the first sub-frame, the upper peak of the justification region is +2UI, and the lower peak region is −1UI. Therefore, when we design discriminator, if buffer is $N=8$, the initial time difference Δt_o takes either 4UI or 3UI: if $N=7$, the Δt_o should be 3UI.

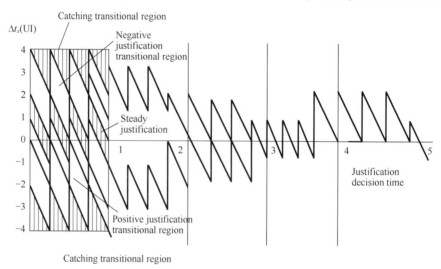

Figure 5-21 Transitional process of positive/negative justification

5.6 PARAMETER DESIGN OF JUSTIFICATION

Formula (5-13) shows stuffing jitter formula of the simplest delta control justification:

$$A_j = \frac{5 f_a^{1/2}(f_1 - f_m)}{F_s^{3/2}}$$

Where,

A_j—effective value of stuffing jitter,

F_s—frame frequency, when justification once in every frame, it is equal to justification Frame frequency (f_s),

$f_1 - f_m$—difference between tributary frequency and synchronous multiplex frequency,

f_a—cut-off frequency of loop filter,

$$f_a = \frac{1}{2\pi RC} \quad (5\text{-}28)$$

To use the formula (5-28) for engineering design, we must modify two points of the formula: the first is to replace the effective value with the peak to peak value, the second is to consider the effect of the load impedance on the loop filter circuit. Because transient wave of stuffing jitter is so complex that it is difficult to find its wave coefficient, hence it is not easy to precisely make these two modifications. Besides, the input impedance of the active circuit following the loop filter is a parameter which changes with the working state, hence it is difficult to find a proper effective value. Thus usually we use two methods to calculate loop parameters: derive an approximate calculation formula from simplified assumption by experience or set parameters by experiment.

(1) approximate formula

Assume that the transient wave of the stuffing jitter is a triangle wave, and let its maximum magnitude be 1; the effective value of the stuffing jitter is the effective value of the base element of triangle wave. Hence, the peak to peak and the effective value of the approximate formula of the stuffing jitter are as follows:

$$A_{jpp} = 2\pi\sqrt{2}A_j$$

For convenience in calculation, we let:

$$A_{jpp} = 10 A_j \quad (5\text{-}29)$$

Assume the load of loop filter be a pure resistance R, then the equivalent cut-off frequency of the loop filter is:

$$f_a = \frac{1}{\pi RC} \quad (5\text{-}30)$$

In this case, the peak to peak value of the simple delta control justification stuffing jitter is:

$$A_{\text{jpp}} = \frac{50 f_a^{1/2} |f_1 - f_m|}{f_s^{3/2}} \quad (5\text{-}31)$$

(2) Set loop parameters by experiment

Reference [62] shows an experimental result of the positive/negative justification: experimental curve of stuffing jitter and loop capacitor; experimental curve of stuffing jitter and justification frequency.

From Figure 5-22 we can see that as the loop capacitor (C) increases, the stuffing jitter monotonously decreases; if the capacitor increases continually, error code will appear in the recovered binary digits. For error free to let the stuffing jitter be as small as possible, the loop capacitor C should take a proper value. In this experiment $A_{\text{jpp}} = 3\% UI$.

When the capacitor is too small, the pass-band of filter is too wide that interfering ripple can not be filtered out, which will convert to big stuffing jitter. Whereas, when capacitor is too high, the pass-band of filter is too narrow that right code can not be recovered, which makes control mistake of the justification so that error code is introduced. Thus, the value of capacitor C is to be limited in a permitted range. Its lower limit of capacitor is changed gradually, the lower is the capacitor the larger is the stuffing jitter. Whereas it burst changes near the higher limit, as long as the value of the capacitor is over the higher limit, error code will appear immediately. Hence, the capacitor is suitable near the higher limit to get the smallest stuffing jitter, but we should take it lower than the margin to avoid error code.

From the experimental curve in Figure 5-23 we can see that as justification

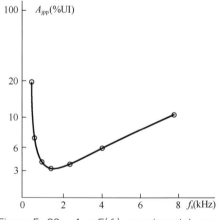

Figure 5-22 $A_{\text{jpp}} = F(C)$ experimental curve

Figure 5-23 $A_{\text{jpp}} = F(f_s)$ experimental curve

rate increases, the stuffing jitter decreases rapidly; after a small region, the stuffing jitter increases slowly. In the experiment, the justification rate is between 500-4000Hz, the stuffing jitter is at the level of 3%UI-4%UI.

In this experiment, corresponding to each justification rate, we justify again the loop circuit capacitor to get the minimum stuffing jitter without error code. Generally speaking, the higher is the justification rate, the lower is the capacitor of ensuring not to cause error code. With premise of error code free, increasing of justification rate causes stuffing jitter to decrease, and decreasing of loop capacitor causes stuffing jitter to increase. These two processes are non-linear. In general, as justification rate increases, the stuffing jitter decreases rapidly, then after keeping a small section of minimum it increases again slowly. Corresponding to the justification rate there exists a plesio-optimum domain. Thus, in engineering design, it is unnecessary to set justification bit in every frame. For example, setting a Justification bit every two flame possibly gets more lower stuffing jitter than the former. Hence, In engineering design we should distinguish the frame frequency (F_s) and the justification rate (f_s) and design them separately. This is the different point between positive/negative justification and positive justification.

5.7 ENVIRONMENT DESIGN OF JUSTIFICATION

The formula of stuffing jitter of sigma delta control code justification is shown in formula (5-24):

$$A_j = \frac{3f_o}{f_1 - f_m}\left(\frac{f_a}{F_s}\right)^{3/2}$$

where, A_j—effective value of stuffing jitter;

F_s — frame frequency, which equals to justification frequency when justification once in every frame;

$f_1 - f_m$ —the difference between tributary frequency and synchronous multiplex frequency;

f_a—out-off frequency of sigma filer;

$$f_a = \frac{1}{2\pi RC} \qquad (5\text{-}32)$$

f_o cut-off frequency of input time difference signal filter

$$f_o = \frac{1}{2\pi R_o C_o} \tag{5-33}$$

As discussed in Section 5.6, the above formulas need be modified for engineering design: Provided that transient wave of stuffing jitter is triangle wave, both these filters are the simplest RC filters, and whose equivalent loads are equal to corresponding resistances. Therefore we can derive out the peak to peak value formula of the stuffing jitter

$$A_{jpp} = \frac{30 f_o}{|f_1 - f_m|} \left(\frac{f_a}{F_s}\right)^{3/2} \tag{5-34}$$

$$f_a = \frac{1}{\pi RC}$$

$$f_o = \frac{1}{\pi R_o C_o}$$

From above formula we can see that the stuffing jitter peak to peak value of the sigma delta control justification is inversely proportional to the frequency difference (f_1–f_m). The experimental result in Figure 5-16 shows such a relationship as well. The results enlighten us: in circuit design, as long as frequency difference $|f_1-f_m|$ is big enough, i.e., it is beyond pass-band of phase-locked loop of code recovery, the stuffing jitter (A_{jpp}) will be small enough. But rather big frequency difference $|f_1-f_m|$ means that more additional slots should be left out in multiplex binary digits. therefore the slot utilization is lower. Of course, the maximum frequency difference $|f_1-f_m|$ should not exceed the justification frequency f_s. In view of single technique, the smaller is stuffing jitter, the better is it. But in view of entire engineering, it is no practical value of pursuing smaller jitter than what is specified by the network system standard. Figure 5-16 shows the actual measuring result: $|f_1-f_m| \geq 120\text{Hz}$, $A_{jpp} \leq 3\%$ UI. It may be an appropriate compromise. The range of frequency difference nominal is summarized as:

$$F_s - |\Delta f_1 + \Delta f_m| \geq |f_1 - f_m| \geq 120\text{Hz} + |\Delta f_1 + \Delta f_m| \tag{5-35}$$

where, F_s—frame frequency of justification;

Δf_1—tolerance of tributary frequency (f_1);

Δf_m—tolerance of synchronous multiplex frequency (f_m);

The following will discuss how to realize the design of the frequency difference $|f_1-f_m|$. Without any prerequisite, the frequency difference design which satisfies above condition is very easy. Detail process is as follows:

(1) Given multiplex frequency $f_h \pm \Delta f_h$ and tributary frequency $f_1 \pm \Delta f_1$;

(2) k demultiply the multiplex frequency to get the synchronous multiplex frequency:

$$f_m = \frac{f_h}{k} \tag{5-36}$$

$$|f_1 - f_m| \geqslant 120\text{Hz} + |\Delta f_1 + \frac{\Delta f_h}{k}| \tag{5-37}$$

(3) Decide the justification frame frequency:

$$F_s = \frac{f_h}{L_s} \tag{5-38}$$

$$F_s \geqslant |f_1 - f_m| + |\Delta f_1 + \frac{\Delta f_h}{k}| \tag{5-39}$$

If prerequisite is given, such as frame length (L_s), the frequency difference $|f_1 - f_m|$ design will be a bit troublesome. Firstly we must note that the design principles of the frequency difference $f_1 - f_m > 0$ and $f_1 - f_m < 0$ are the same. But consider the circuit design, assume $f_1 - f_m > 0$. We Can apply the following process in the circuit design:

(1) Given multiplex frequency $f_h \pm \Delta f_h$, tributary frequency $f_1 \pm \Delta f_1$ and frame length L_s.

(2) Set number of basic information bits (Q) and number of non-information bits (K) in each frame:

$$f_{mo} = f_1 \tag{5-40}$$

$$f_{mo} = \frac{f_h}{L_s} Q, L_s = Q + K \tag{5-41}$$

$$\therefore Q = \frac{f_1}{f_h} \cdot L_s \tag{5-42}$$

$$K = L_s - Q \tag{5-43}$$

(3) Deduct an information slot every M frames (i.e. there is a frame whose information bits are $Q-1$ bits in every M frames), and get the synchronous tributary multiplex frequency f_m:

$$f_m = f_{mo} - \frac{F_s}{M} \quad (5\text{-}44)$$

Require the frequency difference $f_1 - f_m = \frac{F_s}{M}$ satisfy the design condition:

$$F_s - |\Delta f_1 + \Delta f_m| \geq \frac{F_s}{M} \geq 120\text{Hz} + |\Delta f_1 + \Delta f_m| \quad (5\text{-}45)$$

(4) Smooth non-uniform pulse sequence ($f_{mo} - \frac{F_s}{M}$) into a uniform clock f_m through the phase-locked loop, and design usual sigma delta control justification on the basis of uniform f_m. The necessity of such a smooth is to eliminate correlation between structural frequency difference $\frac{F_s}{M}$ and normal justification. To form the frequency difference $\frac{F_s}{M}$, an information slot (Q–1) is deducted in a frame. A negative justification will possibly occur in the next frame (Q+1). They are concealed each other, so that no time difference jitter spectrum introduced by the justification passes on to higher frequency.

(5) Similarly on code recovery side, smooth non-uniform pulse sequence ($f_{mo} - \frac{F_s}{M}$) into a uniform clock f_m through the phase-locked loop, and design normal recovery circuit on the basis of uniform f_m. The necessity of such a smooth is to avoid that the jitter introduced by frequency difference $\frac{F_s}{M}$ and that introduced by normal justification are concealed each other. If these two jitters are concealed each other, a jitter spectrum approaching zero which is hard to filter out. Obviously, the smooth phase-locked loop installed on the justification side plays the same role as the one on the recovery side. Therefore, we use the same circuit shown in Figure 5-24. This circuit is just the same as phase-locked loop of the recovery, except for central frequency of voltage control oscillator.

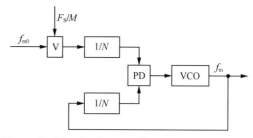

Figure 5-24 Clock smooth phased-locked loop

5.8 THE TECHNICAL APPLICATION OF THE POSITIVE/NEGATIVE JUSTIFICATION

(1) This chapter has discussed two kinds of delta control justification techniques. They all use read/write time difference as a time changing signal which is, through the delta modulation method, transmitted to the other side to trace the phase of binary digits. Although the principles are the same, the methods are different. Therefore, the quantity of equipments and the technical result are different from each other. By comparing with each other, the simple delta control is obviously better than the sigma delta control.

The equipment is simple. The quantity of equipments is almost as same as that of the positive justification. It does not need additional decouple smooth phase-locked loop.

High utilization of channel time slot. Actual design can really realize ($f_1 = f_m$), this is the highest time slot utilization in the justification design. It does not need redundant time slot for deviation.

Because the actual design value is equal to the nominal value ($f_1 = f_m$), it does not need special control to realize compatibility of synchronous/asynchronous.

Without fine justification, stuffing jitter can be smaller than 15% UI, after a little justification it can be smaller than 3% UI, and it does not need any additional cost.

(2) In an plesiochronous digital hierarchy of CCITT recommendation of G.702, all countries do not use positive/negative justification, except Soviet Union. As a national standard our country uses positive justification. Therefore, in the plesiochronous digital hierarchy of the G.702, such a technique is useless. But in the subgroup it is used widely.

(3) In the subgroup of lower than 2048kbit/s, all the tributary binary digits is a simple multiple of multiplex binary digits, therefore the positive justification can not be used. If using asynchronous multiplex system, the choice is only between positive/negative justification (discussed in this chapter) and positive/zero/negative justification (discussed in the next chapter). If the tributary rate is rather high, and second-order phase-locked loop is easy to realize, it is suitable to use positive/negative justification discussed in this chapter.

5.9 EXAMPLE OF POSITIVE/NEGATIVE JUSTIFICATION DESIGN

The multiplexer designed in 1978 which accords with the recommendation of CCITT G.754 consists of 13 plug-in cards (see Figure 5-26). The internal clock is an independent card. The synchronous multiplex uses two cards (T_1,T_2), the electrical principle diagram is shown in Figure 5-27. The synchronous demultiplex part uses two cards (R_1,R_2), the electrical principle diagram is shown in Figure 5-28. Every tributary has independent justification card (G), shown in Figure 5-29. Every tributary has independent recovery card (D) which is common with the positive justification equipment. Here we will not describe it repeatedly.

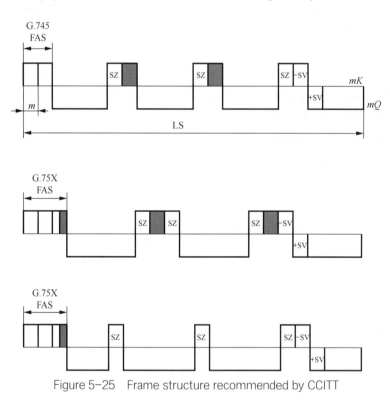

Figure 5-25 Frame structure recommended by CCITT

The main part time wave diagram of the multiplex/demultiplex process is shown in Figure 5-30.

Chapter 5 Positive/Negative Justification

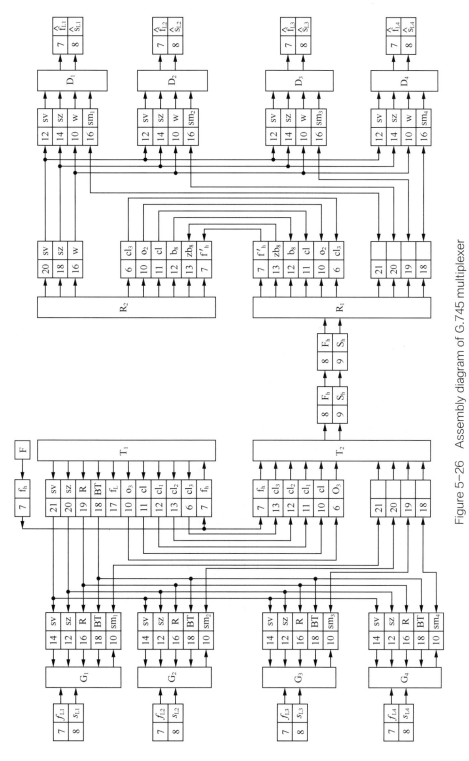

Figure 5–26　Assembly diagram of G.745 multiplexer

– 137 –

PDH for Telecommunication Network

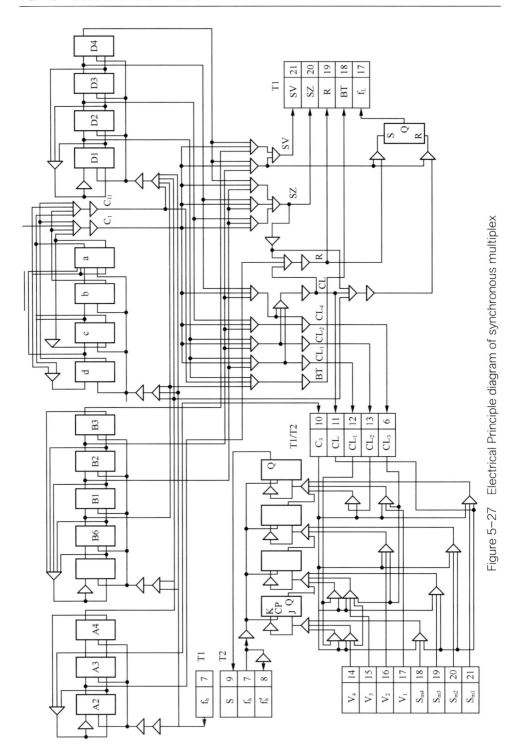

Figure 5-27 Electrical Principle diagram of synchronous multiplex

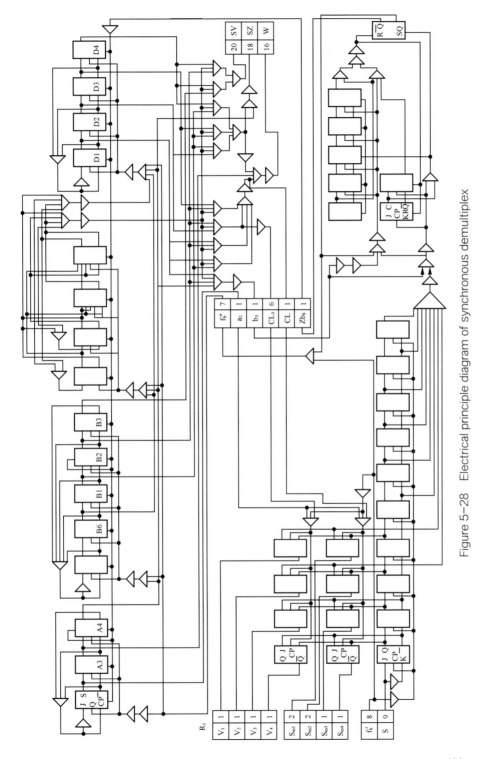

Figure 5-28 Electrical principle diagram of synchronous demultiplex

— 139 —

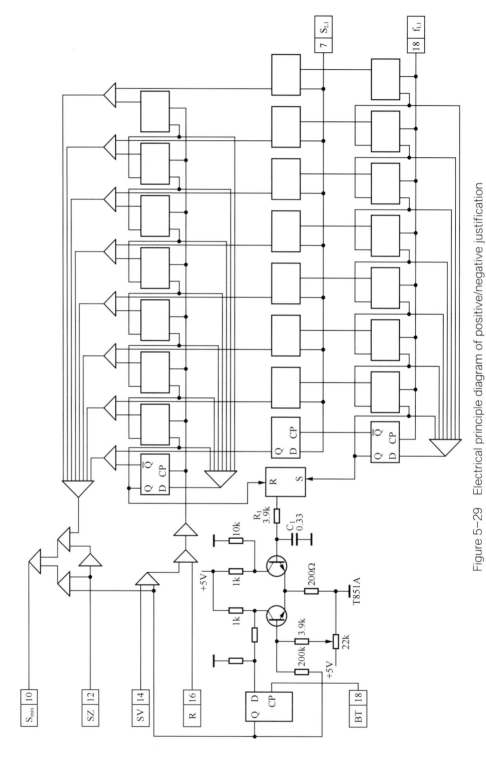

Figure 5-29 Electrical principle diagram of positive/negative justification

Chapter 5 Positive/Negative Justification

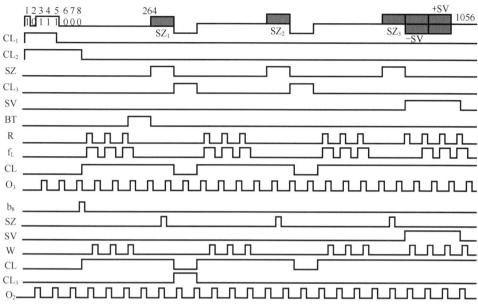

Figure 5-30 Main part wave of multiplex/demultiplex process

The main points of circuit design are discussed as follows.

(1) Sequential chain of multiplex/demultiplex: consider the frame structure requirement and the equipment simplification, the structure of the sequential chain is $4 \times 6 \times 11 \times 4 = 1056$.

(2) The cooperation of the synchronous multiplex and the justification: four kinds of control signal are sent in parallel to four justification cards by synchronous multiplex card, i.e. stuffing indication position (SZ), stuffing position (SV), read clock (R) and justification decision time signal (BT).

(3) The cooperation of synchronous demultiplex and code recovery: three kinds of control signal are sent in parallel to four code recovery cards by synchronous demultiplex card, i.e. stuffing indication position (SZ), stuffing position (SV) and write clock(W).

(4) Synchronous multiplex/demultiplex: the contents of synchronous multiplex/ demultiplex are frame alignment signal, four tributary channel information codes, four channel service information codes. They all have independent multiplex/demultiplex channels, and are controlled by multiplex/demultiplex time sequence. Consider the simplification of the equipments, synchronous multiplex uses four bits as a group (every tributary outputs one bit every time) which is parallel in and serial out, and

synchronous demultiplex is opposite that uses four bits as a group which serial in and parallel out.

(5) Synchronous control: the circuit is as similar as prototype of search circuit of the positive justification. The different point is only that in G.745 the frame synchrconous protection coefficient $\beta = 5$. search protection coefficient $\alpha = 2$. The frame alignment signal uses the pattern (10111000) of G.745 before modification (in may 1980 it is changed into 11101000).

(6) Four cards of multiplex/demultiplex use 129 IC packages altogether: 64 pieces of 7CS43C, 63 pieces of 7MY24D, one piece of 7MY14L and 2 pieces of 7QY14d.

(7) Plug-in card of justification: using simple delta control justification, the gain of differential amplifier is 5, a justification plug-in unit uses 37 IC packages and a pair of TB51A transistors. IC packages include 26 pieces of 7CS43C, 8 pieces of 7MY2D, 3 pieces of 7MY14L.

(8) Entire multiplexer uses one power source whose nominal voltage is 5V. In voltage range of 4.5-5.9 V, the multiplexer works normally. The peak to peak value of jitter is 14.5%UI.

5.10 CHARACTERISTIC COMPAISON OF JUSTIFICATIONS

Compare the positive/negative justification with the positive justification, we get main points of their characteristics as fellows.

(1) Positive/negative justification

A plesiochronous system is easily compatible with a synchronous system if the positive/negative justification is used. Because f_i and f_m take the same nominal, using simple equipment can realize the synchronous compatible. The positive justification also can realize plesiochronization/synchronization compatible but need considerably complex octave multiple frequency phase-locker and necessary control circuits.

(2) For positive/negative justification, design of frequency combination and frame structure is very easy. But for positive justification, to reduce the stuffing jitter need strictly restrict relationship among tributary frequency, multiplex

frequency and frame structure parameter, otherwise the stuffing jitter will be high. For positive/negative justification, stuffing jitter is only related to control method of the justification, and lightly related to $\frac{F_f}{f_1}$. When $\frac{F_f}{f_1}$ changes a little, stuffing jitter will not change rapidly.

(3) The positive/negative justification can fully use residual part of justification, because positive/negative justification is symmetric with nominal value. For reducing jitter, positive justification ratio can not take 0.5, i.e. it is not symmetric. Therefore, in the same condition, the positive/negative justification can permit tributary clock bigger tolerance domain. The bigger is the tributary frequency difference, the lower is the jitter.

(4) For the universal type of frame structure and the ability of anti-frame loss spread, the positive/negative justification has certain benefits also.

(5) For both methods, the maximum value of stuffing jitter is the same. But for positive justification the peaks of stuffing jitter are distributed irregularly within the tolerance. In the outside of the peaks there are very low valley and most of them are lower than 3%UI. For positive/negative justification the peaks are distributed along the nominal values, and ripple is very small. Hence, for the positive/negative justification the average value of stuffing jitter is larger than positive justification.

(6) Quantity of equipment is very close for both methods. But transmission cards of positive justification are digital circuits which are easy to adjust. Whereas that of positive/negative justification has an important delta control circuit which is difficult to adjust.

(7) Other characteristics like stuffing error, utilization of channel and synchronous searching characteristics for both methods are the same.

5.11 RECOMMENDATION OF CCITT

Soviet Union and East Europe use positive/negative justification technique. Now CCITT has recommended three references for positive/negative justification.

Their main parameters is shown in Table 5-1, the frame structure is shown in Figure 5-25.

Table 5-1   Parameters for nositive/negative justification

Recommendation	G.745	G.753	G.754
Multiplex clock frequency f_h (kbit/s)	8448	34 368	139 264
Tributary clock frequency f_L (kbit/s)	2048	8448	34 368
Number of tributaries m(-)	4	4	4
Frame frequency F_s (kHz)	8	16	64
Length of frame Ls (bit)	1056	2148	2176
Bits for tributary message Q(bit)	256	528	537
Bits for non-information mK(bit)	4×8	4×9	4×7
Searching protection frames a(-)	2	2	2
Number synchronous protection frame β(-)	5	3	3
Length of flame alignment signal n(bit)	8	10	10
Length of stuffing indication code n'(bit)	3	3	3

Chapter 6   Positive/0/Negative Justification

6.1   CONCEPT OF POSITIVE/0/NEGATIVE JUSTIFICATION

The positive/0/negative justification is also a justification technique providing a synchronous multiplex environment. The positive/0/negative justification has the same clock values with the positive/negative justification, that is, the synchronous multiplex clock frequency (f_m) is equal to the nominal of the tributary clock frequency (f_l): $f_m = f_l$. Consider that f_m and f_l vary within their own tolerance ranges, hence their transient values may have three cases: $f_m > f_l$, $f_m = f_l$ and $f_m < f_l$. The differences with the positive/negative justification are that these three cases really correspond to the three justification states: positive justification, non-justification and negative justification, and use three justification instructions to inform the recovery end. The justification process is also similar with the previous two techniques: the positive justification does not transmit information code at the positive justification position (+SV), the negative justification does transmit one bit more information code at the negative justification position (−SV); non-justification does transmit information code as usual at the +SV bit slot and does not transmit information code as normal at the −SV bit slot. The main differences of the positive/0/negative justification technique with the previous two justification techniques are: the justification request recognition method and the recovery method are not the same completely. This Chapter will introduce two typical kinds of positive/0/negative justification techniques: the one is a simple positive/0/negative justification of delta controlled justification and phase-lock loop smooth recovery, the another is a positive/0/negative justification of fixed threshold decision controlled justification and digital smooth recovery. According to their main technical characteristics, the former is called a delta controlled positive/0/negative justification technique; the later is called digital smooth positive/0/negative justification technique.

6.2 THE PRINCIPLES OF DELTA CONTROLLED POSITIVE/0/NEGATIVE JUSTIFICATION

The recovery operation of the delta controlled positive/0/negative justification is implemented by a analog two order phase-lock loop. This is the same with the recovery principles of the positive justification and the positive/negative justification, it actually uses the same plug-in unit. Hence, only the justification principles are worth to be described.

The justification recognition circuit of the delta controlled positive/0/negative justification is also the same wim that of the delta controlled positive/negative, both use a simple delta controlled circuit. The difference is the specification of implementing controls and sending the corresponding instructions. The control process of the positive/negative justification is: the request of positive or negative justification is recognized by the previous frame, in the following frame the corresponding positive or negative justification control operation is realized. Although the entire recognition/control process deals with two frames, but the process completes within a duration of a frame long. Because this kind of justification contains only positive or negative two justification control operations, only a group of control instruction position (SZ) is necessary, if specify that all "1" is the positive justification instruction, then all "0" is just the negative justification instruction. The positive and the negative justification controls appear alternately, this equal to non-justification. See Figure 6-1, the positive/0/negative justification control process is a little bit complicated that the recognition results of the previous two frames are used to decide the control operation of the third frame. If the previous two frames both request a positive justification, the third frame will do the positive justification control; if the previous two frames request a negative justification, the third frame will do the negative justification; if the requests of previous two frames are different (the first frame requests a positive justification but the second requests a negative justification or vice versa), the third frame will not do justification. Obviously, the positive/0/negative justification process deals with three frames, but as long as all additional circuit is used to remember the recognition result, the entire control process will complete within a frame period.

Because the positive/0/negative justification contains three kinds of control operations: positive, 0, negative, usually two groups of control instruction positions (+SZ, and −SZ) and two justification positions (+SV and −SV) are need to specified. If +SZ is all "1" code, +SV is idle; if +SZ is all "0" code, +SV transmits information code ad usual. If −SZ is all "1" code, −SV transmits information code, if −SZ is all "0" code, −SV is idle. From the above introduction it is not difficult to see that the specifications of the particular justification control methods of the positive/0/negative justification and the positive/negative justification are different, but the actual circuits are close. The positive/0/negative justification control circuits can be made by supplementing little amount of memory and logical decision circuits into the positive/negative justification control circuits. As for the recovery circuits, only the control instruction decision circuits are different a bit (i.e. for two groups of control instruction positions, set up two groups of the same instruction decision circuits), the smooth phase-lock loop is complete the same. Because the previous chapter already introduced the positive/negative justification in detail, this Chapter will not particularly introduce such delta control positive/0/negative technique.

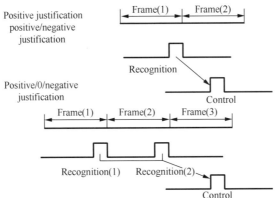

Figure 6-1 Recognition/Control time relationship

The following sections of this chapter will limit only to introduce the digital smooth positive/0/negative justification technique.

6.3 DIGITAL SMOOTH POSITIVE/0/NEGATIVE JUSTIFICATION PRINCIPLES

The digital smooth positive/0/negative justification is already mentioned before, the justification control is very simple. Firstly, set up two fixed thresholds for the tributary buffer: the positive justification threshold and the negative justification threshold. If the read/write time difference is between two thresholds, do not do justification; if it is higher than the positive threshold, request a positive justification, i.e. slow down read clock (f_m) to reduce the read/write time difference; if it is lower than the negative threshold, request a negative justification, i.e. speed up the read clock (f_m) to increase read/write time difference.

This section will emphasize to introduce the digital smooth recovery principles. See Figure 6-2, the digital smooth loop consists of tributary clock generator, buffer and frequency switching circuits. The tributary clock generator generates three kinds of clock frequency, the frequencies are f_1, $f_1 + F_{ms}$, $f_1 - F_{ms}$ respectively, where, f_1 is the tributary nominal frequency; the size of the buffer is N, the write clock frequency is f_m; the read clock frequency is selected from the above three kinds of clock frequencies; the switching circuit is controlled by the timing signal T_{ms} and the stuffing indication signal SZ, where $T_{ms} = \dfrac{1}{F_{ms}}$, which is the minimum interval between two justifications, the tributary clock is switched once in each interval of T_{ms} at the recovery end.

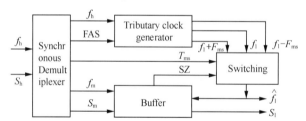

Figure 6-2 Positive/0/Negative justification principles

When non-justification, the buffer write clock frequency is f_m, the read clock frequency is $\hat{f}_1 = f_1$ which is keeping T_{ms}, altogether $N_0 = f_1 T_{ms}$ bit are read out, given $f_m = f_1$. When positive justification, the write clock f_m stops a beat, altogether (N_0-1) bit are written in. The read clock is $\hat{f}_1 = f_1 - F_{ms}$ which is keeping T_{ms}, altogether (f_1-F_{ms}) $T_{ms} = (N_0-1)$ bit are read out. Obviously it still keeps the read/write balance. When negative justification, the write clock is f_m into which a beat is inserted, altogether (N_0+1) codes are written in. The read clock is $\hat{f}_1 = f_1 - F_{ms}$ which keeps T_{ms}, altogether (f_1-F_{ms}) $T_{ms} = (N_0-1)$ codes are read out. Obviously, it also keeps read/write balance. Summarize up the above discussion, all of the three cases insure that the average write frequency is equal to the average read frequency, i.e. $\overline{f}_m = f_1 \cdots$

When positive justification, (N_0-1) codes ale read out during T_{ms}, then the code width is $T_{ms}/(N_0-1)$, consider the $T_{ms}/N_0 = T_1$ (written as UI), when $N_0 \Box T_{ms}/(N_0-1) \approx \left(1 + \frac{1}{N_0}\right) \Box UI$; When negative justification, (N_0+1) codes are read out during T_{ms}, then the code width is $T_{ms}/(N_0-1) \approx \left(1 + \frac{1}{N_0}\right) \Box UI$. Obviously, the longer is the keep time (T_{ms}) or lower is the justification frequency (F_{ms}), the smaller is the code element width. But the lowest justification frequency (F_{ms}) can not be less than the maximum relative variation of the tributary clock frequency, i.e. satisfies the following conditions:

$$F_{ms} \geq |\Delta f_1| + |\Delta f_m|$$
$$T_{ms} \leq 1/(|\Delta f_1| + |\Delta f_m|) \quad (6\text{-}1)$$

In a word, within the maximum limitation, the bigger is T_{ms}, the smaller is the difference of the code elements. The above is just the basic idea or basic principle of such positive/0/negative justification.

From the above discussion it is not difficult to see that the justification control and the recovery control of the positive/0/negative justification and the positive/negative justification have no essential difference, the only difference is the methods to extract the tributary clock; in addition, because the justification process is no longer limited within a frame, there is no problem in the transmission of the justification instructions. Hence, we would rather keep three justification states, transmit three kinds of control instructions. The above two differences will give this new justification some good technical characteristics.

6.4 FRAME STRUCTURE

As it is described already in the previous section, the justification period (T_{ms}) should satisfy the trace condition:

$$T_{ms} \leq \frac{1}{|\Delta f_1| + |\Delta f_m|} \quad (6\text{-}2)$$

For this prerequisite, the bigger is the T_{ms}, the smaller is the difference of the code element width; but the bigger is the T_{ms}, the bigger is the amount of equipment. Generally speaking, do just what is necessary. Except the contents of the primary frame, the following contents should be set within a frame:

—Multiframe alignment signal (MFAS);
—Justification decision position (BT);
—Justification indication signal (SZ);
—Negative justification position (−SV);
—Positive justification position (+SV).

The general principle to form the justification frame structure is that the time sequence link is as simple as possible and easy to generate; the maximum change of the read/write time difference caused by extracting/inserting service bits is as small as possible to reduce the tributary buffer size; give the previous and the following programs enough time, the entire justification control process should complete within a justification frame; fully consider the compatibility with the frame structure of the equipment being used.

[Example 1] Australia 2048/8448kbit/s multiplexer frame structure (Figure 6-3):

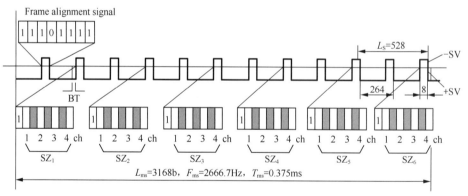

Figure 6-3　Australia 2/8Mbit/s multiframe structure

The justification frame consists of 6 primary frames, the justification cycle $T_{ms}=6 \cdot T_s=528 \times 6T_h$; the frame alignment signal of each primary frame occupies slots 257-264, the pattern is 11101000; the multiframe alignment signal occupies the first slot of the last 8 slots of each primary frames, the pattern is 110010; the stuffing indication signal of each tributary consists of 6 bits code. the slots 2, 4, 6, 8 of the last 8 slots in each primary frame. contral the tributaries 1,2, 3, 4 respectively. The patterns are respectively: positive justification 000111; non-justification 001100; negative justification 11000x, where x is the negative justification position, i.e. when negative justification, transmit the code in slot *x*; the justification decision time is at the second 8 slots group counted backward of the first primary frame; the positive justification position is within the first 4 slots group at the beginning of the next justification frame, i.e. the first slot of each tributary of the next justification frame (Detailed in Figure 6-4).

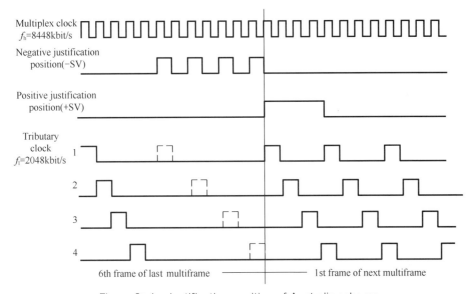

Figure 6-4　Justification position of Australia scheme

[Example 2] West German 64/2048kbit/multiplexer frame structure (Figure 6-5):

A justification frame contains 16 multiframes or 16×16 primary frames, the justification frame period $T_{ms}=16\times256T_h$; the primary frame structure accords with the CCITT recommendation G.732, the frame alignment is at the last 7 slots of the time slot 0 in each odd frame, the pattern is 0011011; the multiframe alignment signal is at the first 4 slots of the 16th time slot of the first primary frame, the pattern is 0000: the justification indication occupies the first slot (slot a) of the 4 slots of the channel associate signaling assigned to the tributary in the 16th time slot. Within each multiframe, each tributary is assigned a justification indication position which occupies 16 slots in a justification time, in the cases of positive, none, and negative justification, their particular patterns are shown in Figure 6-6. Where,

PDH for Telecommunication Network

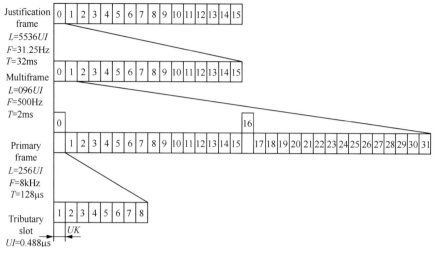

Figure 6-5 [Example 2] 64/2048kbit/s multiframe structure

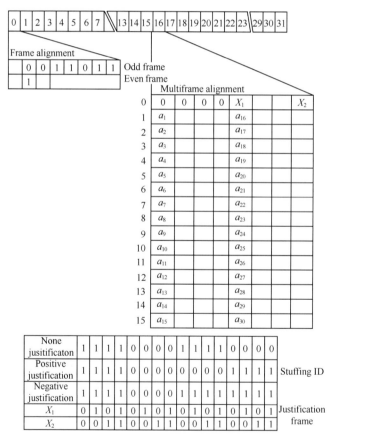

Figure 6-6 West German PCM-30D justification frame structure

the justification frame alignment signal pattern and the slots assigned (slot x_1 and x_2) is also listed; the negative justification position is set at the last slot of the negative justification indication position; the positive justification position is set at the first information code slot of the corresponding tributary in the next justification.

6.5 JUSTIFICATION AND RECOVERY CONTROL

The justification control of the positive/0/negative justification are similar with that of the positive justification and positive/negative justification when consider only the control method that each justification justifies (deduct or insert) a bit. But there are great differences in the justification time selection and its effect on stuffiilg jitter. The positive justification and the positive/negative justification have strict limitation on the justification time, or else the stuffing jitter will be increased obviously. Whereas the positive/0/negative justification is not so, the justification time and the stuffing jitter have no relationship completely; it need only to provide a justification chance in each justification frame, and it is not important when the justification is executed. This is very important that it is not necessary to specify the justification time strictly, then the justification threshold may be properly relaxed. In this case, even though the read/write time difference drifts quite large between the thresholds, it will not request justification, this extends the keep time of non-justification state, in turn further lengthens the non stuffing jitter time (See Figure 6-7). The recovery control are the same with that of the positive justification and the positive/negative justification. All execute the recovery control according to the commands of the justification instructions: if indicate positive justification, deduct the contents of +SV: if indicate negative justification, extract the contents of −SV; if indicate no justification, do not take justification recovery control. Extraction of justification indication (SZ) signal may take correction steps as usual.

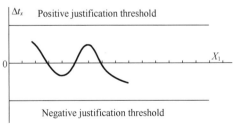

Figure 6-7　Sketch of justification threshold

The particular implementation of the justification control and the recovery control are relative complicated, the particular practical logic and circuits may use various schemes. but the control process and the aim are all the same. The following examples will give further descriptions:

[Example 1] the justification recognition logic of Australia scheme (Figure 6-8):

Status	Positive justfication			Negative justification		
$\Delta t_x(UI)$	2	1	0	5	6	7
	0−6=−6	0−7=−7	0−0=0	0−3=−3	0−2=−2	0−1=−1
	1−7=−6	1−0=1	1−1=0	1−4=−3	1−3=−2	1−2=−1
	2−0=2	2−1=1	2−2=0	2−5=−3	2−4=−2	2−3=−1
	3−1=2	3−2=1	3−3=0	3−6=−3	3−5=−2	3−4=−1
	4−2=2	4−3=1	4−4=0	4−7=−3	4−6=−2	4−5=−1
	5−3=2	5−4=1	5−5=0	5−0=5	5−7=−2	5−6=−1
	6−4=2	6−5=1	6−6=0	6−1=5	6−0=6	6−7=−1
	7−5=2	7−6=1	7−7=0	7−2=5	7−1=6	7−0=7
	−6=1010	−7=1001	0=0000	—	−2=1110	—
	2=0010	1=0001		3=1101	6=0110	1=1111
				5=0101		7=0111
Recognition Logic	$J_+=X\bar{Y}W\bar{Z}+\bar{X}YW\bar{Z}+X\bar{Y}\bar{W}Z+$ $\overline{XY}WZ+\overline{XY}W\bar{Z}=\bar{Y}W\bar{Z}+$ $\overline{XYW}+\bar{Y}W\bar{Z}$			$J_-=XY\bar{W}\bar{Z}+XY\bar{W}\bar{Z}+XYW\bar{Z}+$ $\bar{X}YW\bar{Z}+\bar{X}YW\bar{Z}+\bar{X}YW\bar{Z}+$ $\bar{X}YWZ=XZ+YW$		

Figure 6−8 [Example 1] Justification recognition logic

The justification buffer size is $N=8$; specify: no justification when the read/write time difference Δt_x is between 2-5UI; positive justification when it is equal to or less than 2UI; negative justification when it is equal to or greater than 5UI. Take the corresponding calculation recognition step by step, the digits detected are stored into a four step ($XYWZ$) register, when the state $J_+ = \overline{XYW} + \overline{YWZ} + \overline{YW}\,\overline{Z}$ appears, a positive juistification is decided; when the state $J_-=YZ+YW$ appears a negative justification is decided; when no J_+ and J_- appear, no justification is decided.

The justification instruction code recognition logic of the Australia scheme:

At the recovery end the justification instructions of each tributary are written into the register, realize 1 bit tolerance recognition by the recognition logic.

Chapter 6 Positive/0/Negative Justification

The negative justification instruction code $SZ_-=11000X$, X is negative justification position ($-SV$), write the first 5 bits instruction code into the register abcde, the 1 bit tolerance recognition logic is as following:

$$SZ_- = ab\overline{cde} + \overline{a}b\overline{cde} + a\overline{bcde} + ab\overline{cde} + abc\overline{de} + ab\overline{cd}\overline{e}$$
$$= (a+\overline{a})b\overline{cde} + (b+\overline{b})a\overline{cde} + (c+\overline{c})ab\overline{de} +$$
$$(d+\overline{d})ab\overline{ce} + (e+\overline{e})ab\overline{cd}$$
$$\therefore SZ_- = b\overline{cde} + a\overline{cde} + ab\overline{de} + ab\overline{ce} + ab\overline{cd}$$

The positive justification instruction code $SZ_+=000111$, write it into register abcdef, 1 bit tolerance recognition logic is as following:

$$SZ_+ = \overline{abc}def + \overline{abc}def + \overline{abc}def + \overline{abc}def + \overline{abc}def +$$
$$\overline{abcd}ef + \overline{abcde}f$$
$$= (a+\overline{a})\overline{bc}def + (b+\overline{b})\overline{ac}def + (c+\overline{c})\overline{ab}def +$$
$$(d+\overline{d})\overline{abc}ef + (e+\overline{e})\overline{abc}df + (f+\overline{f})\overline{abc}de$$
$$= \overline{bc}def + \overline{ac}def + \overline{abd}def + \overline{abc}ef + \overline{abc}de$$

The particular justification instruction code recognition circuit is shown in Figure 6-9.

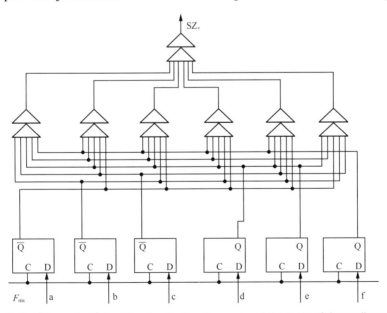

Figure 6-9 [Example 1] Justification instruction recognition logic of Australia scheme

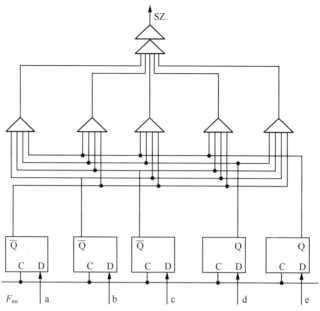

Figure 6-9 [Example 1] Justification instruction recognition logic of Australia scheme (续)

6.6 JUSTIFICATION TRANSITION PROCESS

When the tributary frequency (f_l) and the multiplex frequency (f_h) are both equal to their own nominals, the read/write time difference (Δt_x) is:

$$\Delta t_x = \Delta t_0 + \left[\frac{mf_l}{f_h}(x+g) - x \right] \cdot UI$$

where, Δt_x—write time ahead relative to read time, when read the xth code;

Δt_0—Δt_x when $x=0$, i.e. the initial read/write time difference;

m—number of the tributary multiplexed;

f_l—tributary nominal frequency;

f_h—nultiplex nominal frequency;

g—number of read clock beats stopped;

T_l—tributary code element width, usually written as UI.

Assume $\Delta t_0=0$; the tributary clock nominal (f_l) is equal to the tributary synchronous multiplex clock nominal f_m, here

$$f_m = \frac{Q}{L_s} \cdot f_h \qquad (6\text{-}4)$$

Chapter 6 Positive/0/Negative Justification

where, Q—the number of information codes of each tributary in a frame;
L_s—frame length.
When $x = Q$, $g = K$,

$$\Delta t_x = \left[\frac{mf_1}{f_h}(Q+K) - Q\right] \cdot UI \quad (6\text{-}5)$$

$$= \left[\frac{f_1}{\frac{f_m L_s}{Q}} \cdot m(Q+K) - Q\right] \cdot UI$$

Given $L_s = m(Q+K)$, $f_1 = f_m$;

$$\therefore \Delta t_x = \left[\frac{f_1 Q}{f_m L_s} \cdot L_s - Q\right] \cdot UI = 0 \quad (6\text{-}6)$$

i.e. when the tributary clock and the multiplex clock are both equal to their own nominals, the net variation of the read/write time difference of the positive/0/negative justification is equal to 0 through a entire frame (See Figure 6-10).

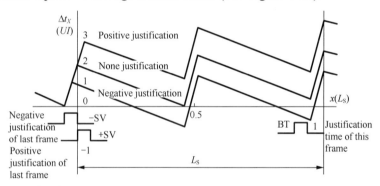

Figure 6-10 Read/Write time difference variation within a primary frame

When both the tributary clock and the multiplex clock deviate from their own nominal, their transient values are respectively:

$$f'_1 = f_1 + \Delta f_1$$
$$f'_h = f_h + \Delta f_h \quad (6\text{-}7)$$

then the read/write time difference is:

$$\Delta t_x = \left[\frac{m(f_1 + \Delta f_1)}{f_h + \Delta f_h}(x+g) - x\right] \cdot UI \quad (6\text{-}8)$$

When $\Delta f_h/f_h \ll 1$, $\Delta f_1/f_1 \ll 1$, let formula (6-8) be first order approximate, get:

– 157 –

$$\Delta t_x \approx \left[\frac{mf_1}{f_h}\left(1+\frac{\Delta f_1}{f_1}-\frac{\Delta f_h}{f_h}\right)(x+g)-x\right]\cdot UI \quad (6\text{-}9)$$

$$= \left[\frac{mf_1}{f_h}(x+g)-x\right]\cdot UI + \left[\frac{mf_1}{f_h}\left(\frac{\Delta f_1}{f_1}-\frac{\Delta f_h}{f_h}\right)(x+g)\right]\cdot UI$$

Let:
$$f_1\left(\frac{\Delta f_1}{f_1}-\frac{\Delta f_h}{f_h}\right)=f_{ms} \quad (6\text{-}10)$$

It expresses the offset of the equivalent transient frequency difference between the tributary frequency and the multiplex frequency relative to the equivalent nominal frequency, this frequency offset is just the frequency difference which the justification mechanism actually need to justify. It is called actual justification rate. Hence

$$\Delta t_x = \left[\frac{mf_1}{f_h}(x+g)-x\right]\cdot UI + \left[\frac{mf_{ms}}{f_h}(x+g)\right]\cdot UI \quad (6\text{-}11)$$

The formula (6-11) is just the general expression of the read/write time difference of the positive/0/negative justification process. Where, m is the number of trilbutaries, f_1 is the tributary frequency, f_h is the multiplex frequency, x is the time to read the xth information code, g is non-information bits between 0 and x, T_1 is the tributary code element width written as UI, f_{ms} is the actual justification rate, accordillg to the difference of symbols and quantity of the frequency difference of the tributaries and the multiplex, it may be positive, negative or zero.

When pass through a frame, and no justification is executed within this frame i.e. $x = Q, g = K$,

$$\Delta t_x = \left[\frac{mf_1}{f_h}(Q+K)-Q\right]\cdot UI + \left[\frac{mf_{ms}}{f_h}(Q+K)\right]\cdot UI$$

Known from formula (6-7),

$$\left[\frac{mf_1}{f_h}(Q+K)-Q\right]=0$$

$$\Delta t_x = \frac{mf_1}{f_h}(Q+K)\cdot UI$$

$$= \frac{m(Q+K)}{f_h}\cdot f_{ms}\cdot UI \quad (6\text{-}12)$$

$$= \frac{f_{ms}}{F_s}\cdot UI$$

The formula (6-12) is the collection of read/write time difference passing through a frame when no justification is executed within the frame. In the positive/0/negative justification system, specify that M frames forma justification frame, there is a chance of justification in a justification frame, and specify that only one information bit is allowed to justified in a chance. Hence, the frame rate of a justification frame:

$$F_{ms} = F_s/M \qquad (6\text{-}13)$$

From the formula (6-12) already get the read/write time difference collection passing through a primary frame, from the formula (6-12) may directly get the read/write time difference collection passing through a justification (no justification):

$$\Delta t_x = \frac{f_{ms}}{F_s} \cdot M \cdot UI$$

$$= \frac{f_{ms}}{F_s/M} \cdot UI$$

$$= \frac{f_{ms}}{F_{ms}} \cdot UI$$

Known that f_{ms} expresses the actual justification frequency; the justification frame rate F_{ms} expresses the maximum possible justification rate. According to the traditional definition of justification ratio (S):

$$S = \frac{f_{ms}}{F_{ms}} \qquad (6\text{-}14)$$

$$\Delta t_1 = S \cdot UI \qquad (6\text{-}15)$$

When $\Delta t_0 \neq 0$ at the beginning of the justification process, passing through y justification frames, and no justification was executed within these y justification frames, the read/write time difference is:

$$\Delta t_y = \Delta t_0 + S \cdot y \cdot UI \qquad (6\text{-}16)$$

The formula (6-16) is the read/write time difference collection passing through y justification frames when no justification is executed. Obviously, the read/write time difference collection (Δt_y) and the justification ratio (S) are proportional. According to different rates of the tributary and the multiplex, S may be positive, negative or zero, hence, Δt_y may also be positive, negative or zero.

When a positive justification is executed while passing through a frame time, i.e. $x = Q-1, g = K+1$,

$$\Delta t_x = \left[\frac{mf_1}{f_h}\{(Q-1)+(K+1)\}-(Q-1)\right]\cdot UI$$
$$+\left[\frac{mf_{ms}}{f_h}\{(Q-1)+(K+1)\}\right]\cdot UI \quad (6\text{-}17)$$
$$=\left[\frac{mf_1}{f_h}(Q+K)-Q\right]\cdot UI + UI + \left[\frac{mf_{ms}}{f_h}(Q+K)\right]\cdot UI$$
$$=\left(1+\frac{f_{ms}}{F_s}\right)\cdot UI$$

When a justification is executed within the period while passing through a justification frame, and

$$\Delta t_1 = \left(1+\frac{f_{ms}}{F_{ms}}\right)\cdot UI = (1+S)\cdot UI \quad (6\text{-}18)$$

When the read/write time difference $\Delta t_0 \neq 0$ at the beginning of the justification process, passing through y justification frames, and K ($K \leq y$) justifications are executed within the period, the read/write time difference is:

$$\Delta t_y = \Delta t_0 + (K+S_y)\cdot UI \quad (6\text{-}19)$$

The formula (6-19) is the general expression of the read/write time difference of the positive/0/negative justification. As a general case, the initial read/write time difference Δt_0 may be positive, negative or zero; the justification times K may also be positive (positive justification times), negative (negative justification times), or zero (no justification); the justification ratio may also be positive, negative or zero. Consider the specification of implementing justification: when $\Delta t_y \geq +\Delta t_s$ (positive justification threshold), the positive justification is executed; when $\Delta t_y \leq -t_s$ (negative justification threshold), the negative justification is executed; when $-\Delta t_s < \Delta t_y < +\Delta t_s$ no justification. Meanwhile consider three cases of $S>0$, $S=0$, $S<0$, may sum up nine particular forms of the general expression (6-19) of the read/write time difference:

Δt_y(UI) \ S \ Δt_y	$S > 0$	$S = 0$	$S < 0$
$\Delta t_y > +\Delta t_s$	1. $\Delta t_0 - (1-S)_y$	2. $\Delta t_0 - y$	3. $\Delta t_0 - (1+S)_y$
$-\Delta t_s < \Delta t_y < +\Delta t_s$	4. $\Delta t_y + S_y$	5. Δt_0	6. $\Delta t_0 - S_y$
$\Delta t_y < -\Delta t_s$	7. $\Delta t_0 + (1+S)_y$	8. $\Delta t_0 + y$	9. $\Delta t_0 + (1-S)_y$

Chapter 6 Positive/0/Negative Justification

Figure 6-11, 6-12 and 6-13 describe the corresponding nine typical justification transition processes with the above formulas. See from the Figures, these nine typical justification processes tend to a steady work states finally.

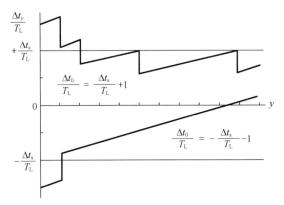

Figure 6-11 Positive/0/Negative justification process ($S > 0$)

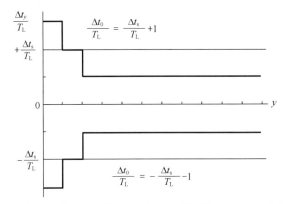

Figure 6-12 Positive/0/Negative justification process ($S = 0$)

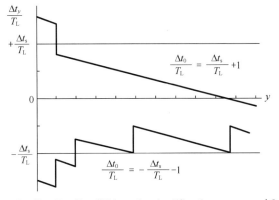

Figure 6-13 Positive/0/Negative justification process ($S < 0$)

6.7 DIGITAL SMOOTH

The digital smooth technique is a method that read clock is provided at the recovery end. The section 3 already introduced its basic work principle. Its work is periodic with the justification frames. When the read/write time difference of the receiving buffer is between two thresholds, the tributary clock nominal (f_1) is used as read clock keeping T_{ms} time, altogether $N_0 = f_1/T_{ms}$ information codes are read; When the read/write time difference is greater than the negative justification threshold, ($f_1 + F_{ms}$) is used as read clock keeping T_{ms} time, the number of information codes read out is:

$$N^- = (f_1 + F_{ms}) T_{ms} = N_0 + 1 \quad (6\text{-}20)$$

When the read/write time difference is less than the positive threshold, ($f_1 - F_{ms}$) is used as read clock keeping T_{ms} time, the number of information codes read out is:

$$N^+ = (f_1 + F_{ms}) T_{ms} = N_0 - 1 \quad (6\text{-}21)$$

See Figure 6-14, repeatedly in this way, the justification and the recovery are balanced.

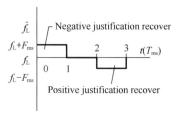

Figure 6-14 Digital phase-lock smooth principle ($T_{ms} \cdot F_{ms} = 1$)

The corresponding output code elements width are respectively:

$$T_0 = T_{ms}/N_0 = T_1$$

$$T_- = T_{ms} N^- \approx \left(1 - \frac{1}{N_0}\right) T_1 \quad (6\text{-}22)$$

$$T_+ = T_{ms} N^- \approx \left(1 + \frac{1}{N_0}\right) T_1 \quad (6\text{-}23)$$

N_0 is usually between $1 \times 10^2 - 1 \times 10^3$, obviously that the difference of code element width is usually less than 2%UI.

The main difficulty to implement this kind of scheme is to generate a group of special clocks for recovery, which have strict phase switch relationship with each other. From the above formulas we may directly get:

$$\frac{f_1}{f_1+F_{ms}} = \frac{N_0}{N_0+1}$$
$$\frac{f_1}{f_1-F_{ms}} = \frac{N_0}{N_0-1}$$
$$\frac{f_1}{N_0} = \frac{f_1+F_{ms}}{N_0+1} \quad (6\text{-}24)$$
$$\frac{f_1}{N_0} = \frac{f_1-F_{ms}}{N_0-1}$$

The principle of the phase-lock corresponding to (f_1+F_{ms}) and (f_1-F_{ms}) is shown in Figure 6-15.

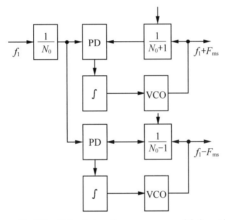

Figure 6-15 Principle of recovery multiplex clock

6.8 PULSE SMOOTHLY PRINCIPLE OF WEST GERMAN PCM30D

PCM30D is a kind of 64/2048kbit/s multiplexer. The recovery part uses pulse smoothly to generate the tributary clock. The common pan generates three kinds of common tributary clocks (f_+, f_0 and f_-), each tributary selects one of them within justification frame (T_{ms}) according to SZ indication. The pulse smoothly control diagram is shown in Figure 6-16).

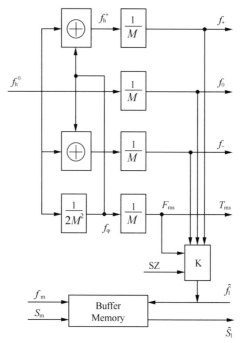

Figure 6-16　Pulse smoothly control

The periodic process to generate the tributary clocks is compled within $T_{ms}=32ms$, the corresponding justification rate F_{ms} is equal to 31.25Hz; partition T_{ms} equally into $2M=64$ subsections, hence each subsection is 32ms/64=0.5ms; in each subsection, f_h^0 and f_φ are modulo N added: at a positive clock tributary, the $\varphi_0 f_h^0$ and the φ_0 of f_φ are modulo 2 added to make f_h^0 delay a half of code element time and get f_h^+; at a negative clock tributary, $\varphi_0 f_h^0$ and the φ_k of f_φ are modulo 2 added to make f_h^0 advance a half of code element time and get f_h^0 (See Figure 6-17); then three clocks (f_h^+, f_h^0, f_h^-) is demultipled by $M=32$. f_h^0 is M demultipled into $f_0=64$kHz which is used as non-justification tributary clock; f_h^0 is M demultipled into $f_+ = f_0 - F_{ms} = 63\ 968$Hz which is used as the positive justification tributary clock; f_h^- is M demultipled into $f_-=f_0-F_{ms}=64\ 031.25$Hz which is used as the negative tributary clock. See Figure 6-18, in each subsection, f_+ is delayed $\frac{1}{2M} \cdot UI$, altogether delayed $1 \cdot UI$ through $2M$ subsections, i.e. recover a positive justification once; in each subsection, f_- is advanced $\frac{1}{2M} \cdot UI$, altogether advanced

1• UI through 2M subsections, i.e. recover a negative justification once.

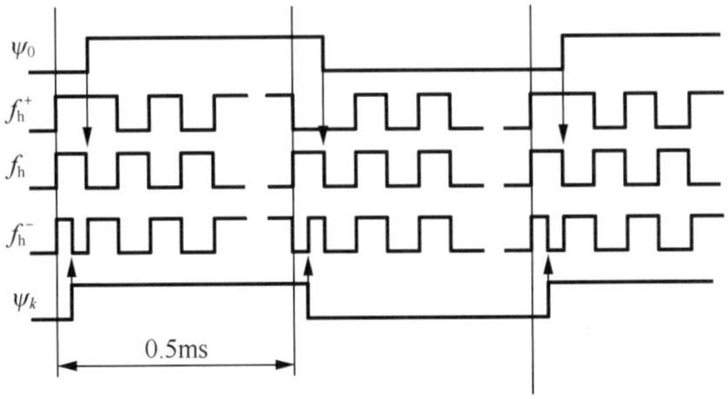

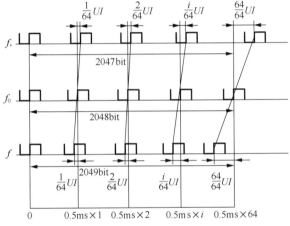

Figure 6-17 Smooth control principle

Figure 6-18 Smooth control time waves

Because a justification is executed once and only once in each subsection, hence there are $2M$ unequal widths of tributary clocks of f_+ and f_-. f_h^0 is M demultiplied into an uniform tributary clock with a period of $T_1 = MT_h^0$; f_h^+ is M demultiplied into a tributary clock of $2M$ beats with a period:

$$T_+ = \left(\frac{1}{2} + M\right)T_h^0 = \left(\frac{1}{2M} + 1\right)T_1 = \left(1 + \frac{1}{64}\right)T_1$$

The periods of the rest beats are all T_1; f_h^- is M demultiplied into a tributary clock of $2M$ beats with period:

$$T_- = \left(-\frac{1}{2} + M\right)T_h^0 = \left(\frac{-1}{2M} + 1\right)T_1 = \left(1 - \frac{1}{64}\right)T_1$$

The periods of the rest beats are all T_1; Thus get the peak-peak value (Δ) of the code element width difference using this kind of pulse smooth technique:

$$\Delta = T_+ - T_- = \frac{1}{M} \cdot UI = 3.1\% UI$$

Sum up the above discussion, the design process using modulo 2 controlled pulse smooth technique:

(1) According to the requirement of rate justification range $F_{ms} > |\Delta f_m| + |\Delta f_1|$, decide the value of F_{ms}, correspondingly get $T_{ms} = 1/F_{ms}$;

(2) According to the tolerance of the code element width difference Δ, decide the value of M: $M \geqslant \dfrac{1 \cdot UI}{\Delta}$;

(3) According to the following formulas to calculate the frequencies:

$$f_h^0 = Mf_1$$
$$f_h^+ = Mf_1 - f_\varphi$$
$$f_h^- = Mf_1 + f_\varphi$$
$$f_\varphi = MF_{ms}$$
$$f_+ = f_h^+ / M = f_1 - F_{ms}$$
$$f_- = f_h^- / M = f_1 + F_{ms}$$
$$f_0 = f_1$$

[Example] PCM-30D: f_h=2048kHz±50ppm, f_1=64kHz±100ppm, $|\Delta f_m| + |\Delta f_1| \approx 10$Hz; let T_{ms}=32ms, F_{ms}=31.25Hz, S=0.32, M=32; get Δ=3.1%UI; f_φ=1kHz, f_h^0=2048kHz, f_h^+=2047kHz, f_h^-=2049kHz, f_0=64kHz, f_+=63 968.7Hz, f_-=64 031.25Hz.

6.9 AUSTRALIA 2/8Mbit/s MULTIPLEXER PULSE SMOOTH PRINCIPLE

The Australia 2/8Mbit/s multiplexer is similar with the West German PCM30D, both use the pulse smooth technique, however the particular methods are different in two points: here T_{ms} is equally partitioned into M subsectinos, each subsection justifies one f_h^0 beat, whereas PCM30D partitions it into $2M$ sutbsections, each subsection justifies $1/2$ f_h^0 beat; here it uses demultiplication control to implement rate justification, whereas PCM30D uses modulo 2 control to implement rate justification. The principle of generation of tributary clock of this kind of multiplexer is shown in Figure 6-19.

Chapter 6 Positive/0/Negative Justification

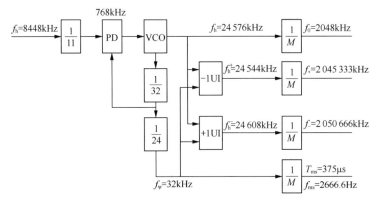

Figure 6-19 Principle of tributary clock generation

The process to implement rate justification using demultiplication control is shown in Figure 6-20. f_h^0 is M demultiplied into uniform clock $f_0 = f_L$; in each subsection, the φ_0 phase pulse of f_φ deducts one beat of f_h^0 to get f_h^+, this is equivalent to delay f_h^0 one beat. it is M demultiplied to get the positive justification tributary clock f_+, the width of its first beat is:

$$T_+ = (M+1)T_h = \left(1 + \frac{1}{M}\right)T_1 \qquad (6-25)$$

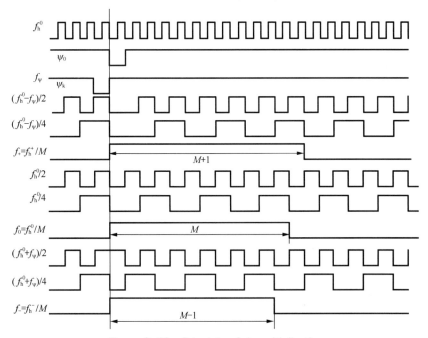

Figure 6-20 Principle of demultiplication

- 167 -

The widths of the rest beats are all equal to T_L; in each subsection, the φ_k phase pulse of f_φ deducts one beat of f_h^0 to get f_h^-, this is equivalent to advance f_h^0 one beat, it is M demultiplied to get the positive justification tributary clock f_-, the width of its first beat is:

$$T_- = (M-1)T_h = \left(1 - \frac{1}{M}\right)T_1 \quad (6\text{-}26)$$

The widths of the rest beats are equal to T_1. Thus get the peak-peak value of the code element width difference using demultiplication control technique:

$$\Delta = T_+ - T_- = \frac{2}{M} \cdot UI \quad (6\text{-}27)$$

Sum up the above discussion, the design process using demultiplication controlled pulse smooth technique:

(1) According to the requirement of rate justification range $F_{ms} > |\Delta f_m| + |\Delta f_1|$, decide the value of F_{ms}, correspondingly get $T_{ms} = 1/F_{ms}$;

(2) According to the tolerance of the code element width difference Δ, decide the value of M: $M \geqslant \dfrac{2 \cdot UI}{\Delta}$;

(3) According to the following formulas to calculate the frequencies:

$$f_h^0 = Mf_1$$
$$f_h^+ = Mf_1 - f_\varphi$$
$$f_h^- = Mf_1 + f_\varphi$$
$$f_\varphi = MF_{ms}$$
$$f_+ = f_h^+ / M = f_1 - F_{ms}$$
$$f_- = f_h^- / M = f_1 + F_{ms}$$
$$f_0 = f_1$$

[Example] Australia 2/8Mbit/s multiplexer: f_h=8448kHz±30ppm, f_1=2048kHz±50ppm, $|\Delta f_m|+|\Delta f_1|$ =164Hz; F_{ms}=2666.6Hz, $S_{max} = (|\Delta f_m|+|\Delta f_1|)/F_{ms}$ =6.1%, T_{ms}=375μs, M=12; $\Delta = \dfrac{2}{12} \cdot UI$ =16.6%UI; f_φ=32kHz, f_h^0 =24 576kHz, f_h^+ =24 608kHz, f_h^- =24 544kHz, f_0=2048kHz, f_+=2 050 666Hz, f_- =2 045 333Hz.

6.10 ENGINEERING APPLICATIONS OF POSITIVE/0/NEGATIVE JUSTIFICATION

The delta control positive/0/negative justification technique may work in the case that the multiplex rate and the tributary rates are related in a simple multiple, it is conveniently to realize synchronization and asynchronization compatible, stuffing jitter is low. Hence, it suits in the applications of the interfaces of a high order synchronous digital system and an asynchronous digital system. CCITT already recommends it in G.709. In addition, this justification technique also suits the application of subgroup synchronization/lasynchronization compatible multiplex, but the work rate may not be over low. Because the recovery of the delta controlled positive/0/negative justification will use analog second order phase-lock loop, the voltage-controlled oscillator is hard to be stable in a relative low frequency. Hence, it is better to use the digital smooth positive/0/negative justification technique in the case of subgroup low rate multiplex.

The digital smooth positive/0/negative justification technique may also in the case that the multiplex rate and the tributary rate are related in a simple multiple, the justification design has no constraints, hence it may be made more reliable, and all use digital circuits which are easy to be integrated. Thus, it suits the subgroup low rate synchronization/asynchronization compatible multiplex. The main differences between this justification technique and the others are: its impair is not phase jitter but the variation of code element widths. This code element width variation in subsections is equivalent to frequency jitter. Through this kind of justification/recovery, if the binary digits participate the other multiplex, or they are observed in a multiplex time coordinate, such a jitter may form 100%UI phase jitter. If the binary digits do not participate any other multiplex after the justification recovery, or they are observed in a time coordinate that uses the code element leading edge as the origin point, such a code element width variation is just equivalent to phase jitter. Obviously, the digital smooth positive/0/negative justification technique suits with low rate tributary synchronous/asynchronous compatible multiplex. avoiding the analog phase-lock loop, it is satisfied that using entire digital circuits to realize compatible multiplex only introduces several percent of code element width offset which is satisfied.

Chapter 7 Measurement Techniques of Justification

Three kind of measurement are related to the justification stuffing jitter measurement,measurement of recovery phase-locked loop and stuffing error measurement.In this chapter we will discuss the particular measurement techniques of these three measurements.

7.1 MEASUREMENT OF STUFFING JITTER

7.1.1 Characteristics of stuffing jitter

The stuffing jitters, which are introduced by various different justification methods to the tributaries, have different characteristics. These different characteristics are the foundation to consider stuffing jitter measurements. The stuffing jitter amplitude of the positive/0/negative justification has its fixed value, which can be simply and strictly calculated (see Figure 7-1). The change of the stuffing jitter amplitude of the positive justification is more complex. But if the equipment of the justification is well designed, its stuffing jitter distribution can be strictly calculated: its peaks have definite values and definite positions. See Figure 7-2, Every jitter peak value steeply rising up. At the top of the peak there is a narrow gap approaching to the ground, and there are valleys up and down between the peak values. What we are interested in is the maximum value of peak in actual measurement, but it is necessary to pay attention to the distribution characteristics of the peaks. The stuffing jitter distribution of the positive/negative justification is monotonous, see Figure 7-3.It only has one peak changing slowly. There is a narrow gap approaching to the ground also at the top of the peak, whose position can be strictly calculated. However, the height of the peak can not be calculated precisely beforehand. As the adjustment conditions of justification circuits and recovery circuits are different, the height of the peak and its slope may be very different, Figure 7-3 shows two kinds of typical distribution.

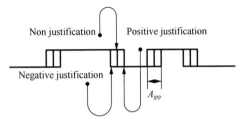

Figure 7-1 Jitter wave form diagram of the positive/zero/negative justification

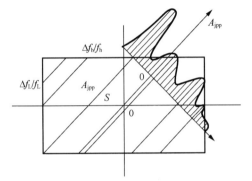

Figure 7-2 Distribution diagram of the positive justification jitter

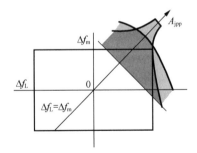

Figure 7-3 Distribution diagram of the positive/negative justification jitter

As mentioned above, we see that the stuffing jitter measurement of the positive/zero/negative justification is quite easy; whereas the corresponding measurements of the positive justification and the positive/negative justification are more difficult. According to the characteristics mentioned above, if we want to measure precisely and quickly, calculation beforehand is very important. By calculation we can find precisely the distribution of every peak position, and the range of their heights.

7.1.2 Measurement range and precision requirement

In the positive/negative justification the maximum value of the stuffing jitter does not exceed 20%UI usually. When equipment is designed correctly, the stuffing jitter peaks are a set of definite discrete values, which are shown in Table 7-1. The jitter peaks of international recommended multiplexers are between 20%UI-14.3%UI. Usually, jitter peaks of practical multiplexes are between 6.7%UI-20%UI. The stuffing jitter peaks of the positive/negative justification are between 10%UI-20%UI. Therefore, the measurement range of the stuffing jitter is between 6.7%UI-20%UI.

Table 7-1   Peak jitter of positive justification

p	A_{jpp}(UI%)	ΔA_{jpp}(UI%)	$\Delta A_{jpp}/A_{jpp}$(%)
5	20.0	0.4	2.0
6	16.7		2.4
7	14.3		2.8
8	12.5		3.2
9	11.0		3.6
10	10.0		4.0
11	9.1		4.4
12	8.4		4.8
13	7.7		5.2
14	7.1		5.6
15	6.7		6.0
16	6.2		6.4
17	5.9		6.8
18	5.5		7.2
19	5.3		7.6
20	5.0		8.0

In the stuffing jitter measurement, it require to distinguish different degree of stuffing jitter in the worst case which is the minimum jitter case in the above measurement range. For example, if we require to distinguish 6.7%UI from its neighbor degree 7.1%UI, the absolute measurement error should not exceed 0.4% UI. If set A_{jpp}=0.4%UI, the relative measurement error $\Delta A_{jpp}/A_{jpp}$ of corresponding possible jitter peak is between 2.0%-6.0%. Such a measurement precision is sufficient in the measurement of the positive justification plesiochronous multiplex.

Experiment shows, such a measurement range and measurement precision may actually be done. But we have to design carefully and adjust correctly.

7.1.3 Measurement equipment

After demultiplex, the stuffing jitter of the tributary clock $\hat{f}_1$ is the phase deviation from multiplex tributary clock f_1. Hence, the stuffing jitter measurement is to measure relative phase deviation of $\hat{f}_1$ from f_1. The block diagram of the jitter measurement is shown in Figure 7-4.

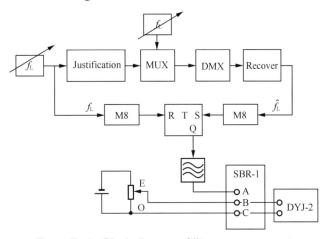

Figure 7-4 Block diagram of jitter measurement

The frequency source (f_l, f_h) uses a frequency synthesizer with resolution of 0.5 Hz. It is necessary to use digital frequency meter to monitor output frequency. To get ideal linearity, We use the RS trigger as a phase discriminator. Two input terminals of the discriminator are connected to two modulo-8 circuits to expand the linear region of phase discrimination. This avoids the influence on discriminator when jitter is bigger, as well as reduces the influence on precision of the phase discriminator, which is caused by non-ideal wave form skip. Low pass filter uses three pieces of LC filter, because the pass-band of the recovery phase-locked loop is 30 Hz, the stuffing jitter spectrum measured is limited within 30 Hz. The other jitter components of the measured tributary clock and the jitter components introduced by the discriminator are shown in Figure 7-5. Hence, the low pass filter requirement is that in the 30 Hz pass band the ripple does not exceed 0.5 dB, and at 1kHz it is suppressed more than 40 dB, there is no specific requirement for decreasing characteristics between 30 Hz-1 kHz. As mentioned above, a complete principle diagram of stuffing jitter measure discriminator is shown in Figure 7-6, its characteristics of the discrimination is shown in Figure 7-7.

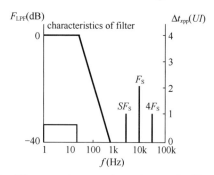

Figure 7-5 Jitter spectrum and requirement of low pass filter

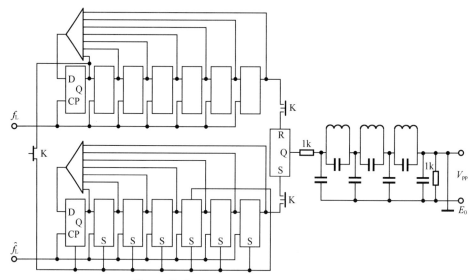

Figure 7-6 jitter measure discriminator

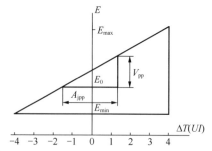

Figure 7-7 Characteristic of discriminator

The output of the discriminator is a rather big DC voltage (E_0) adding a small ripple (V_{pp}), see Figure 7-7. Usually, E_0 is about 2.5 V, V_{pp} is about several decades of mV, i.e. V_{pp}/E_0 is 1/100 in magnitude. Because only V_{pp} is the magnitude of jitter, E_0 have to be rejected before measurement; since V_{pp} is several decades of mV in magnitude, it need to be amplified sufficiently before measurement. See Figure 7-4, these two kinds of works are done by oscilloscope SBR-1.The tested signal is applied at the input End A, a reference DC voltage (E) is applied at the input End B. On the oscilloscope shows the difference of the two Ends A and B. Since the oscilloscope SBR-1 has gain high enough, the signal in mv magnitude can be shown on the entire screen.

In the subtraction and amplification process above, power source ripple interference is a main problem deserved to pay attention to. Considering that measure precision is required not exceeded 0.4%UI (see Figure 7-7, and the output of phase discriminator is usually 4 V equivalent to the peak to peak jitter of 8 UI; corresponding to 0.4%UI, the phase discriminator output is 2 mV. Hence, the interference voltage introduced by subtraction and amplification process, which is equivalent to the voltage at the input end A, should be far less than 2 mV. Therefore, the reference DC voltage source at the end B has to use battery, and the internal interference of oscilloscope is required as low as possible. If the internal interference voltage can not be neglected, a correct method has to be used. The reason why we use oscilloscope SBR-1 is that it has differential input, high gain and low internal interference.

Finally, it is worth to mention that the number of stuffing jitter measurement points is quite large, values near the peak point are changing rapidly, sometimes we need measure again and again. Hence, convenient regulation of measurement equipment is particularly important. It is required that the rough value of the clock source can be conveniently regulated, the tail value of the clock can be regulated continuously and precisely. Thus we use high steady frequency synthesizer usually. Its regulation step unit is 10 Hz, we use continuous regulation dial at lower than 10Hz. Besides, the reference DC voltage (E) should be regulated precisely, its maximum range is about 2.5 V, and the precision is 1 mv in magnitude. Therefore, we may only use precise multi-coil resistor. The output reference voltage is monitored and read out by a digital volt-meter. Because every time when we change clock source (i.e. change measure point), the multiplex system will rebuild synchronization, Which introduces the change of DC level (E_0) of the phase discriminator and the loss of the wave form V_{pp} in the oscilloscope. For the stability of E_0, we set a switch (K) to synchronize two modulo-8 circuits.

7.1.4 Measurement method

The measurement process of stuffing jitter includes five steps: distributed calculation, clock adjustment, peak value measurement, correction measurement and data arrangement.

(1) Distributed calculation

For the stuffing jitter measurement of the positive justification we need to calculate all peak measurement points (f_h, f_l) and its corresponding jitter peak values (A_{jpp}) beforehand. For the positive/negative justification we only need calculate corresponding point $f_l = f_m$, where $f_m = \dfrac{Q}{L_s} \cdot f_h$, and find the corresponding measurement point (f_l, f_h). Usually, the measurement of jitter distribution is to cut the jitter distribution into several cross-section area ($f_{h1}, f_{h2} \cdots$), then measure every cross-section area one by one. While measuring the ith cross-section area, f_{hi} is keeping constant, and only f_l need to be adjusted.

(2) Clock adjustment

While measuring the ith cross-section area, the multiplex clock (f_{hi}) is keeping constant, and only the tributary clock f_l need to be adjusted. From the basic relationship formula of the positive justification we can derive out:

$$S = Q - L_s \cdot \dfrac{f_l}{f_{hi}}$$

Let
$$S = S_o + \Delta s, f_l = f_{lo} + \Delta f_l, S_o = \dfrac{q}{P}, (p,q) = 1;$$

$$f_{lo} = \dfrac{f_{hi}}{L_s}(Q - S_o),$$

$$\therefore \quad \Delta S = -\dfrac{L_s}{f_{hi}} \cdot \Delta f_l$$

Where, the frame length (L_s) is definite, the multiplex clock f_{hi} is selected. Therefore, the value (ΔS) that the justification ratio deviates from a simple rational number (i.e. a peak point), depends only on the tributary clock deviation (Δf_l). Given the calculation formula of the stuffing jitter frequency:

Chapter 7 Measurement Techniques of Justification

$$f_j \approx \Delta S \cdot p \cdot F_s$$

$$= -\frac{L_s}{f_{hi}} \cdot PF_s \cdot \Delta f_1$$

$$= -P\left(\frac{F_s}{F_s}\right) \Delta f_1$$

$$f_j \approx -p \cdot \Delta f_1$$

That is, the jitter frequency f_j is equal to the product of the measured peak value P and the frequency difference of the tributary clock from that peak value. In Figure 7-8, when $\Delta f_1=0$, $\Delta S=0$, $S = \frac{q}{p} S_o$, the peak to peak jitter value $A_{jpp}=0$; when Δf_1 approaches zero, ΔS approaches zero, and f_j approaches zero also, the peak to peak jitter value A_{jpp} approaches $\frac{1}{p}$ (UI), i.e. it approaches its maximum value. In whatever direction it approaches the peak point, the jitter size is all equal to $\frac{1}{p}$ (UI). But in different direction approaching peak point, the phase of jitter is opposite. According to this characteristic, we may precisely find out the peak points.

Besides, from Figure 7-8 we can find that the more the jitter approaches its peak, the more its amplitude approaches the maximum value, but the lower the jitter frequency is. The more the jitter amplitude approaches its maximum value, the higher the measurement precision is. But the lower the jitter frequency is, the more inconvenient the measurement is. They are contradictory with each other. In actual adjustment, we hope when the jitter amplitude approaches its maximum value in an enough precision, at the same time the jitter frequency is as high as possible to measure conveniently.

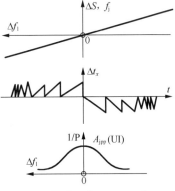

Figure 7-8 Corresponding relationship of clock adjustment

The function relationship between the jitter frequency (f_j) and the jitter amplitude ($A_{jpp}^{(k)}$) can be strictly calculated. This is a problem what is its peak to peak amplitude $A_{jpp}^{(k)}$ after a ideal indent wave with peak to peak amplitude A_o passes a ideal filter with band pass $f_o = Kf_j$:

$$A_j(t) = \frac{A_o}{\pi}\sum_{n=1}^{\infty}\frac{(-1)^{n-1}}{n}\cdot\sin(n2\pi f_j t), f_j = \frac{1}{T_j},$$

$$f_o = kf_j$$

$$\therefore A_j^{(k)}(t) = \frac{A_o}{\pi}\sum_{n=1}^{k}\frac{(-1)^{n-1}}{n}\sin(n\cdot 2\pi f_j t)$$

from

$$\frac{dA_j^{(k)}(t)}{dt} = 0$$

get

$$t = t'$$

$$\therefore A_j^{(k)} = \frac{A_o}{\pi}\sum_{n=1}^{k}\frac{(-1)^{n-1}}{n}\sin(2\pi f_j nt')$$

$$= A_o F\left(\frac{f_o}{f_j}\right)$$

According to the precision requirement $\left[A_o - A_{jpp}^{(k)}\right]/A_o = m\%$, may get $A_j^{(k)}$; and according to $a_j^{(k)} = A_o F\left(\frac{f_o}{f_j}\right)$, may calculate $k = \frac{f_o}{f_j}$. The particular is shown in Figure 7-9 in detail.

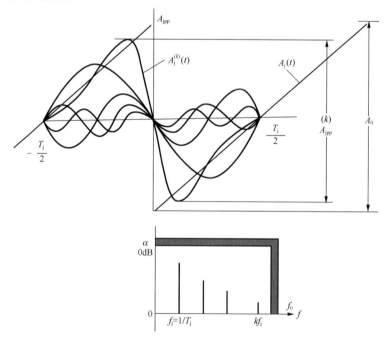

Figure 7-9 Adjustment deviation of stuffing jitter

Usually, we may use an experimental method to measure the peak value of jitter directly. See Figure 7-10, if the absolute regulation error is required not exceed 0.2%UI, as long as the jitter frequency is regulated to about 1Hz, the measurement may be carried out.

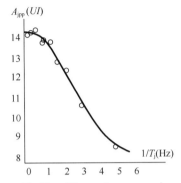

Figure 7-10 Jitter value near q/p point

(3) Peak value measurement and correction measurement

The following method is to measure the peak value, see Figure 7-11. Firstly, we select a mechanical scale line in the center of the screen, or select another idle wave line as a reference line; secondly, we regulate precise resistor (R) so that the upside peak of jitter wave can coincide with the reference line, at the same time we write down the digital voltage meter value (E_1); thirdly, we continue regulating R so that the downside peak of jitter wave coincide with the reference line, then write down the digital voltage meter value (E_2); finally, using the method as the same as the above mentioned, we can measure interference wave peak to peak value E_1' and E_2'. Then get:

$$V_{pp} = |E_1 - E_2|$$
$$S_{pp} = |E_1' - E_2'|$$

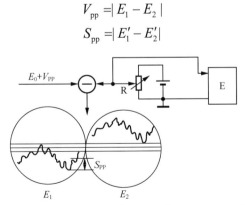

Figure 7-11 Demonstration diagram of measurement method

See Figure 7-6, push down the key K_1, at that time the oscilloscope line changes into a direct line, regulate R to coincide with the reference line, write down the digital voltmeter value E_{max}; then push down the key K_2 and measure the voltage E_{min}, then get the correction voltage:

$$\Delta E = |E_{max} - E_{min}|$$

See Figure 7-7, the jitter value can be calculated according to the following formula:

$$A_{jpp} = \frac{8(V_{pp} - S_{pp})}{\Delta E} \cdot UI$$

From the above measure process we can see that although the measure errors can be caused by a lot of factors, such as stability of clock source, characteristics of discriminator and filter, reference powor interference and reading error, stability of the oscilloscope, precision of correction measurement, and various external interferences, but after a series of the above steps, there are two kinds of main error sources of justification error $\Delta A'_{jpp}$ and observation error $\Delta A''_{jpp}$. The experimental results show that if the jitter frequency is about 1 Hz, error range caused by justification $\Delta A'_{jpp}$ does not exceed 0.2%UI. Observation error is limited by width of the oscilloscope wave and the resolution of human vision. Considering that the oscilloscope usually take 5 mV per grade in one range, and the resolution of human vision is 0.2 grade, hence the voltage measurement deviation is 1 mV. If the correction voltage difference $\Delta E = 2V$, the error range introduced by observation is:

$$\Delta A''_{jpp} = \frac{\pm 1mV}{2V} \cdot 8UI = \pm 0.4\% UI$$

Thus, the total measurement error range is:

$$\Delta A_{jpp} = \Delta A'_{jpp} + \Delta A''_{jpp} = \pm 0.6\% UI$$

But considering that both the justification error and the observation error are random, the measurement error obviously decreases by repeated measurement. In this way, taking standard deviation as measurement error is suitable. As estimated error, rather large value is taken when standard deviation σ_{jpp} is estimated:

$$3\sigma'_{jpp} = \Delta A'_{jpp}$$
$$3\sigma''_{jpp} = \Delta A''_{jpp}$$

The maximum error bound is used as the boundary not exceeded within 99.7%

Chapter 7 Measurement Techniques of Justification

time. At that time:

$$\sigma'_{jpp} = \frac{\Delta A'_{jpp}}{3} \approx 0.07\% UI$$

$$\sigma''_{jpp} = \frac{\Delta A'_{jpp}}{3} \approx 0.14\% UI$$

$$\sigma_{jpp} = \sqrt{(\sigma'_{jpp})^2 + (\sigma''_{jpp})^2} = 0.16\% UI$$

$$3\sigma_{jpp} \approx 0.5\% UI$$

That is, the standard deviation of absolute measurement error caused by this measurement method is 0.16%UI. The estimated standard deviation in 99.7% of the data does not exceed 0.5%UI.

7.1.5 Example of measurement

A 2nd group multiplexer which accords with the recommendation of CCITT G.742 and uses the positive justification is measured. The measurement range is: $f_h = 8\,448\,000 \pm 250$Hz; $f_l = 2\,048\,000 \pm 100$Hz. The measurement result is shown in Table 7-2 and Figure 7-12.

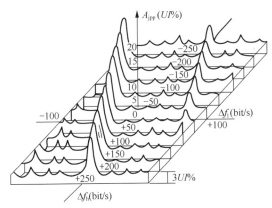

Figure 7-12 Jitter measurement result of multiplexer 2/8Mbit/s

The maximum peak is at $S=0.428\,57$. The theoretical calculation peak value is 14.3%UI, the result of measurement is 13.9%UI $\pm$ 0.2%UI. The arithmetical average measurement result is 2.8% lower than the theoretical value. The second peak appears at $S=0.416\,67$. The theoretical calculation peak value is 8.3%UI, the result of measurement is 7.8%UI $\pm$ 0.4%UI. The arithmetical average measure result is 6.0% lower than the theoretical value.

Table 7-2

A_{jpp} (UI%) \ S / f_h(bit/s)	0.409 09	0.411 76	0.413 79	0.416 67	0.419 35	0.421 05	0.423 08	0.424 24	0.428 57	0.433 33	0.434 78	0.437 50	0.440 00
8 448 250						4.8	3.9	3.8	13.7	4.4	5.1	6.0	5.4
8 448 200					3.4	4.8	3.8	3.2	13.9	4.4	5.0	6.7	
8 448 150					3.8	4.8	4.2	3.3	14.0	6.0	6.1	6.7	
8 448 100				7.1	4.2	5.9	4.2	3.2	14.1	4.9	5.9		
8 448 050				6.3	3.5	5.3	4.0	3.5	13.5	4.2	5.4		
8 448 000				8.1	4.2	5.4	3.8	4.4	14.0	5.5			
8 447 950			3.4	8.1	3.9	5.0	4.0	4.6	14.3	4.2			
8 447 900		5.5	3.7	8.3	3.9	5.0	4.2	4.0	13.6				
8 447 850		4.7	3.7	7.9	4.2	5.3	4.4	3.4	13.8				
8 447 800	5.3	5.2	4.2	7.5	4.2	5.2	4.6	4.0	14.2				
8 447 750	3.9	5.0	3.5	7.8	4.0	5.8	3.9	4.2					

7.2 MEASUREMENT OF MULTIPLEX CODE ERROR

7.2.1 Characteristics of multiplex code error

Multiplex code error includes frame loss code error (P_{ef}) and stuffing code error (P_{es}). In an actual measurement process, data measured include average code error of stable channel (P_e), P_{ef} and P_{es}. To distinguish them is difficult and unnecessary. In the 2nd group multiplex system using positive justification, the transfer relationship of the multiplex code error is shown in the Figure 7-13.

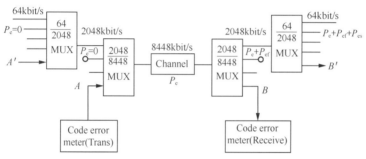

Figure 7-13 Multiplex code error transfer

The frame loss code error is: the code error (P_e) in the 8Mbit/s binary digits cause frame loss of the 2/8Mbit/s demultiplexer, then it introduces into the multiplex code error (P_{ef}) in the 2Mbit/s. In the same way, this kind of error (P_e+P_{ef}) passes into the 64kbit/s binary digits. The stuffing code error is: code error (P_{ef}) in the 8Mbit/s binary digits cause the stuffing indication code error, therefore code recovery operation in the 2/8Mbit/s demultiplexer is out of order. The result is that the 2Mbit/s code gets slip, it will lose or insert a code element in the binary digits (except for this no error increased in the 2Mbit/s). The 64kbit/s/2Mbit/s demultiplexer frame loss is caused by slip of 2Mbit/s binary digits. Hence, the 64kbit/s binary digits will get stuffing error.

Therefore. the multiplex code error is more complex than the code error in a stable channel. In the multiplex code error meaasurement we not only consider stable average code error measurement, but also consider code error measurement in frame loss and slip condition. Thus, the error meter has corresponding special requirement.

7.2.2 Requirements of code error meter

In usual stable average code error measurement it need not any special search requirement. The important thing is to ensure enough lower loss lock probability in a high code error measurement. Loss lock probability P_L will decrease to a necessary degree if the detection period (T_e) is suitably increased. Generally speaking, increasing detection period (for example using step search method) will increase average search time, but it not very important in usual stable error measurement.

In the flame loss error measurement, we hope in the period of multiplex frame loss, the error meter will not be loss locked. Actually this can be done, because in the frame loss period the binary digits of the tributary have no any structural damage. For example, in the frame loss period a special signal is sent to the error meter, which forces the meter not to do the search even if the error code exceeds the specified limit. For example, we can fully increase the detection period that is far bigger than the extent of frame loss time of the multiplexer. Even if the code error is as high as 50% in the frame loss period of the multiplexer, the integral level of tributary search circuit of the error meter can not get loss lock.

In the stuffing code error measurement, there is complete different requirement for the error meter. For example, in the case of Figure 7-13, even if it has slip, the 2Mbit/s binary digits do not add any code error. In this case, it is required that the error meter can find slip as soon as possible, then search as quick as possible and set new lock. The time interval from slip happening to the new lock having been set, the error meter will introduce an additional measurement error for the 2Mbit/s binary digits. This additional code error is the error of the code error measurement. To reduce this kind of error the best way is that the decision time of the meter is as short as possible and the search is as soon as possible. That is, except using good search method, the detection period is not too long also.

From mentioned above, we know that the requirements of the code error meters are different or even contradictory when measuring three kinds of code error. The usual solution is that if frame loss or stuffing error happen, let the meter to search immediately and as soon as possible.

Firstly, let us discuss the multiplex code error measurement of ($n-1$)-ary group (for example, 2048kbit/s): in the frame loss period of the n-ary group multiplexer, half of the ($n-1$)-ary group binary digits cause code errors. Hence, the average code error ratio is:

$$P_{ef} = \frac{1}{2} P_f t_r$$

After the synchronization recovery of *n*-ary group demultiplexer, the code error meter can begin the search process (Δt_r). Therefore, after the frame loss the difference of the measurement error ratio of the code error meter is:

$$\Delta P_{ef} = \frac{1}{2} \rho_f \Delta t_r$$

When the (*n*−1)-ary group appears slip, the difference of the code error measurement caused by rebuilding the lock of the meter is:

$$\Delta P_{es} = \frac{1}{2} \rho_s \cdot \Delta t_r$$

If the relative measurement error of multiplex code error is required not to exceed 5%, then get

$$(\Delta P_{ef} + \Delta P_{es})/P_{es} \leqslant 5\%$$

$$\Delta t_r \leqslant \frac{5}{100} \cdot t_r / \left(1 + \frac{\rho_s}{\rho_f}\right)$$

Where, the frame loss frequency of *n*-ary group ρ_f, the stuffing error frequency ρ_s and the average frame loss time t_r are respectively:

$$\rho_f = (nP_e)^\beta \cdot F_s$$

$$\rho_s = C_n^{d+1} \cdot P_e^{d+1} \cdot F_s$$

$$t_r = \left(a - \frac{1}{2}\right) T_s$$

$$\Delta t_r \leqslant 0.05 \left(a - \frac{1}{2}\right) T_s / 1 + \frac{C_n^{d+1} \cdot P_e^{d+1}}{(nP_e)^\beta}$$

Where, the parameters of the *n*-ary group are:

 n—frame alignment length;
 η—stuffing indication code length;
 d—the number of stuffing jitter indication tolerance bits;
 α—the number of search protection frames;
 β—the number of synchronous protection frames;
 T_s—frame period;

P_e — channel average error rate.

For example: Recommendation of CCITT G.742 (8448kbit/s): $n=10$, $\eta=3$, $d=1$, $\alpha=3$, $\beta=4$, $T_s=848$UI. In the extreme case, $P_e=1\times 10^{-2}$,

$$\Delta t_r \leqslant 27\text{UI}$$

Even if in the above condition, to ensure the relative measurement error not exceeding 5%, the average search time can not exceed 32 code elements.

Below, let us discuss the multiplex code error measurement and the stuffing error measurement of the ($n-2$)-ary group (for example, 64kbit/s): usually, if the ($n-1$)-ary group has flame loss, "1" will be sent out to the ($n-2$)-ary group to tell them not to search, thus the frame loss will not spread. At that time, the working state of the code error meter is as the same as the code error of stable channel state. The another part is the stuffing error which introduces the stuffing code error into the ($n-2$)-ary binary digits, the process is: the n-ary group stuffing error causes the ($n-1$)-ary group binary digits to slip, then the ($n-1$)-ary group multiplexer loses frame, in turn, introduces the stuffing code error into the ($n-2$)-ary group binary digits. Therefore, the character of stuffing code error of the ($n-2$)-ary group is as the same as that of the ($n-1$) -ary group frame loss. Their code errors both are introduced by the frame loss, and the structures of the binary digits both are not damaged. At that time, if the multiplexer send "1" to the tributary in the frame loss period to tell the meter not to do search and to keep in the original stable state, the meter will not introduce any additional error. Then the meter will be working as the usual stable measurement state. The measurement error only depends on the selection of the code form of pseudo random code, the timing precision and the counting error. If the selection of code form accord with the recommendation of CCITT, the measurement errors introduced by other two factors are very easily limited within 1×10^{-3}. Hence, it is neglectable in the multiplex code error measurement.

7.2.3 Design of Special code error meter

The requirement of the average search time is $\Delta t_r < 32$UI.

The usual fast search of the delayed lock loop has two methods: the step by step method and the sequential forecast method. The search time ratio of them is:

$$\frac{\Delta t_r(\text{step})}{\Delta t_r(\text{forecast})} = (1-P_e)^l \cdot 2^{l-1}$$

Where,

P_e—average code error ration of binary digits,

I—stages of the pseudo-random-code generator.

When 1=4, $P_e<1\times10^{-2}$, then Δt_r (forecast) $<\Delta t_r$ (step). Therefore, we use forecast fast search method here.

Using the sequential forecast fast search method, the average search time of delayed lock loop:

$$\Delta t_r = T_e/(1-P_e)^l(1-P_1')(1-P_y')$$

Where,

T_e—detection period,

P_1'—virtual loss probability, i.e. probability of decision error after correct forecast,

P_y'—virtual alarm probability, i.e. probability of decision correct after wrong forecast.

Assume: there are k code elements in a detection period; if a code elements are different in k code elements, the decision is forecast error; if b code elements are the same code, the decision is forecast correct, then:

$$T_e = k \cdot UI, P_1' \approx (P_e)^a, P_y' \approx \left(\frac{1}{2}\right)^b, a+b = k-2.$$

Usually, $P_1' = P_y' = 1\times10^{-8}$ is enough, then,

$$a = 8/\lg(1/P_e)$$
$$b = a \cdot \lg(1/P_e)/0.3$$

When $P_e=1\times10^{-2}$, get $a=4$, $b=27$, $k=29$, hence

$$T_e = 29 \cdot UI$$

In a practical design, to simplify the circuit, the detection period is taken $32 \cdot UI$.

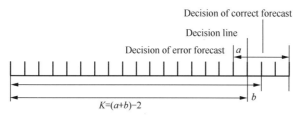

Figure 7-14 Detection period and range of forecast decision

The principle diagram of a practical error meter is shown in Figure 7-15. The

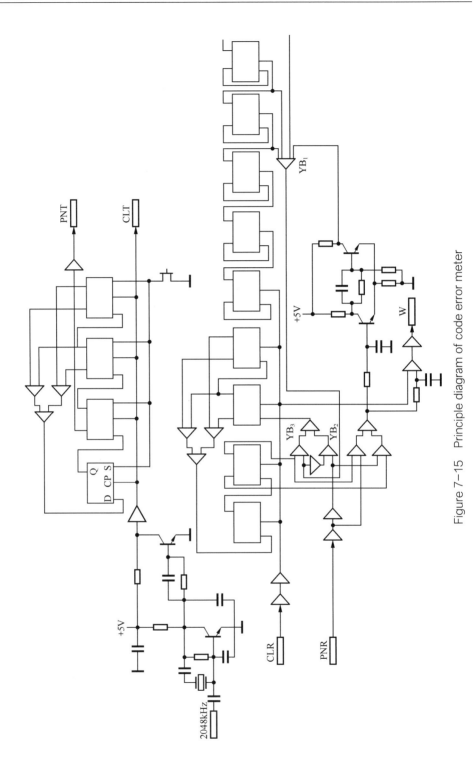

Figure 7-15　Principle diagram of code error meter

transmission side of the error meter is a general pseudo-random-code generator, which is made up of 4 stages shift register and a modulo-2 circuit. The code length M=15, and the particular code sequence is 111100010110110. the main part of the error meter is a delayed lock loop, which uses sequential forecast fast search, and whose detection period is T_e=32 • UI. The principle diagram of the meter is shown in Figure 7-16. In the search process, the control pulse with width 4 • UI is output every 32 • UI by YB_1. Open the YB_2, to let the input pseudo-random-code directly inject into the local pseudo-random-code generator, and just make the generator full. The control signal closes the YB_2 and opens the YB_3 at the same time, to let the local code generator to operate automatically. If the forecast (inject) is completely correct, then through the modulo-2 addition circuit are compared the local code and the received code with each other. If identical digits reach or exceed 27/32, the recognition circuit outputs the low voltage level to close YB_1. The YB_1 stops outputting negative control pulse, then the delayed lock loop is getting into the locked state. If the local code and received code are compared with each other by the modulo-2 addition circuit, when different digits reach or exceed 5/32, then the recognition circuit outputs the high voltage level that the YB_1 is opened again and sends out the control pulse. At that time the delayed lock loop is getting into search state. The local code and received code are compared bit by bit, if they are different, the error code pulses will be sent out. In a definite time interval the number of the error code pulses divided by the number of total information code pulses is the average error ratio.

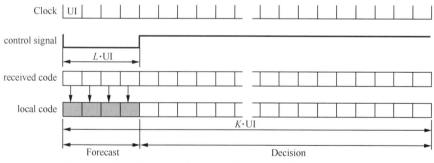

Figure 7-16   Principle diagram of sequential search forecast

7.2.4   Design of analog channel

The function of the analog channel is to inject random noise signals into the binary digits to let it get random code error. The average code error ratio (P_e) can be regulated according to the requirement and indicate the average error ratio. The principle diagram of the analog channel is shown in Figure 7-17.

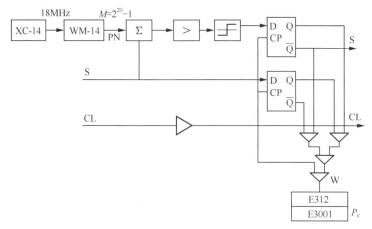

Figure 7-17   Principle diagram of analog channel

The signal source XC-14 supplies 18MHz clock. The general code error meter WM-14 is used as a noise signal generator whose highest working rate limit is 18MHz, the code length is $2^{20}-1$; the clock of the channel signal is 8448kHz, the code length is 2^4-1. In such case, the noise signal spectrum density ratio is $2^{20}-1/2^4-1 \approx 1 \times 10^{16}$; the ratio of clock is $18/8.448 \approx 2.1$. Therefore, the density of noise spectrum is high enough, but the spectrum amplitude is not flat. Generally speaking, it is a white noise signal. The noise and signal are added by a resistor adder, by regulating which, the signal noise ratio can be regulated, then the average code error ratio is changed. The added signal after being amplified is fixed decided by a smith circuit. Finally, binary digits with average code error ratio P_e is reformed, and it is output by the trigger circuit D. The error free signal code and signal code with error are compared with each other by a modulo-2 addition circuit to detect out the number of the error code, counted by E312 and printed out by E3001 printer.

7.2.5   Example of measurement

The diagram of measurement is shown in Figure 7-13;the diagram of the analog channel is shown in Figure 7-17 and the principle diagram of the code error meter is shown in Figure 7-15. The variable is the average code ratio (P_e) of binary digits 8448kbit/s; the dependent variable is the average code ratio (P_e+P_{ef}) of binary digits 2048kbit/s. Each point is measured 10s. The measurment data are shown in Table 7-3. The comparison between the measured curve and calculated curve is shown in Figure 7-18.

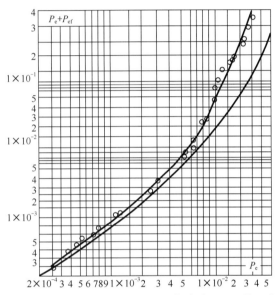

Figure 7-18　Diagram of comparison between calculation result and measuring result

From the comparison, we can see that the tendency of measurement result coincides with that of calculation result. If $P_e=1\times10^{-3}$, the measurement result is 10% larger than the calculation result in magnitude; If $P_e=1\times10^{-2}$, the measurement result is 15% larger than the calculation result in magnitude; If $P_e\geqslant 1\times10^{-2}$, the deviation of the measurement result from the calculation result increases rapidly. The main reason is that the calculation formula uses small approximation, when P_e approaches the 1×10^{-1} in magnitude, the error of the approximate calculation formula becomes bigger and bigger. In the actual measurement, when P_e approaches the 1×10^{-1} in magnitude, the slip frequency increases more and more, which lets the measurement error, introduced by research, of the error meter increases also. The theoretical calculation causes smaller error; whereas the actual measurement causes bigger error; Therefore, as P_e increases, the deviation between the calculation value and the measurement value increases also.

Table 7-3 Data of loss frame code error

P_e	4.1×10^{-2}	3.6×10^{-2}	3.4×10^{-2}	2.8×10^{-2}	2.4×10^{-2}
P_e+P_{ef}	2.2×10^{-1}	1.6×10^{-1}	1.4×10^{-1}	1.1×10^{-1}	6.7×10^{-2}

1.8×10^{-2}	1.4×10^{-2}	8.6×10^{-3}	6.2×10^{-3}	6.0×10^{-3}	3.0×10^{-3}
2.7×10^{-2}	1.8×10^{-2}	9.7×10^{-3}	6.8×10^{-3}	6.7×10^{-3}	3.3×10^{-3}

1.5×10^{-3}	8.0×10^{-3}	5.8×10^{-4}	3.8×10^{-4}	2.4×10^{-4}	4.5×10^{-2}
1.8×10^{-3}	1.1×10^{-3}	7.4×10^{-4}	2.4×10^{-4}	2.6×10^{-4}	3.3×10^{-1}

4.2×10^{-2}	4.0×10^{-2}	3.6×10^{-2}	3.2×10^{-2}	2.6×10^{-2}	2.3×10^{-2}
2.6×10^{-1}	2.1×10^{-1}	1.6×10^{-1}	1.3×10^{-2}	8.9×10^{-2}	4.9×10^{-2}

1.7×10^{-2}	1.3×10^{-2}	7.7×10^{-3}	5.3×10^{-3}	2.7×10^{-3}	1.4×10^{-3}
2.4×10^{-2}	1.5×10^{-2}	7.7×10^{-3}	5.5×10^{-3}	2.9×10^{-3}	1.8×10^{-3}

7.2×10^{-4}	5.1×10^{-4}	3.3×10^{-4}			
9.4×10^{-4}	6.5×10^{-4}	3.4×10^{-4}			

7.3 MEASUREMENT CHARACTERISTICS OF THE PHASE-LOCKED LOOP OF THE CODE RECOVERY[14]

The phase-locked loop used by code recovery is a common analog phase-locked loop, and the measurement method is also a usual method. Therefore, we only give a simple introduction here.

7.3.1 Tracing error measurement

The plug-in cards of the code recovery, either positive or positive/negative justification, are commonly used. The diagram of the measurement tracing characteristics is shown in Figure 7-19. The minimum scale of frequency dial of the frequency synthesizer MG525C is 0.2Hz. The last digit of the voltage indication of the digital volt-meter PZ-5 is 0.001 V. an oscilloscope SBE-7 is used to monitor loss lock.

Chapter 7 Measurement Techniques of Justification

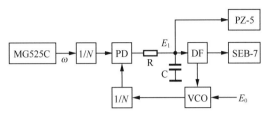

Figure 7-19 Block diagram of tracing characteristic measurement

Before measurement, firstly disconnect the input signal, regulate E_0 to let the output frequency be equal to the nominal frequency. Then, connect the input signal. Whenever change a frequency (ω), then write down the output voltage of integrator (E_1). The measurement will go on till the loop loses lock. After the test is finished, do the correction measurement: set the end S of the discriminator trigger to ground, test $E_1=E_{1\min}$; then set the end R to the ground, test $E_1=E_{1\max}$.

In Figure 7-20, according to the following formula, the measurement data (E_1) is converted into tracing phase error (φ_ε):

$$\varphi'_\varepsilon = \frac{E'_1 - E_{10}}{E_{1\max} - E_{1\min}} \cdot 8 \cdot UI$$

The measurement result of tracing error of the phase-locked loop, which coincide with recommendation of CCITT G.742, is shown in Figure 7-21. The frequency tolerance is in the range $2\,048\,000 \pm 100$Hz. The tracing error does not exceed 1rad.

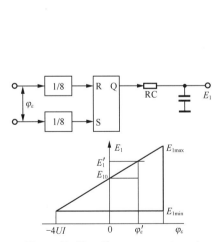

Figure 7-20 E_1-φ_ε conversion of phase discriminator

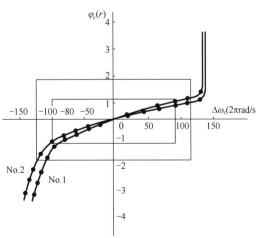

Figure 7-21 Result of tracing error

7.3.2 Measurement of jitter suppression characteristics

The diagram of suppression characteristics of the code recovery phase-locked loop is shown in Figure 7-22. The input end is connected to a frequency source with fixed frequency 2 048 000Hz. The initial frequency of wave analysis meter TF455E is 10Hz, and by correction the frequency range can be expanded to 5Hz. The oscilloscope SBE-7 monitors amplifier not to limit the amplitude. The audio frequency synthesizer XD-1 is used to regulate the voltage control oscillator to modulate frequency.

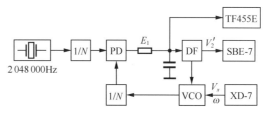

Figure 7-22 diagram of depressing characteristic measurement

Regulate amplitude V_x to as high as possible, but ensure amplifier not to limit the amplitude, then keep amplitude V_x constant. Whenever XD-1 changes frequency, a corresponding voltage $E_1'(\omega)$ is tested by TF455E. We can prove the following relationship between the jitter suppression characteristics and the measurement value:

$$|H(j\omega)| = \frac{E_1(\omega)}{E_0(0)}$$

The actual $E_1(0)$ can not be tested conveniently, but we can use value tested at the frequency as low as possible to replace it approximately: $E_1(0) \approx E_1(\omega)|\omega \to 0$.

The phase-locked loop of the code recovery, which accords with the recommendation of CCITT G.742, whose characteristics of the suppression jitter tested is shown in Figure 7-23, where, f_0=30Hz. If f<30Hz, the maximum jitter gain is less then 0.5dB. If f>30Hz, the slop of jitter suppression is 12dB per double frequency range.

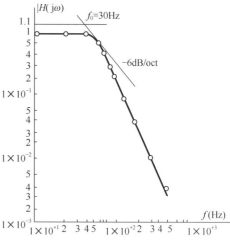

Figure 7-23 Measurement result of jitter suppression characteristics

Chapter 8 Frame Adjustment Principle

8.1 GENERAL

As stated in the first chapter, in order to realize synchronous multiplexing, the synchronization environment must be provided in advance, in some concrete applications it is relative easy to supply certain synchronization environment but in other concrete applications it is relative difficulty to create such synchronization environment. As mentioned in the following chapter, at present, it is internationally recognized that adopting the code rate adjustment technology to create synchronization environment in the trunk (group) transmission is a relevant proper method. But the code rate adjustment technology leaves over an very important unsolved problems, which is the signal wander. The code rate adjustment can only track the wander but can not remove the original wander in the code stream, and even introducing new wander into the participated multiplexing/demultiplexing code stream.

Usually, in a digital switching network, the channel switching takes the voice channel as one unit. As a matter of fact, the access network of a digital switch is a synchronization digital multiplexer under special control, it may be either a channel synchronous multiplexer or a group synchronous multiplexer, which depends on the concrete system of the exchange. The group code streams from all directions are divided into voice channels for switching, then the new groups are multipldxed again to flow to all directions. Thus, it is not allowed for all the digital streams to have different wander. Therefore, the prerequisite to realize the switch multiplexing is to cancel the digital stream wander in the voice channels and set up the synchronization with local clock and establish a certain phase relation with local frame alignment signal. As mentioned in the first chapter, the group synchronous multiplexing requires to put the frame alignment signals from various tributaries into the defined locations of the multiplexing frame. At this time, the bit rate adjusting technology is not applicable, and the frame regulating technology is put forward against this background.

Practically, the frame regulator is an positive/zero/negative frame regulator without restoring capability. Its main body is a frame buffer and it is quite similar to that of bit rate adjuster, whose main body is a code element buffer, those two only having different control approaches. when the time difference between the reading and writing in the frame buffer is small to a certain extent, the positive adjustment will be done once reread one frame; when the time difference is big to a certain extent, the negative adjustment will be done once-skip reading one frame. The difference between this and code adjustment is that the frame, being regulated and controlled at 'transmitting end' need not to be restored and controlled at "receiving end". Thus it is not necessary to allocate the regulation flag. and provide negative adjustment message code channel too. So, there is no such concept as "transmitting end" or "receiving end". It is clear that the frame regulation means only the regulation-positive one is to reread one frame and negative one is to skip one frame-such phenomenon is called slip. It is a structure damage of a code stream. Having gone through such a damage once the code stream is reborn-completely removing the wander, setting up the synchronization with local clock, maintaining a stipulated phase relation with the regulation signal of local frame.

The frame regulator is usually installed between the group transmission link and digital switch, i.e. at the input of various network nodes of the switching network (see Figure 8-1). The remote terminal generally is a multiplexer equipped with independent clock source, a concentrator or digital switch; local network node usually is a digital switch equipped with independent clock source or synchronous group multiplexer. Generally speaking, such frame regulator is required as long as the wander exists in the group transmission between the nodes, no matter the various network noes have independent clock source or an unified clock source.

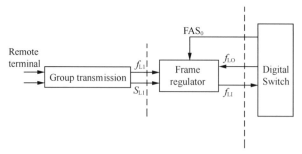

Figure 8-1 The environment of frame regulator

Chapter 8 Frame Adjustment Principle

The general frame regulation takes local clock and local frame alignment signals as the basement. The input code stream will have following differences or impairs relative to local clock and local frame alignment signal: frequency difference of the clock, relative time delay, wander and jitter.

For instance, primary group (2048kbit/s) [28] code stream's typical data is as follows:

— Frequency difference: According to CCITT Recommendations, the transmission code stream tolerance is $\pm 50\times 10^{-6}$; office clock tolerance (for international use) $\pm 1\times 10^{-11}$, therefore, the scope of clock frequency difference may be $\Delta f/f = 2\times 10^{-11} \sim 1\times 10^{-4}$;

—Relative time delay and its variation scope: typical data is as follows:

Transmission media	Time delay (μs/km)	Temperature variation ratio (ns/km·°C)
Non-inductance: paper dielectric cable	5.1	3
coaxial cable	3.5	0.033
microwave relay	3.3	0.0033

—To be calculated in accordance with the shortest hypothetical reference connection (1000km), the time delay of coaxial cable is 33ms while the frame cycle of 2048kbit/s code stream is 125μs. It is clear that the time delay is much bigger than the frame length.

—Wandering and jitter: CCITT has recommended the wander and jitter sample of 2048kbit/s code stream (Figure 8-2).

—The environment mentioned above is a typical one where tile frame regulator is located in under such concrete environment, the frame regulator, on condition that the slip is allowed to occur for certain times, has to set up the synchronization between input code stream and local timing sequence. The following concrete functions need to be completed.

—Use local clock instead of input

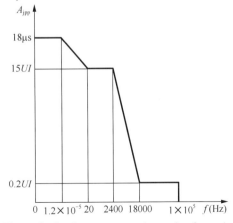

Figure 8-2 Recommended sample of wander

code stream clock to realize bit synchronization.

—Remove the jitter and wander in the input code stream by means of slip.

—Make use of frame time delay adjustment to realize frame synchronization between the input code streams from all directions with local timing sequence.

—No more losses of transmitted messages except the complete frame slip.

—It is not difficulty to imagine that the frame regulator to realize above-said concrete functions is a suitable-controlled buffer with the capacity of one frame at lest. Therefore, sometimes, such kind of technology is called reservoir buffeting technology.

8.2 FRAME REGULATION CATEGORIES

It can be seen from the statement in the previous chapter that the frame regulation is a buffer provided with special control, and its writing-in clock (W) and reading-out clock (R) are separately. When time difference between reading and writing is less than one frame, the code stream will be transmitted without error; when time difference between reading and writing is decreased to zero or less than zero, one frame in the code stream is reread the transmission is done without error; when the time difference between writing and reading reaches to one or more than one frame, after one frame in the code stream skip, the transmission is done without error. Obviously the equipment to finish such function is a buffer whose capacity is equal to one frame (see Figure 8-3). In case the writing-in clock and reading-out clock are constant and without jitter and wander, then this buffer can realize above said functions without any control.

In fact, the jitter and wander exists in the writing-in clock, whose frequency difference relative to local clock is not constant, so, when R catches up with W or W catches up with R, it is possible to have slip back and forth frequently (Figure 8-4), which is not allowable. For such purpose, some necessary control must be added. On the basis of different control modes being added, the different kinds of frame regulation technologies appear.

In addition, in consideration of selecting the devices, both RAM and SR can be used as the buffer. Generally, RAM is used to read and write same address simultaneously while SR is used for synchronous reading and writing, Therefore, some control circuits need to be added if they are used as the buffer. With the different circuit added, different frame regulator are designed.

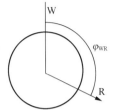

Figure 8-3 Basic principle of frame regulation (1)

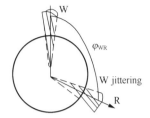

Figure 8-4 Basic principle of frame regulation (2)

With different control modes and circuits, the different frame regulation schemes and equipment are configured, and the relevant performances, quality of equipment and the applicable cases are different. The reference paper [41] shows a simplified result.

Categories		Working rate	Memory capacity	Control circuit	Circuit design	Frame synchronization	Suitability
Use RAM	Timing-selection control	High	Small	Small	Easy	No	Low rate high capacity
	Memory-selection control	Low	Big	Middle	Easy	Yes	High rate
	Pre-buffer control	Middle	Middle	Middle	Relative difficulty	Yes	
Use SR		Low	Big	Middle	Easy	Yes	

Among them, the pre-buffer control scheme using RAM can meet the requirement in respect of its performance, the other conditions is also suitable, applicable for primary group frame regulator.

8.3 THE WORKING PRINCIPLE OF PRE-BUFFER FRAME REGULATOR[42][43]

The pre-buffer frame regulator is composed of pre-buffer memory, frame memory, frame synchronization circuit and time-difference control unit. The principle is shown in Figure 8-5.

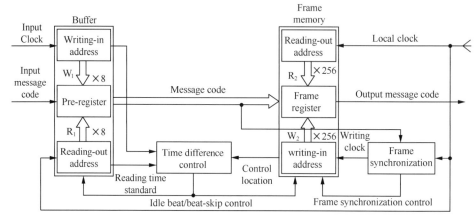

Figure 8-5   Pre-buffer frame regulation principle

(1) Pre-buffer

This is a typical small-capacity buffer, taking input clock as writing-in clock (W_1); and the local clock is idle beat/beat-skip controlled to be reading-out clock (R_1). if $W_1=R_1$, then the time difference of reading relative to writing ($\varphi_{W_1R_1}$) maintains unchanged and the written-in message codes are read out in sequence to realize bit synchronization; if $W_1<R_1$, as time goes by, the time difference between writing and reading will decrease gradually. When $\varphi_{W_1R_1}$ decreases to some certain lower limitation φ_{min}, R_1 will hesitate for one beat, thus, the time difference between writing and reading $\varphi_{W_1R_1}$ will in crease by one bit time (UI). Such repeat will not lose the message but maintain the bit synchronization; if $W_1>R_1$, as time goes by the $\varphi_{W_1R_1}$ will increase gradually, when $\varphi_{W_1R_1}$ increases up to some certain limitation, R_1 will insert one beat, then, the tim0e difference between reading and writing $\varphi_{W_1R_1}$ will decrease by one bit time. Such repeat will not lose message but maintain bit synchronization (see Figure 8-6). It is thus clear that the pre-buffer can complete the functions to replace clock, track wander and absorb jitter. The buffer inputs message codes and outputs message codes and there is only one difference that the gradual-change of phase in the input code stream and the time difference accumulated in the course of clock change will be converted to the transient beat-hesitation and beat insertion of the output code stream. this is identical with positive/zero/negative adjustment procedure.

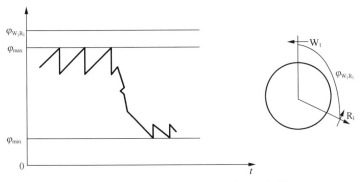

Figure 8-6　Working principle of pre-buffer

(2) Frame elastic memory

The capacity of frame memory is just equal to the bit numbers of one frame. Its reading-out clock (R_2) is the local original clock while its writing-in clock (W_2) is same local clock, but its writing-in address generator is controlled simultaneously with input frame alignment signal of the frame memory and the beat-hesitation/beat-skip control signals. The result being controlled with frame alignment signals can ensure one complete frame signals put separately on the dedicated locations of frame memory. In a network node, if the frame regulator which receives the code streams from all directions can put message codes of various frames of the input code stream into stipulated locations of frame memory. but all the reading-out clock is a same clock, then the output frames from different regulators are naturally synchronized each other. i.e. having realized frame synchronization. The result being controlled with beat-hesitation/beat-skip makes R_1 and W_2 completely united, in this way, there will be no message codes loss from buffer into frame memory. There is no any other control except those mentioned above. Therefore, W_2 is indirectly controlled with input clock (W_1), and there is only one difference that W_2, taking UI as a units, is of sudden-change and it is complete equal to R_2 when sudden change does not happen because they are all from local clock.

If W_2 has neither beat-hesitation nor beat-skip then $W_2=R_2$, at this moment, the time difference between reading and writing of the frame memory $\varphi_{W_2R_2}$ will maintain unchanged, i.e. the message is not inserted or lost to maintain the frame synchronous operation; each beat hesitation of W_2 makes the time difference $\varphi_{W_2R_2}$ decrease by one UI time, when $\varphi_{W_2R_2}<0$, i.e. when R_2 overtakes W_2, the phenomenon of reading (inserting) one frame message code will occur, i.e. one slip happens but the frame synchronization is still maintained, every time when W_2 has a beat-skip, $\varphi_{W_2R_2}$ will increase by one UI time. When $\varphi_{W_2R_2}$ exceeds the frame memory capacity. i.e. W_2 overtakes R_2, the phenomenon to lose one frame, message codes will occur, i.e. slip happens once, at this moment the frame synchronization is still maintained (see Figure 8-7).

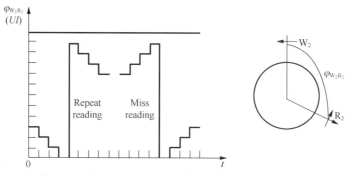

Figure 8-7   Operating principle of frame regulation

(3) Frame synchronization circuit

Its task is to collect frame synchronization signals from output message codes of the buffer, it is used to synchronize with writing-in address of frame memory so as to guarantee the message codes to be put in the stipulated address of frame memory, taking a frame as a unit. Its working principle is identical to that of frame synchronization of the multiplexer (see Figure 8-8).

Figure 8-8   Frame synchronization working principle

(4) Time difference control unit

The time difference control unit is controlled with writing-in and reading-out pulses of the buffer, the output values of those two being discriminated presents the size of time difference between reading and writing ($\varphi_{W_1R_1}$). When $\varphi_{W_1R_1} \leq \varphi_{min}$, it gives beat-hesitation signal; when $\varphi_{W_1R_1} \geq \varphi_{max}$, it provides beat-skip signal, in which, φ_{min} and φ_{max} is protection margin for time difference between reading and writing, also named as magnetical stagnation margin (Refer to Figure 8-9).

Chapter 8 Frame Adjustment Principle

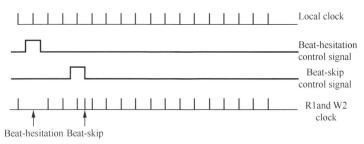

Figure 8-9 Time difference control principle

Because the equipment has the frame synchronization circuit already, which can be used to select a proper beat-hesitation/beat-skip locations to simplify the control circuit. For example, beat-skip control is usually relative troublesome, but the beat-skip location is selected in the location of frame alignment signals, the beat-skip control can be simplified (see Figure 8-10), i.e. W2 will write the first bit into the first address followed by the third bit into third address, it is not necessary to write the second bit into the second address because this bit is useless. Such processing has one UI time in advance and it is equal to insert one bit but easy to be realized.

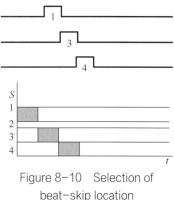

Figure 8-10 Selection of beat-skip location

8.4 2048kbit/s FRAME REGULATOR DESIGN

The material of this section comes from Reference paper [42] and [43].

(1) Frame structure

The frame structure is in compliance with CCITT Rec. G.732, the nominal rate is 2048kbit/s, tolerance±50ppm, each frame has 32 time slots and each time slot has 8bit. That is to say, frame length is 256bit, frame frequency 8kHz and the frame alignment signals are located in the following 7bit of zero time slot in odd frames (see Figure 8-11).

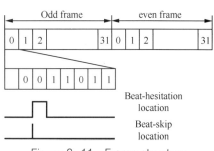

Figure 8-11 Frame structure

In each two frames one possible beat-hesitation location and one possible beat-skip location are arranged: the beat-hesitation location is at the second bit position of the frame alignment signal; while beat-skip location is selected between the first bit and second bit of the frame alignment.

(2) Frame synchronization unit

The schematic drawing of frame synchronization is shown in Figure 8-12.

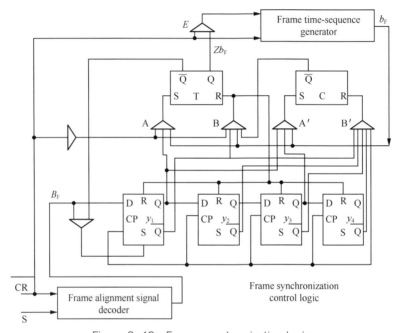

Figure 8-12　Frame synchronization logic

The frame alignment signal decoder provides the frame synchronization control logic with decoding location signal (B_F); the frame time sequence generator provides the frame synchronization control logic with frame time sequence location signal (b_F); depending on the relative time location relation between B_F and b_F, the frame synchronization control logic controls the frame time sequence generator or set the frame time sequence generation onto b_F location (i.e. the eighth beat of frame time sequence), or feeds the clock to drive the frame time sequence generator to operate synchronously with the input frame code stream.

The acquisition logic of Figure 8-12 is designed according to the flow chart of Figure 8-13. The main advantage of such acquisition logic is that the acquisition checking counter is unified with synchronization protective counter so as to simplify the equipment[37], the time waveform drawing is shown in Figure 8-14.

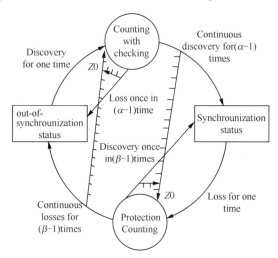

Figure 8-13 The flow chart of synchronization acquisition logic

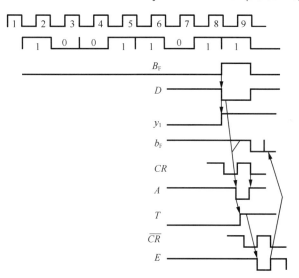

Figure 8-14 Acquisition procedure for the first time

Acquisition procedure for the first time: as acquisition logic is at the acknowledgment of out-of-synchronization status, the frame time sequence generator will be set to the eighth beat status; at this moment, the output status signal of frame time sequence generator b_F will be at high electric potential; when the frame alignment signal is not decoded from the input code stream, the output B_F will be low electric potential, at this moment, the whole system is at the waiting status. When frame alignment decoder decodes the frame alignment signals, B_F will become high potential to make frame synchronization control logic take place the control procedure according to Figure 8-14 until the clock control gate (E) feeding clock to drive frame time sequence generator; the initiation of frame time sequence makes b_F become low. potential at once, and B_F returns to low potential, system gets into synchronization status.

Acknowledgment of synchronization procedure: the system will be driven into synchronization status when B_F and b_F is aligned for one time, but it is not yet acknowledged.It will return to acknowledgment of out-of-synchronization status only if non-alignment occurs for one time, only when continuous alignments are reached for three times, the flip-flop C can be converted to get into acknowledging synchronization status, then, non-alignment phenomenon occurs for less than four times by accident it will not be recognized as acknowledgment of out-of-synchronization (see Figure 8-15).

Acknowledgment of out-of-synchronization: once upon the synchronization is acknowledged, if the non-alignment between B_F and b_F is discovered for one time. the system will counts the errors but the control measure will not yet be used; as long as the non-alignment is less than four times followed by the alignment for one time, then the acknowledgment of synchronization status will be restored again; the system will acknowledge the out-of-synchronization and adopt acquisition control so long as the non-alignments occur up to four times (see Figure 8-16).

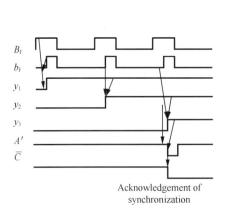

Acknowledgement of synchronization

Figure 8-15 Acknowledgment of synchronization procedure

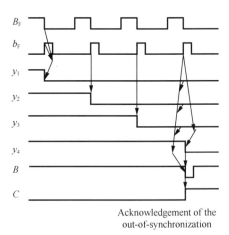

Acknowledgement of the out-of-synchronization

Figure 8-16 Acknowledgment of the out-of-synchronization procedure

Chapter 8 Frame Adjustment Principle

(3) Checking and controlling the time difference between reading and writing

There is no specific time to check the time difference between reading and writing. As long as the front edge of W_{1-3} enters into R_{1-4} section, the flip-flop G will get into high potential, i.e. apply for beat-skip control; so long as the front edge of R_{1-4} enters into W_{1-5} section, the flip-flop will get into high potential, i.e. apply for beat-pause control. Its circuitry is shown in Figure 8-17. The relevant relation of time difference between reading and writing when the control application takes place (see Figure 8-18).

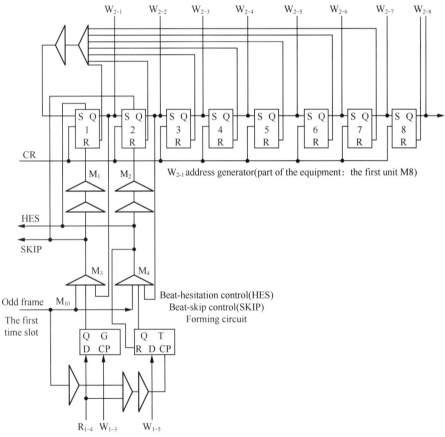

Figure 8-17　Checking and control principle of the time difference between reading and writing

As the front edge of W_{1-3} pulse begins to overlap with R_{1-4}, it expresses that

W_{1-4} differs from R_{1-4} by two UI time and requires R_1 faster; As R_{1-4} front edge begins to overlap with W_{1-5}, it expresses that R_{1-4} differs from W_{1-4} by two UI time and requires R slower. That is to say, the buffering margins are two UI time respectively.

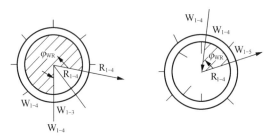

Figure 8-18 The control application of time difference relation between reading and writing

The time to employ control is specified in the first time slot of odd frame (controlled by M_{10}); the beat-skip control is at the beginning of the first bit, i.e. the first address pulses (R_{1-1} and W_{2-1}) are omitted. Thus the first time slot only leaves seven UI time, and R_1 and W_2 is one UI ahead of time; the beat-hesitation control is at the second bit time, i.e. the width of the first address pulses (R_{1-1} and W_{2-1}) are doubled, so as to make the first time slot occupy nine UI times and make W_2 and R_1 be postponed by one UI time. The control procedure is given in Figure 8-19 and Figure 8-20.

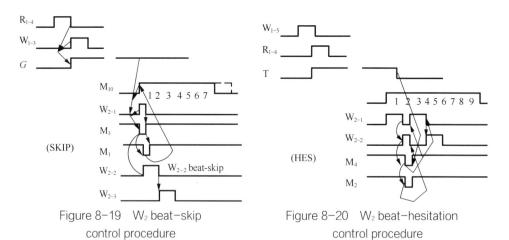

Figure 8-19 W_2 beat-skip control procedure

Figure 8-20 W_2 beat-hesitation control procedure

(4) Writing-in (W_2) control of the frame memory

The beat-skip control is that the beat-skip control signal (SKIP) sets the first level of module-8 circuit of W_2 generator to zero, and the second level to one so as to make the first circle of the first time slot in the odd frame generate W_{2-2}, then normally generate W_{2-3}, W_{2-4} etc.; the beat-hesitation control is that the beat-hesitation control signal (HES) sets the second level to zero but the first level to one so as to double W_{2-1} i.e. occupy two bits circle time, then generate normally W_{2-2}, W_{2-3} etc.

(5) Reading-out (R_1) control of the buffer

Reading-out address control and its generating circuit is shown in Figure 8-21.

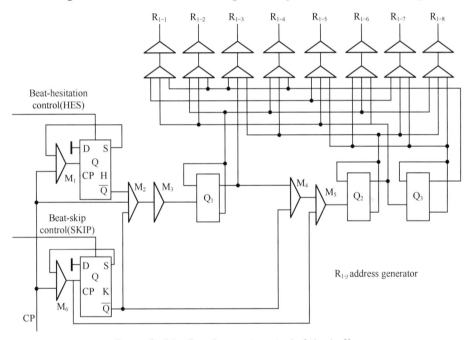

Figure 8-21　Reading-out control of the buffer

The control procedure is shown in the relevant time waveform drawing. The circuit differ from W_8 generator. Here the third level divided by 2 circuit plus decoder circuit generates eight circulating address signals but its function is exactly same as the former. The beat-skip control causes the first bit position of the first time slot in the odd frame generate directly R_{1-2} beat-hesitation control doubles the address pulse R_{1-1}. Their activity time is exactly corresponding to W_2, the detail procedures are shown in Figure 8-22 and Figure 8-23.

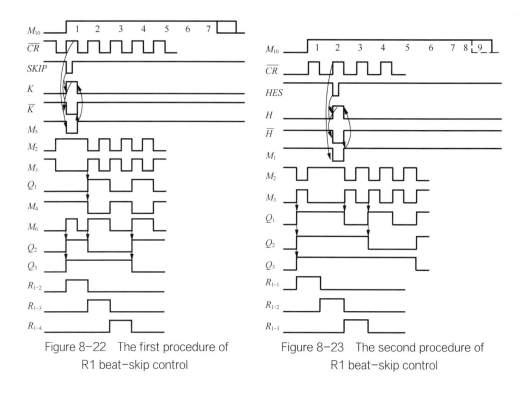

Figure 8-22 The first procedure of R1 beat-skip control

Figure 8-23 The second procedure of R1 beat-skip control

8.5 THE ADDITIONAL FUNCTION OF FRAME REGULATOR

The functions mentioned-above are the basic ones of the frame regulator, i.e. bit and frame regulating function. The frame regulator is convenient to complete more task with a little more relative circuit added. Such items includes: the inserting and collecting of the frame alignment signals, inserting and collecting of the control signals as well as the inserting and collecting of various service signals etc. These functions, in fact, are the multiplexing functions. Because the frame regulator has the frame structure sequence generator and frame synchronization acquisition and maintaining circuit already, only small amount of the multiplexing position shift links are added, the above said multiplexing functions can be completed. The regulating function plus multiplexing function to the frame regulator will finish the trunk terminal functions of the digital switching system. In such case it is not proper to be called frame regulator, normally called switching terminal.

Chapter 9 PDH/SDH Interface

9.1 GENERAL

In order to meet the demand of high rate digital transmission via optic fiber, CCITT No. 18 working group studied and worked out three new Recommendations in June of 1988, i.e. Rec. G.707 (Synchronous Digital Hierarchy Bit Rates), Rec. 708 (Network Node Interface for the Synchronous Multiplexing Structure). and Rec. 709 (Synchronous Multiplexing Structure). These three coherent Recommendations jointly specify a new synchronous digital hierarchy (SDH)of higher order group and also stipulate the interface between SDH and plesiochronous digital hierarchy (PDH) described in Rec. 702 (Digital Hierarchy Bit Rates)[73][74][75].

This chapter is related to SDH and PDH/SDH interfaces respectively.

9.2 HIGHER ORDER GROUP SDH

(1) Higher order group SDH bit rates

Rec. 707 specifies that the lowest level bit rate of SDH shall be 155 520kbit/s, and the higher SDH bit rates shall be obtained as $N=4X$ of the lowest rate (among which, X is non-zero integer multiples) and SDH higher levels should be denoted by the corresponding multiplication factor N. The concrete value stipulation is shown in the following table. It can be seen from the table that the multiplexing rate is exactly equal to the sum of various tributary rates, i.e. no any other bits are contained in the multiplexing code stream except the tributary code streams. Obviously, this differs from the plesiochronous digital multiplexing hierarchy.

SDH lowest level is called Synchronous Transport Module level 1, abbreviated as STM-1. SDH level N is expressed as Synchronous Transport Module level N, abbreviated as STM-N. The multiplexing schematics from STM-1 to STM-N is given in Figure 9-1.

Study status	Synchronous digital hierarchy module(*N*)	Bit Rates(kbit/s)
CCITT Rec G.709	1	155 520
	4	622 080
(To be stuidied)	8	1 244 160
	12	1 688 240
	16	2 488 320

It is clear in Figure 9-1 that realization of such synchronous multiplexing is quite simple, as matter of fact, it is a high speed parallel to serial conversion circuit which takes the code word as unit, once it is multiplexed into multiplexing code stream, the occurred moment of various tributary code can not be recognized again. Therefore, the simple serial to parallel conversion

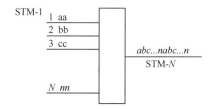

Figure 9-1   The synchronous multiplexing from STM-1 to STM-*N*

circuit can not be used to realize the demultiplexing correctly, there will be some troubles for the synchronous demultiplexing. The numbers of the tributaries are known and it is also known that the time slot sequence of the adj acent code word occupied by the tributary code streams in the multiplexing code stream while such multiplexing is adopted. So, after being divided into various tributary code streams by means of the serial to parallel conversion circuit taking code word as unit, once a tributary code stream can prove what is its sequencing number, then the original sequence numbers can be identified for all the tributaries. In this way, the correct demultiplexing will be realized. Of course, if each tributary can mark out its own tributary sequence number, it will be more sure to realize correct demultiplexing. Practically, we do it in this way, because each tributary has its own flag, it is possible to realize the correct recognition quicker by using a relative simple recognition circuit.

From the rate table of high order group SDH. We can see that STM-1 code stream is most basic one in the whole SDH, which can be multiplexed into STM-*N* directly. Thus, as long as STM-1 code stream is able to give its own tributary sequence number, that is enough to implement correct tributary dividing. For such flag, there is a concrete arrangement in the STM-1 frame structure.

(2) STM-1 frame structure

STM-1 frame structure is shown in Figure 9-2. STM-1 frame structure is also called the basic frame structure for high order SDH, it consists of digital section formed with 270 rows (one byte per row) times 9 lines, its basic parameters are concluded as follows:

- Bit rate f_h=155 520kbit/s;
- Frame length L_s=23 430kBytes=19 440kbit;
- Frame frequency F_s=8kHz;
- Frame circle T_s=125μs.

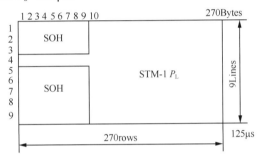

Figure 9-2 STM-1 frame structure

The whole frame is called a section, each section can be divided into two parts: section overhead (SOH) occupy in lines 1-3 and lines 5-9 among the row 1-9; STM-1 payload occupying the other parts.

Figure 9-3 shows the concrete arrangement of STM-1 section overhead: A1, A2: frame alignment signal. which occupies 6 codes time slots in the STM-1 frame overhead with the pattern A1 A1 A1 A2 A2 A2, in which, A1=11110110, A2=00101000;

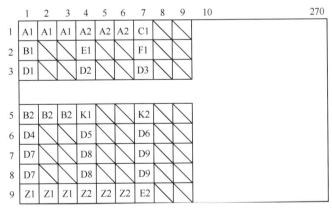

Figure 9-3 STM-1 section overhead arrangement

D1-D2: data communication link. The data link occupies, in total, 12Bytes with total rate of 768kbit/s. It is specified that only first channel STM-1 has such data communication link among STM-N.

C1: STM-1 flag signal. It is a unified number assigned to STM-1 tributaries before N pieces of STM-1 tributary code streams are multiplexed into one STM-N, then its own sequence number can be identified in accordance with this flag signal after the demultiplexing so as to realize correct demultiplexing.

E1, E2: telephone private link. This two bytes provide 128kbit/s dispatch circuit link. In STM-N, only the first STM-1 has such link arrangement.

F_1: subscriber link. Assigned 64kbit/s link used by subscriber. In STM-N, only the first STM-1 has such link.

B1, B2: bit error supervisory link. This system uses bit interleaving parity check code (BIP-N) to supervise the bit errors. B1 is used to transmit BIP-8 code for supervising the bit errors in the regeneration section, three B2 types are used to transmit BIP-24 code for supervising the bit errors in the section. In STM-N, all of STM-1 provides such links.

K1, K2: auto-protection switch link. Used to transmit automatic protection switch message. it is specified that only first STM-1 has such link in STM-N.

Z1, Z2: idle bytes.

The other bytes arranged for domestic use.

9.3 THE GENERAL ARRANGEMENT OF PDH/SDH INTERFACE

At present, the existing digital network in the world is on the basis of PDH. Therefore, STM-1 must be able to interface with all the plesiochronous digital rates stipulated in Rec. G.702. That is a relative complicated technical arrangement. In order to make these interfaces arrangements as unified as possible, it is necessary to have a related careful system arrangement.

(1) Basic Definition

Container: C-n (n=1-4).

It is an effective payload unit with the stipulated capacity. Such unit not only has the scale capable to carry any existing levels stipulated in Rec. G.702, but also provides the capability to carry broadband signals to be specified. The payload stipulated in Rec. G.702 herein means H4 (139 264kbit/s), H32 (44 736kbit/s), H31 (34 368kbit/s), H22 (8448kbit/s), H21 (6312kbit/s), H12 (2048kbit/s) and H11 (1544kbit/s). Correspondingly, the containers capable to carry correspondent effective payload are listed below:

Asynchronous rate (kbit/s)	Payload Hn	Container(C-n)
139 264	H04	C-04
44 763	H32	C-32
34 368	H31	C-31
8448	H22	C-22
6312	H21	C-21
2048	H12	C-12
1544	H11	C-11

Virtual Container: VC-n (n=1-4).

The VC-n consists of the C-n and path overhead (POH) of VC-n appropriate to its own level. The POH provides the communication between the assembly point and disassembly point in the VC. The concrete communication contents include path function supervisory signal, maintenance function signal and alarm status indicating signal. The VC has two types, i.e. basic VC (VC-n, n=3,4).

Tributary Unit: TU-n (n=1-3).

The TU-n consists of the VC-n and tributary unit pointer (TUPTR). The TUPTR indicates VC-n phase offset relative to the POH of higher order VC-n which contains such TUPTR. The TUPTR is fixedly allocated corresponding to the POH of last higher level.

Tributary Unit Group: TUG-2.

TUG is composed of several equal TUs.

Administrative Unit: AU-n(n=3, 4).

The AU-n consists of high order VC-n and Administrative Unit Pointer (AUPTR). The AUPTR indicates high order VC phase offset relative to STM-1 frame. The AUPTR location is fixed relative to STM-1.

The first level of synchronous transmission Mode: STM-1.

STM-1 is a basic assembly module of SDH, which consists of STM-1 payload and section overhead (SOH).

The definitions said above is shown in Figure 9-4. It is clear that, for the H4 and H3 (139 264kbit/s, 44 736kbit/s and 34 368kbit/s), the high order container G4 or G3 can be entered directly; the high order VC-4 or VC-3 can be formed directly, by adding POH; followed by adding AUPTR, the administrative unit AU-4 or AU-3 can be formed directly to contain the payload area of STM-1 at last. For H2 and H1 (8448kbit/s, 6312kbit/s, 2048kbit/s and 1544kbit/s), it can enter the basic container C-2 or C-1 directly; and the basic VC-2 or VC-1 can be formed directly by adding POH; followed by adding TUPTR the TU-2 or TU-1 can be formed directly; several TUs can form tributary unit group (TUG-2). TUG-2, as effective payload of high order container, at last, enters the payload area of STM-1 through the high order assembly procedure.

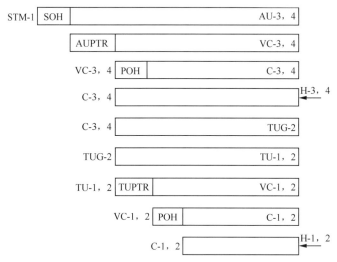

Figure 9-4　The relationship between the basic definitions

(2) Interface multiplexing structure

In fact, the containing relationships between the definitions herein above have stipulated the interface relationship between PDH/SDH. Such interface relation is the multiplexing relation in fact. In order to provide the most flexibility for the engineering application and the networking between the networks, the possibility of different multiplexing has been considered in such interface multiplexing structure.

It can be seen from the Figure 9-5 that there are seven effective payloads in STM-1 in total, i.e. H4, H32, H31, H22, H21, H12 and H11; all this effective payload directly enter the corresponding containers (C-4, C-32, C-31, C-22, C-21, C-12 and C-11); all the containers will firstly form the corresponding virtual containers (VC-4, VC-32, VC-31, VC-22, VC-21, VC-12 and VC-11); all the high order VCs get into corresponding AU or TU (AU-4, AU-32, AU-31, TU-31, TU-32 and TU-31); all the basic VC get into corresponding TUG (TUG-22 and TUG-21), TUG and high order container (C-4, C-32 and C-31) are in the equal positions and can get into high order VC (VC-4, VC-32 and VC-31).

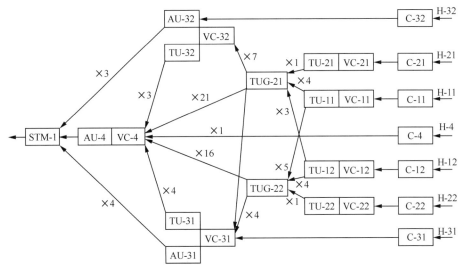

Figure 9-5 Interface multiplexing structure

It is clear that the internal multiplexing relation is quite complicated in such interface multiplexing structure, but the effective payload access is rather flexible. Such interface multiplexing design just make use of such complexity of internal multiplexing relation to obtain the flexibility of integrated equipment's application. All the interface relation will be described as follows:

9.4 THE ARRANGEMENT DESIGN FOR C-4 TO ENTER STM-1

The arrangement design for C-4 to enter STM-1 is in three steps: The first step is the arrangement for AU-4 to enter the STM-1 payload section, as a matter of fact, it gives the concrete stipulation for AU-4 PTR, the second step is the arrangement for VC-4 to enter the AU-4 payload section, in fact, it gives the concrete stipulation for VC-4 POH; at last, it is the concrete arrangement for C-4 to enter the VC-4 payload section. Obviously, the arrangement design for C-4 to enter STM-1 is, in fact, the concrete arrangement design of AU-4 PTR, VC-4 POH and C-4 in the STM-1 payload section. The concrete plan recommended by CCITT is shown in Figure 9-6.

PDH for Telecommunication Network

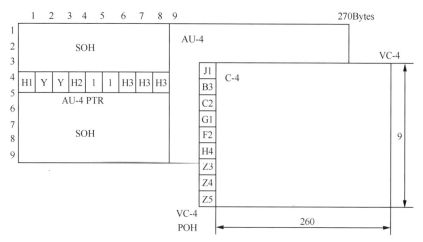

Figure 9-6 The arrangement for C-4 to enter STM-1

(1) AU-4 PTR

AU-4 PTR function is to express the phase difference of the VC-4 relative to the SOH of STM-1; and the frequency justification indication of VC-4 frame frequency relative to AU-4 frame frequency. The AU-4 frame frequency. The AU-4 PTR makes VC-4 being arranged in a flexible and random way.

Figure 9-7 shows the AU-4 PTR arrangement relation. the AU-4 PTR occupies 5 bytes of H1, H2 and H3 etc. Among them, the first 4 bits (N) in H1 expresses New Data Flag (NDF), 1001 means effective and 0110 means ineffective; the 5th and 6th bits (S) of H1 expresses the type of AU, when AU-4 is among the multiplexing, it is specified as 10; the last two bits of H1 and all the bits of H2 mean the PTR value in the locking status, which is a binary value from 0 to 782 and expresses the distance value of the first byte of AU-4 relative to the last byte (H3) of AU-4 PTR, the distance value takes three bytes as one unit and among the three the first is taken as the flag, e.g. if AU-4 PTR value is 0, it means that the end of last H3 of AU-4 PTR is followed closely by the first word of VC-4; if the frame frequency of VC-4 is not equal to the frame frequency of STM-1 (i.e. the frame frequency of SOH), it is necessary to do the rate justification on VC-4 frame frequency. In the course of frame frequency justification, a length of four basic frames is taken as justified frame length. If VC-4 frame frequency is lower than SOH frame frequency, the positive frame frequency justification will be done, Five I-bits are inverted as the positive frame frequency justification indication so as to get three idle bytes in the positive justification position, i.e., VC-4 has been postponed for three bytes time; If VC-4 frame frequency is higher than SOH frame frequency, the negative frame frequency

Chapter 9 PDH/SDH Interface

justification will be done. Five D-bits are inverted as the negative frame frequency justification indication so as to transmit three bytes of VC-4 in the negative justification position (i.e. three H3 bytes), i.e. VC-4 has been advanced for three bytes time. As the positive justification being implemented, the AU-4 PTR value is increased by one; as the negative justification being implemented, AU-4 PTR value is decreased by one. Then, the PTR value will be maintained onto the new values.

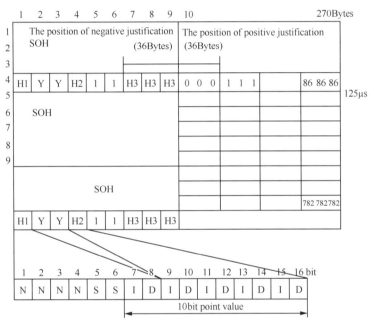

Figure 9-7 AU-4 pointer allocation

The regulation to form AU-4 PTR is summarized as follows:

In normal operation, VC-4 is located in the beginning position of AU-4 and regulates the PTR value and keeps the PTR unchanged. In the following cases, the PTR value will be changed: When positive frame frequency justification is done, the PTR value is increased by one; When negative frame frequency justification is done, the PTR value is increased by one; When the relative location of VC-4 is changed because of other reasons, the PTR value will be changed accordingly. But no matter what kind of reasons, the PTR value will not be changed again within the following three frames after the PTR change. Accordingly, the regulation to decode AU-4 PTR is summarized as follows: In normal operation, the PTR value regulates the beginning position of VC-4 in the AU-4 and keeps it unchanged; if I-bits in most of PTR words are inverted, which indicates that the positive frame frequency justification has occurred, and 1 needs to be added to the following PTR value; If D-bits in most of PTR words are inverted, which indicates that the negative frame frequency justification has occurred, and 1 need to be reduced from the following PTR value; When new flag (1001) occurs in N bit, the new PTR value will be used to replace the original PTR value.

(2) VC-4 POH

VC-4 POH is assigned to specific payload unit from the beginning of multiplexing to demultiplexing and it fulfills specific function necessary for transmitting VC-4. The POH is only stipulated by payload unit independently, nothing to do with other specific operation and do not obstruct other specific operation. The arrangement of VC-4 POH in VC-4 is shown in Figure 9-6.

VC-4 POH occupies the time slots of nine words in total (J1, B3, C2, G1, F2, H4, Z3, Z4 and Z5), The concrete functions are assigned as follows.

J1: It is used as path flag and expresses the start of VC-4, and its relative location is indicated with AU-4 PTR. The transmission of fixed word series of 64 Bytes is repeated in J1, by which the continuous connection with assigned transmitting end is verified at the receiving end. The concrete contents in J1 will be agreed by the subscribers at both transmitting and receiving ends.

B3: It transmits path BIP to realize the BER supervisory function. BIP-8 depends on the calculation of all VC-4 bits before the scramble.

C2: Signal flag, this byte is used to indicate the integrity of VC-4 code stream, it may have 256 types of binary values, among which, 2 have been stipulated: value 0 expresses that VC-4 is not equipped and value 1 expresses that VC-4 is equipped but without the effective payload stipulated.

G1: path status. The byte is used to feed back the path status and performances to path source unit so as to supervise the status and performances of all duplex paths. The concrete arrangement is that bits 1-4 are used to indicate the numbers of bit errors (0-8), the bit 5 is used for remote alarm (transmitting all "1" code).

F2: Path subscriber link. This byte is assigned to be used by subscribers.

H4: Multiframe indication. It is used to indicate the multiframe category of effective payload.

Z3-Z5: Spare bytes.

(3) C-4 container arrangement

It can be seen from Figure 9-6 that C-4 container occupies the 11-270 rows in

STM-1 basic frame structure with total capacities of 260×9 Bytes. So, C-4 container's basic parameters are listed below:
- Frame frequency F_s=8kHz;
- Frame length L_s=2340Bytes=18 720bit;
- Rate f_h=18 720kBytes/s=149 760kbit/s.

9.5 THE ARRANGEMENT DESIGN FOR C-3 ENTERING STM-1

(1) The arrangement format for C-32 entering STM-1

There are two ways for C-32 to enter STM-1: C32 enters STM-1 directly; C-32 firstly enters VC-4, then enters STM-1. No matter what kind of method will be used, each STM-1 frame can contain three C-32 containers in maximum (see Figure 9-8, 9-9).

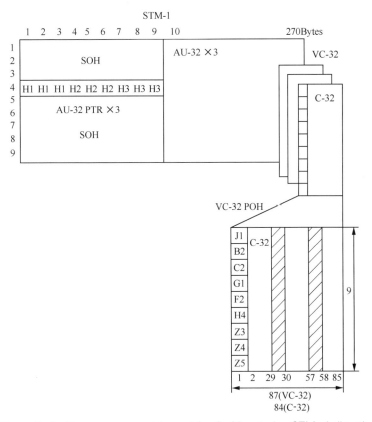

Figure 9-8 The arrangement format for C-32 entering STM-1 directly

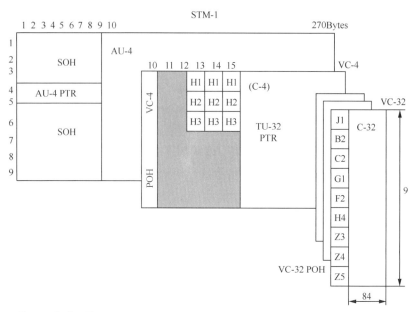

Figure 9-9 The arrangement format for C-32 entering STM-1 indirectly

In case the C-32 enters STM-1 directly, three AU-32 PTRs occupy the 4th columns of the rows 1-9 of STM-1, others occupy the AU-32 payload area; AU-32 payload accommodates three VC-32, each VC-32 capacity is 87×9 Bytes; the 1st row of each VC-32 is assigned to VC-32 POH, the two rows of idle area are distributed between 29th-30th rows and 57th-58th rows of VC-32, others are VC-32 payload area, i.e. C-32; So, C-32 capacity is 84×9 Bytes.

In case the C-32 enters STM-1 indirectly, three TU-32 PTRS occupy 1st-3rd columns of the 13th-15th rows in VC-4, the others of 11th-15th rows are idle area, and the 16th-270th rows are VC-4 payload area; VC-32, each VC-32 capacity is 85×9 bytes, the 1st row of each VC-32 is VC-32 POH, others are VC-32 payload area, i.e., container C-32; So, C-32 capacity is 84×9 Bytes.

Therefore, no matter what kind of access method is used, the C-32 effective capacities are identical. The basic parameters of the C-32 are summarized as follows:

- Frame frequency F_s=8kHz;
- Frame length L_s=756kBytes=6048kbit;
- Rate f_h=6048kByte/s=48 384kbit/s.

(2) The arrangement format for C-31 entering STM-1

There are two approaches for C-31 to enter STM-1: C-31 enters STM-1 directly; and C-31 firstly enters VC-4, then enters STM-1. No matter what kind of method will be used, each STM-1 frame can accommodates four C-31 in maximum (see Figure 9-10, 9-11).

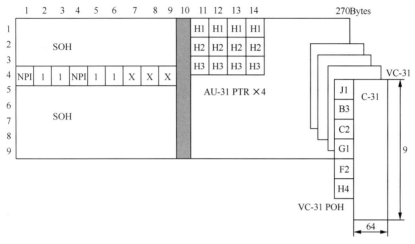

Figure 9-10 The arrangement format for C-31 entering STM-1 directly

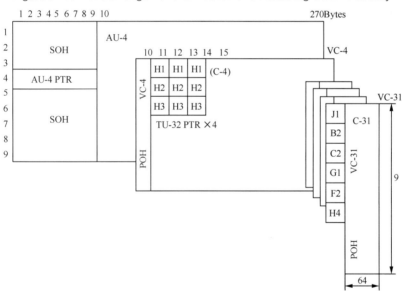

Figure 9-11 The arrangement format for C-31 entering STM-1 indirectly

In case C-31 enters STM-1 directly, four AU-31 PTRs occupy 1st-3rd columns of 11th-14th rows in STM-1, meanwhile, AU-4 PTR Null Pointer Indication (NPI) is set in the 4th columns of 1st-9th rows to ensure AU-31 PTR having a certain position relative to STM-1 SOH. Two NPI bytes have 16 bits in total to form the fixed NPI code: 1001ss1111100000, among them, ss contents are not regulated.

The 10th row of STM-1 is idle, the others are AU-31 payload area; AU-31 payload area accommodates four VC-31, each VC-31 capacity is 64×9+6 Bytes; 6 Bytes of 1st row in each VC-31 are assigned to VC-31 POH, the others are VC-31 payload area, i.e. accommodating C-31. So, C-31 capacity is 64×9 Bytes.

In case the C-31 enters STM-1 indirectly, four TU-31 PTRs occupy 1st-3rd columns of 11th-14th rows of VC-4, the others are VC-4 payload area; VC-4 payload area accommodates four VC-31, each VC-31 capacity is 64×9 Bytes; the 6 word time slots of 1st row in each VC-31 are VC-31 POH, the others are VC-31 payload area, i.e. container C-31, So, C-31 capacity is 64×9 Bytes.

Therefore, no matter what kind of access approach is used, C-31 effective capacities are identical. C-31 basic parameters are summarized as follows:

- Frame frequency F_s=8kHz;
- Frame length L_s=576kBytes=4608kbits;
- Rate f_h=4608kBytes/s=36 864kbit/s.

(3) AU-3 PTR

The arrangement of AU-3 PTR is illustrated in Figure 9-8 and Figure 9-10. The PTR value range of AU-32 is 0-782; the value range of AU-31 is 0-518. The others such as PTR function, frequency justification, NDF, PTR formation and decoding principle are as same as that of AU-4.

(4) TU-3 PTR

TU-3 PTR arrangement is illustrated in Figure 9-9 and Figure 9-11. The PTR value range is 0-764 for TU-32; TU-31 value range is 0-581. AU-32/TU-32 category flag is ss=10; AU-3/TU-31 category flag is ss=01. The others such as PTR function, frequency justification, NDF, PTR formation and decoding principle are as same as the that of AU-4 PTR.

(5) VC-3 POH

VC-3 POH arrangement is illustrated in Figure 9-8, Figure 9-9, Figure 9-10 and Figure 9-11. VC-32 POH occupies 9 Bytes; but VC-31 POH occupies 6 Bytes. The corresponding bytes application regulation is as same as VC-4 POH.

9.6 THE ARRANGEMENT DESIGN FOR TUG ENTERING HIGH ORDER VC

TUG has preassigned fixed position when it enters high order VC, So, no PTR is arranged in VC; TUG makes the groups accommodating several tributary units and it has not the POH. So, the arrangement design for TUG to enter high order VC is only to work out the concrete specifications on the interleaving position.

(1) TUG-22 entering C-4

The arrangement format for TUG-22 to enter C-4 is illustrated in Figure 9-12. 1st-4th bytes of C-4 is idle, the other 256 bytes are occupied with cyclic interleaving of 16 channels of TUG-22 according to their sequence numbers and taking the byte as a unit. Each TUG-22 Capacity is 16×9 Bytes.

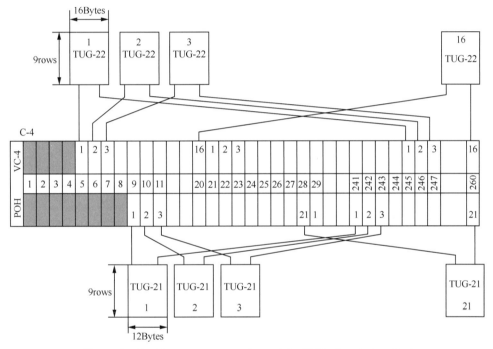

Figure 9-12 The arrangement format for TUG-2 entering C-4

(2) TUG-21 entering C-4

The arrangement format for TUG-21 to enter C-4 is illustrated in Figure 9-12. The 1st-8th Bytes of C-4 are idle, the other 252 Bytes are occupied with cyclic interleaving of 21 channels of TUG-21 according to their sequence numbers and taking the byte as a unit. Each TUG-21 capacity is 12×9 Bytes.

(3) TUG-21 entering C-32

The arrangement format for TUG-21 to enter C-32 is illustrated in Figure 9-13. All 84 Bytes of C-32 are occupied with cyclic interleaving of 7 channels of TUG-21 respectively according to their sequence numbers and taking the byte as a unit.

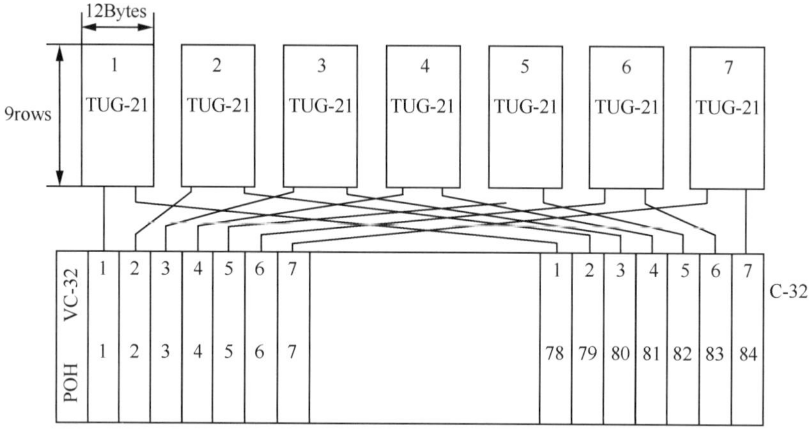

Figure 9-13 The arrangement format for TUG-21 entering C-32

(4) TUG-22 entering C-31

The arrangement format for TUG-22 entering C-31 is illustrated in Figure 9-14. All 64 Bytes of C-31 are occupied with cyclic interleaving of 4 channels of TUG-22 respectively according to their sequence numbers and taking the byte as a unit.

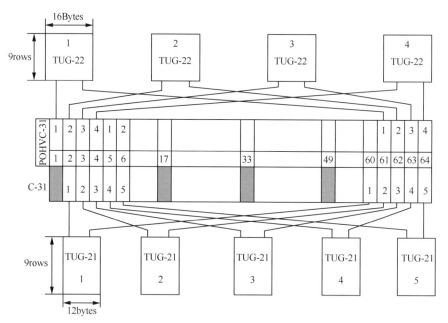

Figure 9-14 The arrangement format for TUG entering C-31

(5) TUG-21 entering C-31

The arrangement format for TUG-21 entering C-31 is shown in Figure 9-14. Firstly, the 1st, 17th, 33rd and 49th bytes are deducted from C-31, the others are occupied with cyclic interleaving of 5 channels of TUG-21 according to its sequence numbers and taking the byte as a unit.

9.7 BASIC CONTAINER ENTERING TUG

(1) The frame structure of basic container

The position for TU entering TUG is fixed, i.e. according to the tributary sequence numbers, the interleaving multiplex is carried out, taking the byte as one unit. The concrete arrangement is given in Figure 9-15.

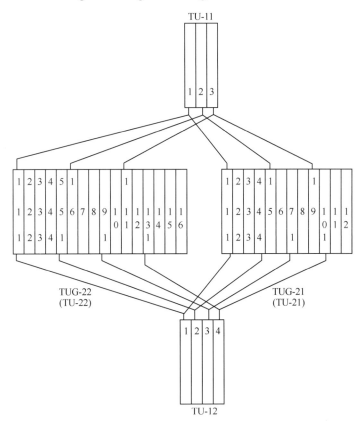

Figure 9-15 The arrangement format for TU entering TUG

There are two possible modes for basic VC to enter the TU, i.e. floating mode and locked mode. The floating mode means that the VC does not have the relative locked position in the TU, a VC PTR is designed at a stipulated position of each TU, and the relative position of VC is indicated with PTR. The locked mode means that the VC has the relative locked position in the TU, so the arrangement of VC PTR is not required.

It is considered that the basic TU is too short popularly, in order to improve effective payload efficiency and signalling arrangement, all kinds of multiframes are practically used. Now, there are three basic tributary multiframes:

500μs multiframe: formed with 4 basic frames, applicable for TU-1 and TU-2 floating mode arrangement;

2ms multiframe: formed with 16 basic frames. applicable for out-of-channel time slot signalling arrangement for 2048kbit/s word synchronization of TU-12;

3ms multiframe: formed with 24 basic frames, applicable for out-of-channel time slot signalling arrangement for 1544kbit/s word synchronization of TU-11;

The category of multiframe is expressed with H4 bytes in the high order VC POH. All coding array of TU multiplexing indication (H4) is given in Figure 9-16.

H4 bit sequence number 12345678	Frame No.	Time	Description
		0	
00000000	0		
00000001	1		
00010010	2		
00010011	3	500μs	TU multiframe
00100100	4		
00100101	5		
01000110	6		
01000111	7		
01011000	8		
01011001	9		
01101010	10		
01101011	11		
10001100	12		
10001101	13		
10011110	14		2048kbit/s signalling cycle
10011111	15	2ms	
10100000	16		
10100001	17		
11000010	18		
11000011	19		
11010100	20		
11010101	21		
11100110	22		1544kbit/s signalling cycle
11100111	23	3ms	

Figure 9-16 H4 coding array

The relative arrangement in 500μs multiframe is illustrated in Figure 9-17. One multiframe is formed with 4 basic frames. In TU 500μs multiframe, 4 Bytes (V1,

V2, V3, V4) are remained as TU PTR or for other application. In VC 500μs multiframe, one byte(V5) is left as the VC POH position. The related basic parameters of TU, VC and C are summarized in Table 9-1.

Table 9-1　　　　　　　　TU, VC, C Basic parameters

Number		22	21	12	11
Basic frame (125μs)Byte	TU	144	108	36	27
	VC	143	107	35	26
500μs multiframe Byte	TU	576	432	144	108
	VC	572	428	140	104
	C	571	427	139	103
C parameters	F_s(kHz)	2	2	2	2
	L_s(bit)	4568	3416	1112	824
	f_h(kbit/s)	9136	6832	2224	1648

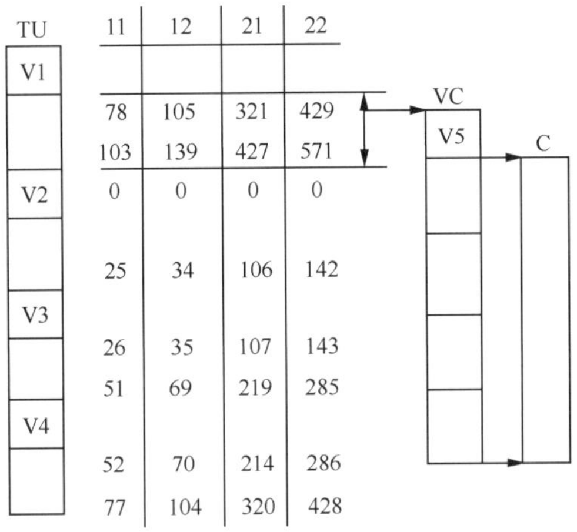

Figure 9-17　Basic container 500μs multiframe structure

(2) Basic VC POH

The first byte (V5) of basic VC (VC-2 and VC-1) is arranged to be VC POH. When basic VC works in locked mode, V5 is not arranged for the concrete application and filled with non-message bit R; When VC works in floating mode, The concrete applicable application of V5 is stipulated as follows.

The 1st and 2nd bits of V5 are used for bit error supervisory to transmit bit interleaving parity check code (BIP-2); The 3rd bit of V5 is used as bit error indication for remote terminal. If any bit error is checked by BIP-2, a series of "1" as bit error indication will be sent to VC path source unit; if no bit error is checked, a series of "0" will be sent; the 4th bit of V5 will not be used; the 5th-7th bits of V5 transmit one type of signal flag. The value 000 expresses the VC path not being equipped; the value 001 means the VC path being equipped, but the effective payload not being equipped; the 8th bit of V5 is used as remote alarm for VC path, in the case of alarm, a series of "1" is transmitted, otherwise, a series of "0" is transmitted.

(3) Basic tributary unit PTR

Only when the VC floating mode arrangement is used, the TUPTR setting is required. TUPTR can make VC to realize a flexible and random arrangement in TU multiframe and have on relationship with concrete contents.

Refer to Figure 9-17, TUPTR is arranged among V1 and V2. The concrete application design of TUPTR is shown in Figure 9-18. The new data flag (N), TU category indication (S) and PTR value (I and D) bits are arranged in 18bit time slots of V1 and V2. When N value is 0110, it means maintaining the original PTR value; When N value is 1001, it means new value. S value stipulates the tributary unit category: TU-22(01), TU-21(00), TU-12 (10), TU-11(11). 7th-16th bits are used to transmit PTR value, the maxim justification range is TU-22(571), TU-21(427), TU-12(139), TU-11(103), 0 PTR deviation position is located at the beginning of 1st byte after V2.

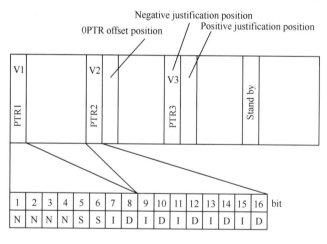

TU	N	N	N	N	S	S	Norminal range of PTR value
22	0	1	1	1	0	1	0 ~ 571
21	0	1	1	1	0	0	0 ~ 427
12	0	1	1	1	1	0	0 ~ 139
11	0	1	1	1	1	1	0 ~ 103

Figure 9-18 TU-1/TU-2 PTR design

When VC rate is not equal to TU rate, the justification of VC rate can be employed relative to TU rate. In each 500μs multiframe, D-bit of PTR value is used as negative justification indication, I-bit as positive justification one; V3 as negative justification position and one word time slot after V3 as positive justification position. The justification procedure and related stipulation is totally in accordance with that of AU-4.

The formation and decoding principle of basic TU PTR is basically as same as that of high order TU PTR. The difference only depends on the change of the number of the order; in addition, if TU capacity in TUG-21 has a change, the new data flag (NDF:1001) will be generated simultaneously in all the TU which has the new capacity in the group; the corresponding additional PTR decoding is stipulated as follows: Once upon all the TU of TUG-21 receives NDF (content: 1001) and any new capacity simultaneously, the corresponding PTR and capacity will be used to replace original PTR and capacity at once.

9.8 STM-N MULTIPLEXING

(1) The frame format of STM-N

STM-N signal is formed by interleaving N pieces of STM-1 signal according to the order of tributary number, taking byte as one unit. The 1st byte of STM-N signal should be from 1st byte (i.e. A1 byte) in the 1st STM-1, in this way the rank is interleaved according to the tributary number.

The STM-N frame structure is given in Figure 9-19. In the implementation of interleaved multiplexing, it is required that various STM-1 SOH must be put on the stipulated positions in STM-N, in detail, all the SOH will be centralized in the first $9N$ rows of STM-N. Therefore. it is required that the delay adjustment in the range of 0-125μs must be carried out before various STM-1 is multiplexed. It is rather difficult to implement practically the delay adjustment at the rate of 155 520kbit/s. The equivalent method used herein is rather simple: the SOH of various channels of STM-1 are set in the stipulated positions in the first $N9$ rows of STM-N, then, the AUPTR value in various STM-1 will be reset, i.e. various AUs are corresponding to their own SOH time (or phase) differences, In this way the equivalent delay adjustment is fulfilled.

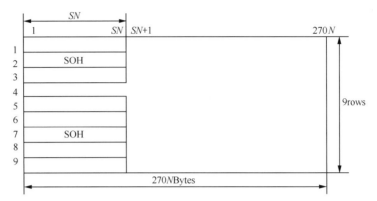

Figure 9-19   STM-N frame structure

(2) STM-N interleaving

STM-N, through frame interleaving is multiplexed to form higher signal group STM-M (M>N), see Figure 9-20. For example, the three TUGs STM-X, STM-Y, STM-Z with their frame lengths X, Y, Z, the STM-M multiplexing group is formed through frame interleaving, among which, STM frame length M=X+Y+Z. For such multiplexing, there is no any other additional stipulation except the frame interleaving on the basis of time sequence.

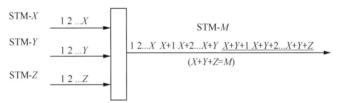

Figure 9-20   The interleaving schematic of STM-N

(3) STM-1 concatenation

Several STM-1 can form a STM-NC with a longer frame structure so as to accommodate the effective payload whose capacity is larger than that of C-4. In the implementation of STM-1 concatenation, the stipulation is as follows: the AU-4 PTR position of 1st STM-1 in STM-NC expresses the PTR value, all the other AU-4 PTR positions (H1, H2) transmit concatenation indication (CI): 1001ss111111111111 (among them, no stipulation for ss), so as to express these STM-1 are the part of STM-NC. Corresponding PTR decoding procedure: if PTR is found to accommodate CI, the operation implemented in this STM-1 frame is same as that of 1st STM-1 in STM-NC.

(4) High order group SDH features

Through the description hereinabove, it is not difficult to see some features of high order group SDH.

To simplify multiplexing/demultiplexing technology: That is the main advantage of such SDH. It is known to all that the main difficulty of high order group multiplexing/demultiplexing technology is due to the difficulty caused by high rate. It includes that the devices working rate are not high enough (or the price is too high), the pass-band of instrument is not wide enough (or the price is too high) and the other troubles to adjust the test method. Therefore, when the multiplexing experts design high order group multiplexing/demultiplexing equipment,they always try to carry out the processing on tributary rate to simplify the operation on the multiplexing rate. Thus, it can avoid or reduce the difficulties in respect to the devices, instruments and the operation due to the high rate. In the design from STM-1 to STM-N, firstly, the respective identification flag (C1) will be added to various STM-1 tributaries, then, the most simple parallel/series operation will be used to complete the synchronous multiplexing procedure, at the receiving end, the synchronization separation procedures are fulfilled through the most simple series/parallel operation, at last the correct demultiplexing will be realized through the identification sequence number of various tributaries. To simplify the high rate operation as much as possible is the basic consideration to put forward the high order group SDH in the earlier days.

To enforce supervisory and administrative function: No doubt, it is proper to have a rather large SOH section in the STM-1 frame structure being remained and distributed to all kinds of supervisory and administrative function. As the communication network is developing so far, the people recognize the importance of network supervisory and administration gradually and understand the difficulty to improve the network supervisory and administration due to insufficient service path capacity of asynchronous digital hierarchy. Such arrangement of STM-1 frame structure lays the foundation for the modernization of network administration.

To enforce the flexibility of multiplexing/demultiplexing: It is known that the multiplexing/demultiplexing in the asynchronous digital hierarchy are implemented level by level. So you can feel how trouble to skip the level in multiplexing/demultiplexing. It is stipulated herein that the STM-1 can be multiplexed into STM-N directly, there is no any concrete restriction on N value. It is clear that no matter what value will be taken for N, it is possible to realize multiplexing/demultiplexing directly between STM-1 and STM-N. Because of such flexibility of multiplexing/demultiplexing, the multiplexing/demultiplexing procedure is simplified and it will be easier to develop to higher bit rate gradually along with the development of technology.

9.9 PAYLOAD CONTAINER INTERFACE IN CCITT REC

(1) Container's parameters

From the description hereinabove, the container's parameters are obtained and listed in Table 9-2.

Table 9-2　　　　　　　　　　Container's parameters

Container	C-4	C-32	C-31	C-22	C-21	C-12	C-11
Cycle (μs)	125	125	125	500	500	500	500
Structure (Bytes)	260×9	84×9	64×9	(16×9−1) 4−1	(12×9−1) 4−1	(4×9−1) 4−1	(3×9−1) 4−1
Capacity C(Bytes)	2340	756	576	571	427	139	103
Frame frequence F_s (kHz)	8	8	8	2	2	2	2
Frame length L_s(bit)	18 720	6048	4608	4568	3416	1112	824
Rate f (kbit/s)	149 760	48 384	36 864	9136	6832	2224	1648

The payload rate stipulated in G.702 is shown in Table 9-3.

Table 9-3　　　　　　　　　　Payload rate

Payload	H-4	H-32	H-31	H-22	H-21	H-12	H-11
Rate (kbit/s)	139264	44736	34368	8448	6312	2048	1544

(2) Contain justification frame structure

The Container justification frame structure in CCITT Rec. is shown in Figure 9-21 to Figure 9-27.

The symbol description in Figure 9-21 to Figure 9-27:
- I—message bit;
- R—fixed-inserting non-message bit;
- C—control bit in positive justification;
- C1—negative justification control bit in positive/zero/negative justification;

- C2—positive justification control bit in positive/zero/negative justifications;
- S—justification position in the positive justification;
- S1—negative justification position in the positive/zero/negative justification;
- S2—positive justification position in the positive/zero/negative justification;
- O—overhead bit.

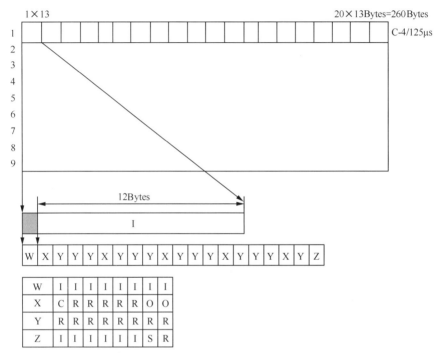

Figure 9-21   C-4 justification frame structure

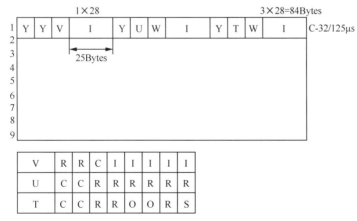

Figure 9-22   C-32 justification frame structure

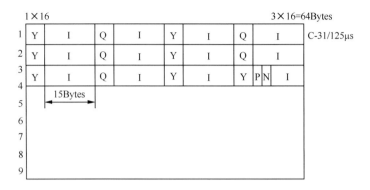

Figure 9-23 C-31 justification frame structure

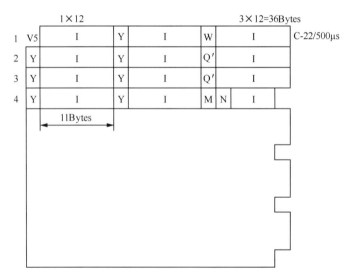

Figure 9-24 C-22 justification frame structure

Chapter 9 PDH/SDH Interface

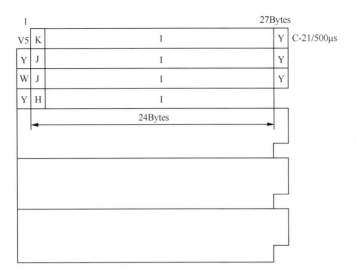

Figure 9-25 C-21 justification frame structure

Figure 9-26 C-12 justification frame structure

Figure 9-27 C-11 justification frame structure

In the parameters calculation of C-22, C-21, C-12 and C-11 for the sake of convenience to have an unified calculation, V5, the VC POH, is counted into the justification of frame length and treated as non-message bits.

According to the container's parameters and the justification of frame structure design mentioned-above, the basic parameters of justification frame is summarized in Table 9-4.

Table 9-4 Justification frame parameters

Container	C-4	C-32	C-31	C-22	C-21	C-12	C-11
Justification frame structure	$\dfrac{260\times 9}{9}$	$\dfrac{84\times 9}{9}$	$\dfrac{64\times 9}{3}$	$\dfrac{4(16\times 9-1)}{4}$	$\dfrac{4(12\times 9-1)}{4}$	$4(4\times 9-1)$	$4(3\times 9-1)$
Justification frame frequensy F_s(kHz)	72	72	24	8	8	2	2
Justification frame cycle T_s(kHz)	125/9	125/9	125/3	125	125	125×4	125×4
Justification frame length L_s(bit)	2080	672	1536	1144	856	1120	832
Container capacity f_h(kbit/s)	149 760	48 384	36 864	9152	6848	2240	1664
No.of message bits Q(bit)	1935	622	1432	1056	789	1024	772
No.of no message bits K(bit)	145	50	104	88	67	96	60
Payload rate f_i(kbit/s)	139 264	44 736	34 368	8448	6312	2048	1544

(3) Check calculation of justification design

When positive justification technology is used to realize plesiochronous interface, stuffing jitter value should not be higher than the recommended value in CCITT Rec. Stuffing jitter calculation formula is:

$$S = \left(1 - \dfrac{f_1}{f_h}\right)L_s - K$$

in which, S is positive justification rate, S can determine the stuffing jitter value A_{jpp} of the positive justification directly:

$$A_{jpp} = \frac{1}{P}$$

$$S = \frac{q}{P}, (P,q) = 1$$

To use positive/zero/negative justifications technology to realize plesiochronous/synchronous compatible interface is subject to payload frame frequency (F_{SH}) being equal to the container frame frequency (F_{SC}):

Interface(C/H)	C-4/H-4	C-32/H-32
f_h(kbit/s)	149 760	48 384
f_L(kbit/s)	139 264	44 736
L_s(bit)	2080	672
K(bit)	145	50
S	0.7777	0.6666
A_{jpp}(%UI)	11.1	33.3

$$F_{SH} = F_{SC}$$
$$F_{SC} = f_h / L_s$$
$$F_{SH} = f_l / Q$$

Interface(C/H)	C-31/H-31	C-22/H-22	C-21/H-21	C-12/H-12	C-11/H-11
f_h(kbit/s)	36 864	9152	6848	2240	1664
L_s(bit)	1536	1144	856	1120	832
F_{SC}(kHz)	24	8	8	2	2
f_L(kbit/s)	34 368	8448	6312	2048	1544
Q(bit)	1432	1056	789	1024	772
F_{SH}(kHz)	24	8	8	2	2

The check calculation result of the justification design shows that the design of interfaces 31, 22, 21, 12 and 11 are in compliance with the necessary conditions under which positive/zero/negative justifications are used to realize plesiochronous/synchronous interfaces.

As long as the proper positive/zero/negative justification technology is selected, it is possible to make the stuffing jitter small enough. The positive justification technology is used to realize plesiochronous interface for interface 4 and 32. The stuffing jitter is too big for interface 32.

9.10 THE IMPROVED DESIGN OF PAYLOAD CONTAINER INTERFACE

(1) The question being put forward

It is clear from the discussion hereinbefore that, in seven types of payload container interfaces design recommended by CCITT, the interface 4 and 32 only have plesiochronous interface and use positive justification technology, interface 31 also only has plesiochronous interface but uses positive/zero/negative justifications technology, and interfaces 22, 21, 12 and 11 have plesiochronous/synchronous compatible interfaces and also use positive/zero/negative justification technology. So, we have such question: is it reasonable to adopt the different technologies to obtain the different function for same group of interfaces design?

Usually, in the interfaces of same group, it is required that the same functions are provided and the same technical methods are adopted. For example, between the digital streams stipulated in G.702, there is only plesiochronous multiplexing relationship and all of them adopt the positive justification technology. The different technical method has to be adopted unless the engineering application requires the different function, or there are technical difficulty.

Talking about engineering practical experience most of them are tributary digital streams with higher rate to interconnect with the trunk having high rate of 155 520kbit/s. CCITT works in this mode for many years: for example, in the plesiochronous rate series, the multiplexing series from N to $(N+1)$ was formed firstly, then, the multiplexing series from N to $(N+2)$ was put into consideration, i e. the lower order group digital streams are always multiplexed level by level (or skip one level) and from low level to high level. That is to say, lower order group digital stream is less connected with high rate digital stream directly. Thus, the inference can be made in this way, if the lower order group needs to do direct synchronous interface with the 155 520kbit/s, the higher order group is much required to do direct synchronous interface with it; if we say that such kind of synchronous interface is not required, firstly. the synchronous interface for the lower order group is not required.

The other question is if two justification technologies must be used in the same group of plesiochronous interfaces. First of all, CCITT traditional approach is that, in the same series, identical technical method is used. For example, the positive justification technology is used for the whole series of PDH. There is an exception, the whole series of Russia adopts positive/zero/negative justification technology. There is no any precedent to use two technologies together. In addition, same justification technology used in this same group of plesiochronous interface is not difficult technically. In detail, there is no any technical difficult to totally adopt positive justification technology or adopt positive/zero/negative justification technology.

(2) To use positive justification technology to realize plesiochronous interface

One design plan is provided hereinafter, in which, for seven kinds of interfaces between SDH and PDH, positive justification technology is unitedly used to realize the plesiochronous interface function. It is considered that the complexity of seven related frame structure in CCITT Rec.G.709, the design plan is subject to the defined frame structure being unchanged.

After STM-1 and its interface frame structure is determined, the defined multiple (K_j) relationship between various container capacity (f_{hj}) and STM-1 basic rate (f_h=155520kbit/s) is:

$$f_{hj} = k_j f_h$$
$$\therefore \Delta f_{hj} = k_j f_h$$

Under the certain premise of frame structure parameters (L_{sj}, K_j), the ratio between container capacity (f_{hj}) and payload rate (f_{lj}) can determine the interface justification ratio:

$$S_j = \left(1 - \frac{f_{lj}}{f_{hj}}\right) L_{sj} - K_j$$

From which, the normal formula between interface justification ratio deviation (ΔS_j) and STM-1 basic rate deviation (Δf_h) near f_h=155 520kbit/s is derived:

$$\Delta S_j \approx \frac{f_{lj} L_{sj}}{k_j f_h^2} \cdot \Delta f_h$$

The calculating result is given in Table 9-5 and the curve is shown in Figure 9-28.

Table 9-5　　　　　　　　$\Delta S_j = F(\Delta f_h)$ calculation

Interface	H-4/C-4	H-32/C-32	H-31/C-31	H-22/C-22	H-21/C-21	H-12/C-12	H-11/C-11
$k_j(-)$	260/270	84/270	64×3/ 270×3	143/ 270×9	107/ 270×9	140/ 270×9×4	104/ 270×9×4
f_h(kbit/s)	155 520						
f_l(kbit/s)	139 264	44 736	34 368	8448	6312	2048	1544
L_{sj}(bit)	2080	672	1536	1144	856	1120	832
$\Delta S_j/\Delta f_h$ $(1\times10^{-3}/\text{kHz})$	12.485	4.053	9.352	6.790	5.073	6.584	4.693

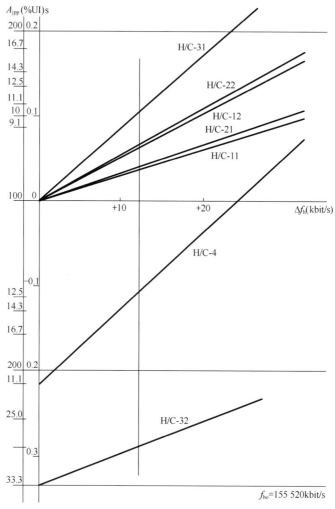

Figure 9-28　A_{jpp}, $S_j = F(\Delta f_h)$ calculating curve

By means of Figure 9-28, lots of compromise plan can be obtained. When

STM-1 basic rate deviation takes Δf_h=+12kbit/s (i.e. at f_h=155 532kbit/s), the design result is as follows:

$$\Delta f_h / f_h = +15 \times 10^{-6}$$

$$f_{hj} = k_j f_h$$

$$Q_j = L_{sj} - K_j$$

$$S_j = \left(1 - \frac{f_{lj}}{f_{hj}}\right) L_{sj} - K_j$$

$$\Delta S_j = \left(\left|\frac{\Delta f_{lj}}{f_{lj}}\right|\right) + \left(\left|\frac{\Delta f_{hj}}{f_{hj}}\right|\right) Q_j$$

$$S_{j\max} = S_j + \Delta S_j$$

$$S_{j\min} = S_j - \Delta S_j$$

$$A_{jpp} = \frac{1}{p_j}$$

$$S_j = \frac{q_j}{p_j}, \quad (p_j, q_j) = 1$$

The calculating result is given in Table 9-6.

Table 9-6   Payload container interface jitter calculation

Interface	H4/C4	H32/C32	H31/C31	H22/C22	H21/C21	H12/C12	H11/C11
f_{hj}(kbit/s)	149 771.54	48 387.73	36 866.84	9152.71	6848.53	2240.17	1664.13
f_{lj}(kbit/s)	139 264	44 736	34368	8448	6312	2048	1544
L_{sj}(bit)	2080	672	1536	1144	856	1120	832
K_j(bit)	145	50	104	88	67	96	60
Q_j(bit)	1935	622	1432	1056	789	1024	772
$\Delta f_{lj}/f_{lj}$ (1×10^{-6})	15	20	15	30	30	50	50
S_j	0.9270	0.7146	0.1104	0.0819	0.0611	0.0777	0.0603
ΔS_j	0.0581	0.0218	0.0430	0.0475	0.0355	0.0666	0.0502
A_{jpp}(%UI)	12.5	14.3	14.3	12.5	9.1	14.3	11.1

This kind of design plan uses unified positive justification technology to keep the frame structure unchanged. Only STM-1 clock frequency needs to be increased by 12kbit/s, the stuffing jitter of seven plesiochronous interfaces can be decreased below 15%UI popularly; and the master clock frequency is easy to be generated, for example, for 6.4805MHz×2^3×3 or 5.184MHz×2×3×5, they are convenient to generate 155 532kHz master clock.

(3) Using positive/zero/negative justification technology to realize plesiochronous/synchronous compatible interface

One design plan is provided hereinafter, in which, for seven kinds of interfaces between SDH/PDH, positive/zero/negative justification technology is unitedly used to realize plesiochronous/synchronous compatible interface. The plan is subject to STM-1 basic rate (f_h=155 520kbit/s) being unchanged. The premise to realize the synchronous interface is that the payload frame frequency (F_{SH}) is required to be equal to the instantaneous value of the container frame frequency (F_{SC}) on the synchronous interface:

$$F_{SH} = F_{SC}$$

$$F_{SH} = \frac{f_{lj}}{Q_j}$$

$$F_{SC} = \frac{f_{hj}}{L_{sj}}$$

$$\therefore Q_j = \frac{f_{lj}}{f_{hj}} \Box L_{sj}$$

If positive/zero/negative justification technology is used to realize the plesiochronous/synchronous compatible interface, the payload frame frequency (F_{SH}) is required to be equal to container frame frequency (F_{SC}) normally within their tolerance domain ($F_{SH} \pm \Delta F_{SH}$, $F_{SC} \pm \Delta F_{SC}$). That means. that the interface is required to provide enough justification capability for compensation of frequency instantaneous deviation. Assuming that there is only one justification opportunity in each justification frame and only one bit is justified each time, then, the justification rate (f_{sj}) is equal to frame frequency:

$$f_{sj} = F_s$$

$$f_{sj} \geq \left|\frac{\Delta f_{hj}}{f_{hj}}\right| \Box f_{hj} + \left|\frac{\Delta f_{lj}}{f_{lj}}\right| \Box f_{lj}$$

$$\therefore F_s \geq \left|\frac{\Delta f_{hj}}{f_{hj}}\right| \Box f_{hj} + \left|\frac{\Delta f_{lj}}{f_{lj}}\right| \Box f_{lj}$$

For the seven interfaces in Rec.G.709, interfaces 31, 22, 21, 12 and 11 are already in compliance with the compatible condition. Only the design on interface

4 and 32 need to be modified.

The modified design of H-4/C-4 interface:

To redesign the justification frame structure of container C-4: to take one fourth of C-4 capacity (260×96Bytes) as justification frame length:

L_{sj}=260×9/4=585Bytes=4680bit;

$f_{hj}=L_{sj} \cdot F_{SC}$=149 760kbit/s;

f_{lj}=139 264kbit/s;

$Q_j=f_{lj}/F_{SC}$=544Bytes=4352bit;

$K_j=L_{sj}-Q_j$=41Bytes.

The concrete arrangement of interface frame is illustrated in Figure 9-29.

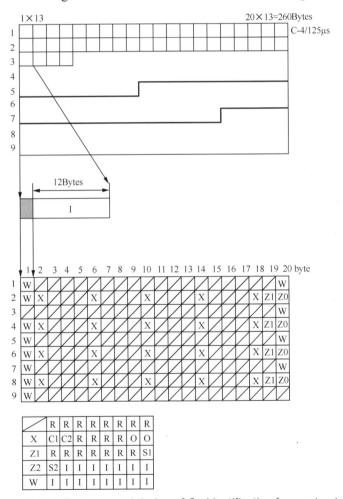

Figure 9-29 The improved design of C-4 justification frame structure

The improved design of H-32/C-32 interface:

To redesign the justification frame structure of container C-32: to take one third of C-32 capacity (84×9Bytes) as justification frame length:

L_{sj}=84×9/4=252Bytes=2016bit;

F_{SC}=3×8kHz=24kHz;

F_{hj}=L_{sj} • F_{SC}=48 384kbit/s;

F_{lj}=44 736kbit/s;

Q_j=f_{lj}/F_{SC}=233Bytes =1864bit;

K_j=L_{sj}−Q_j=19Bytes.

The detail arrangement of interface frame is illustrated in Figure 9-30.

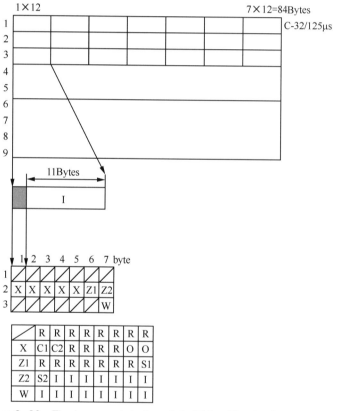

Figure 9-30 The improved design of C-32 justification frame structure

The design result to use positive/zero/negative justification for realization of plesiochronous/synchronous compatible interfaces is listed in Table 9-7.

Table 9-7 The design data of plesiochronous/synchronous compatible interfaces with positive/zero/negative justification

Interface	H4/C4	H32/C32	H31/C31	H22/C22	H21/C21	H12/C12	H11/C11
Frame Structure	$\dfrac{260 \times 9}{4}$	$\dfrac{84 \times 9}{3}$	$\dfrac{64 \times 9}{3}$	$(16 \times 9) - 1$	$(12 \times 9) - 1$	$(4 \times 9 - 1) \times 4$	$(3 \times 9 - 1) \times 4$
L_{sj} (bit)	4680	2016	1536	1144	856	1120	832
f_s (kHz)	32	24	24	8	8	2	2
f_{hj} (kbit/s)	149 760	48 384	36 864	9152	6848	2240	1664
F_{lj} (kbit/s)	139 264	44 736	34 368	8448	6312	2048	1544
Q_j (bit)	4352	1864	1432	1056	789	1024	772
K_j (bit)	328	152	104	88	67	96	60

On the basis of such frame structure design, it is possible to do the interface having stuffing jitter less than 10%UI by using very simple justification technology and without fine justification, To do fine justification can reduce the stuffing jitter of plesiochronous interface to less than 3%UI.

Chapter 10 Anti Fading Frame Synchronization

10.1 THE PRINCIPLES OF ANTI FADING FRAME SYNCHRONIZATION

Frame synchronization is the basic problem of digital multiplex. Either synchronous multiplex or plesiochronous multiplex must use frame synchronization technique. The practice in theory and engineering has proven that the techniques of frame synchronization systems decide the technical characteristics of the whole digital multiplex devices to a great extent. This problem has been paid close attention widely these years.

The digital multiplexer using the frame synchronization scheme recommended by CCITT may be used together with a digital transmission system of constant parameter channels but may not be used together with that of fading channels; even with the former it is not a good scheme. The major reason is that the average keeping time of frame synchronization with the scheme recommended by CCITT in the case of burst errors is too short, it is unacceptable in engineering; even in the case of uniform errors, the time is shorter than the other schemes.

The above defect of the CCITT scheme is from its design idea and the working mechanism. The basic design idea is to "keep" the frame synchronous state; hence its working mechanism is to detect continuously if the frame syhchronization signals are lost at the synchronization bits. Suppose a synchronization system is in the synchronous state, if all the synchronization signals are lost at the consecutive β synchronization positions in the case of errors, the frame synchronization system is decided as frame loss (it is call a virtual missing event) and the synchronization search begins immediately. This search operation iust causes the real frame loss. Thus such a virtual missing frame loss is relevant to errors but the code structures.

The basic design idea of anti fading frame synchronization scheme is to "refresh" frame synchronous state; hence its working mechanism is to detect continuously if a new frame alignment signal occurs at a position. Suppose frame synchronization is in the synchronous state, if the frame synchronization pattern is detected using the code combination among the β consecutive non-synchronization positions which are a frame distant from each other the synchronization system decides a new frame synchronization (it is called virtual alarm event) and resets time sequential generator of the local frame, it in turn causes a real frame loss. Hence such a virtual alarm frame loss is relevant with code combination but errors.

The above two kinds of schemes are based on different design ideas and use different working mechanisms, hence they show absolute different technical characteristics. But neither of these two schemes make full use of the time resource from the synchronization deciding time to the loss deciding time, hence the frame synchronization keeping time of these two kinds of schemes usually shorter. To counter such weakness a synthetical synchronization scheme is proposed. The basic design idea of the synthetical synchronization scheme is to fully use of the time resource from the synchronization deciding time (t_0) to the frame loss deciding time (t_n); it synthetically uses a strategy of "keeping" and "refreshing". Appearing on the working principles, the keeping strategy is used at the non-synchronization positions and the refreshing strategy is used at the non-synchronization positions. According to sequential combination of refreshing and keeping strategies such a kind of synthetical frame synchronization scheme may form four different particular synthetical schemes. The above two schemes plus these four schemes constitutes six typical frame synchronization schemes.

It is not difficult to understand physically that in the case of continuous high error rate, all the frame synchronization scheme will have mechanical difficulties. The working environment discussed in this chapter is existence of the voltage level fading or burst error codes. There are high error code time intervals and also low error code time intervals, the so-called anti fading frame synchronization scheme is a scheme that is able to adapt such a working environment, that is, to set frame synchronization quickly at a low error rate and to keep fairly the frame synchronization at a high error rate.

The frame loss of a frame synchronization system is divided into two kinds: real frame loss and false frame loss. The real frame loss is meant that the frame loss really occurs in the system, for example, because of input bit stream slips the receive frame synchronization system loses frames; the false frame loss is meant that frame loss does not occur, the frame synchronization system makes wrong decisions only because of error code in the input bit stream, it misunderstands there are loss frames in the system and takes search or reset operation which in turn makes a real frame loss.

Figure 10-1 shows the characteristics of the above six kinds of typical schemes

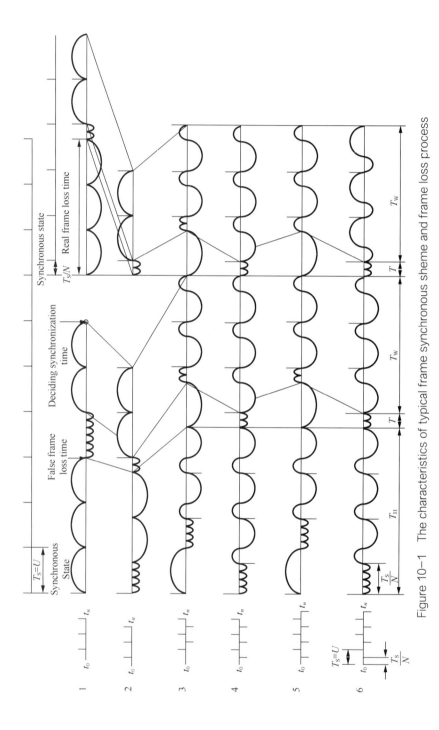

Figure 10-1 The characteristics of typical frame synchronous sheme and frame loss process

($\beta=3$), and the frame synchronization keeping processes, frame synchronization search processes and frame synchronization deciding processes in the cases of real and false frame losses. where the abscissa stands for time, the arc lines above the abscissa stand for the processes which are affected by the virtual missing probability (P_s); the arc lines below the abscissa stand for the processes which are affected by the virtual alarm probability (P_c) Real frame loss process:

There exists frame loss deciding process only in the first scheme, the synchronization search process will start only after frame loss is decided. After the synchronization search reaches at the synchronization position the frame synchronization deciding process starts. when frame synchronization is decided it then ends the frame loss process and enters frame synchronization keeping process. In fact, there are no concept of frame loss deciding in the rest five schemes, but only the frame synchronization concepts. It continuously searches the frame synchronization pattern, as long as the pattern matching the frame synchronization deciding criterion is found, it decides frame synchronization and implements a new synchronization reset to the local sequential circuits. The search processes in the schemes (2), (4), and (6) are carried out bit by bit, whereas the search processes in the schemes (3) and (5) are carried out frame by frame. In the schemes (2), (5) and (6), a new reset is carried out as long as frame synchronization is decided, whereas in the schemes (2) and (4), it is delayed ($N-i$)UI after frame synchronization is decided.

False frame loss process:

During synchronization keeping cycle, only the first scheme is in keeping detecting state. As long as the frame synchronization deciding pattern is disappeared a frame loss is decided immediately and the synchronization search is started, hence the frame synchronization process is ended. So far, in the case of false frame loss, in the scheme (1) the frame loss deciding process is carried out during the synchronization keeping time. In the rest five schemes, the frame synchronization search and deciding processes are continuously repeated during the synchronization keeping and loss periods: a correct result causes frame synchronization whereas a wrong result causes false frame loss.

The duration from the frame synchronization deciding time to frame loss time is called the frame synchronization keeping time (T_H). The duration from the frame loss time to the frame synchronization deciding time is called the frame loss keeping time (T_L). The frame loss keeping time of the scheme (1), in the case of real frame loss, includes the frame loss deciding time (T_D), the synchronization searching time (T_A) and the frame synchronization deciding time (T_W); the frame loss keeping time of the scheme (1) (in the case of false frame loss), (2), (5), and (6), includes the synchronization search time (T_A) and the frame synchronization deciding time (T_W); the frame loss keeping time of the schemes (3) and (4) include the synchronization search time (T_A), the frame synchronization deciding time (T_H) and the frame synchronization reset delay time(T_Z).

In the function of the combination of error codes and information codes, a frame synchronization system has finite countable states (A, B, C, D, ···)which are called frame states. Where the beginning letter (e.g. A) stands for a initial state(e.g. the frame synchronization initial state, the frame loss initial state); the ending letter (e.g. D) stands for a final state (e.g. the frame synchronization flnal state, the frame loss final state). For a deterministic system, these states are all defined precisely by recognizable criterion. hence the transition probabilities among the frame states are all constants. In the function of the virtual missing factor (P_s), the frame synchronization system transits from the initial synchronization state to the final synchronization state, which is called occurrence of virtual leak frame loss; in the function of the virtual alarm factor (P_c), the frame synchronization system transits from the initial synchronization state to the final synchronization state which is called occurrence of the virtual alarm frame loss. So far, the virtual alarm frame loss process and the virtual leak frame loss process are all discrete Markovian Processes. Hence, the Z transform of the state transition diagram and the state transition probability can be used to calculate the average keeping times of the various processes of a frame synchronization system (the frame synchronization average keeping time T_H, frame synchronization average search time T_A, the frame synchronization average deciding time T_W). It has been proved that the frame synchronization average keeping time (T_H) of the above six typical schemes are different greatly, however, their frame loss average keeping times (T_L), i.e. the sum of the frame synchronization search times (T_A) and the frame synchronization deciding times (T_W), are very close. Thus, based on the frame synchronization average keeping time, the frame synchronization schemes can be evaluated.

Chapter 10 Anti Fading Frame Synchronization

10.2 FRAME SYNCHRONIZATION AVERAGE KEEPING TIME

10.2.1 Scheme (1)

The first frame synchronization scheme is the scheme recommended by CCITT. Figure 10-2 shows the state-time diagram of the scheme (1). Where A stands for the synchronization deciding state, D stands for the loss deciding state, the rest B, C stand for intermediate transition states; t_0 is synchronization deciding time, t_n is loss deciding time, t_n-t_0 is frame synchronization average keeping time (T_H). Figure 10-3 shows the state transition diagram of the transition process from the synchronization deciding state to loss deciding state and a simple example.

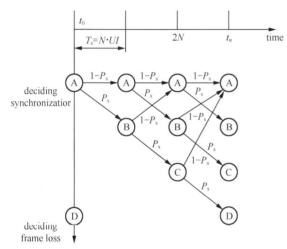
Figure 10-2 The State-Time diagram of scheme (1)

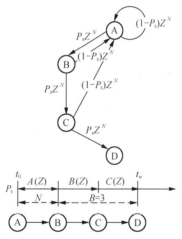
Figure 10-3 Synchronization keeping state transition diagram of scheme (1)

Where, P_s—Frame synchronization signal virtual missing probability;
$P_s = 1 - (1 - P_e)^n$;
P_e—Multiplex receive code stream average error rate;
n—Frame synchronization signal code length;
N—Frame length;
β—Frame synchronization protection coefficient;
t_0—Frame synchronization deciding time;

t_n—Frame loss deciding time;
T_H—Frame synchronizatton average keeping time;
$T_H = t_n - t_0$;
T_s—Frame cycle;
$T_s = N \cdot \text{UI}$.

The Z transform of the transition probability from state A to state D at time t (Case of $\beta=3$):

$$F_{AD}(Z) = A(Z) \cdot B(Z) \cdot C(Z)$$

$$A(Z) = \frac{P_s Z^N}{1-(1-P_s)Z^N}$$

$$B(Z) = \frac{P_s Z^N}{1 - A(Z)(1-P_s)Z^N}$$

$$C(Z) = \frac{P_s Z^N}{1 - A(Z) \cdot BI(1-P_s)Z^N}$$

Frame synchronization average keeping time (Case of $\beta=3$) solved:

$$T_H = \left.\frac{dF_{AD}(Z)}{dZ}\right|_{Z=1}$$

$$= \left[\left.\frac{dA(Z)}{A(Z)dZ}\right|_{Z=1} + \left.\frac{dB(Z)}{B(Z)dZ}\right|_{Z=1} + \left.\frac{dC(Z)}{C(Z)dZ}\right|_{Z=1}\right] F_{AD}{}^{(Z)}\big|Z=1$$

Q $A(Z)|_{Z=1} = 1$

$B(Z)|_{Z=1} = 1$

$C(Z)|_{Z=1} = 1$

$F_{AD}(Z)|_{Z=1} = 1$

$$\therefore T_H = \left.\frac{dA(Z)}{dZ}\right|_{Z=1} + \left.\frac{dB(Z)}{dZ}\right|_{Z=1} + \left.\frac{dC(Z)}{dZ}\right|_{Z=1}$$

$$\left.\frac{dA(Z)}{dZ}\right|_{Z=1} = \frac{N}{P_s}$$

$$\left.\frac{dB(Z)}{dZ}\right|_{Z=1} = \frac{N}{P_s^2}$$

$$\left.\frac{dC(Z)}{dZ}\right|_{Z=1} = \frac{N}{P_s^3}$$

$$T_H = \frac{N}{P_s^3}(1+P_s+P_s^2) \cdot UI$$

$$\frac{T_H}{T_S} = \frac{1}{P_s^3}(1+P_s+P_s^2)$$

This is the first frame synchronization average keeping time expression when $p=3$. The frame synchronization keeping time of the first scheme when $\beta \geq 1$ is general expression is:

$$\frac{T_H}{T_S} = \frac{1}{P_s^\beta} \sum_{j=0}^{\beta-1} P_s^j \qquad (10\text{-}1)$$

10.2.2 Scheme (2)

The second frame synchronization scheme is the scheme whose frame synchronization keeping time is unrelevant to error codes, which is called basic anti fading frame synchronization scheme. Figure 10-4 shows the primitive state-time diagram and the simplified state-time diagram of the scheme (2). Where A stands for synchronization deciding state, D stands for loss deciding state. Figure 10-5 shows the process state transition diagram from state A to state D and the simplest example.

Where,

P_c—Frame synchronization virtual alarm probability,

$$P_c = \left(\frac{1}{2}\right)^n$$

α—Frame synchronization detection coefficient.

The Z transform of the transition probability at time t from state A to state D (Case of $\alpha=3$):

$$F_{AD}(t) = A(Z) \cdot B(Z) \cdot C(Z)$$

$$A(Z) = \frac{P_c Z^i}{1-(1-P_c)Z^N}$$

$$B(Z) = \frac{P_c Z^N}{1-A(Z)(1-P_c)Z^{2N-i}}$$

$$C(Z) = \frac{P_c Z^N}{1-A(Z) \cdot B(Z)(1-P_c)Z^{2N-i}}$$

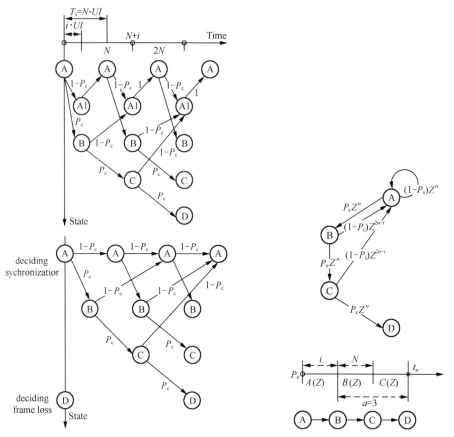

Figure 10-4 state-time diagram of scheme (2)　　Figure 10-5 synchronization keeping state transition diagram of scheme (2)

Frame synchronization average keeping time (Case of $\alpha=3$) is solved:

$$T_{\mathrm{H}} = \left.\frac{\mathrm{d}F_{\mathrm{AD}}(Z)}{\mathrm{d}Z}\right|_{Z=1}$$

$$= \left.\frac{\mathrm{d}A(Z)}{\mathrm{d}Z}\right|_{Z=1} + \left.\frac{\mathrm{d}B(Z)}{\mathrm{d}Z}\right|_{Z=1} + \left.\frac{\mathrm{d}C(Z)}{\mathrm{d}Z}\right|_{Z=1}$$

$$\left.\frac{\mathrm{d}A(Z)}{\mathrm{d}Z}\right|_{Z=1} = i + \frac{1+P_{\mathrm{c}}}{P_{\mathrm{c}}}N$$

$$\left.\frac{\mathrm{d}B(Z)}{\mathrm{d}Z}\right|_{Z=1} = N + \frac{1-P_{\mathrm{c}}}{P_{\mathrm{c}}}\left[\left.\frac{\mathrm{d}A(Z)}{\mathrm{d}Z}\right|_{Z=1} + (2N-i)\right]$$

$$\left.\frac{\mathrm{d}C(Z)}{\mathrm{d}Z}\right|_{Z=1} = N + \frac{1-P_{\mathrm{c}}}{P_{\mathrm{c}}}\left[\left.\frac{\mathrm{d}A(Z)}{\mathrm{d}Z}\right|_{Z=1} + \left.\frac{\mathrm{d}B(Z)}{\mathrm{d}Z}\right|_{Z=1} + (2N-i)\right]$$

$$T_H = \frac{N}{P_c^3}(1+P_c+P_c^2)-(N-i)\cdot UI$$

$$\frac{T_H}{T_S} = \frac{1}{P_c^3}(1+P_c+P_c^2)-\left(1-\frac{i}{N}\right)$$

This is the second frame synchronization average keeping time expression when $\alpha=3$. When $\alpha \geq 1$ the general expression of second frame synchronization average keeping time is:

$$\frac{T_H}{T_S} = \frac{1}{P_c^\alpha}\sum_{j=0}^{\alpha-1} P_c^j - (1-i/N) \qquad (10\text{-}2)$$

[Case 1] $i = 0$

$$\frac{T_H}{T_S} = \frac{1}{P_c^\alpha}\sum_{j=0}^{\alpha-1} P_c^j - 1 \qquad (10\text{-}2\text{-}1)$$

[Case 2] $i = N$

$$\frac{T_H}{T_S} = \frac{1}{P_c^\alpha}\sum_{j=0}^{\alpha-1} P_c^j \qquad (10\text{-}2\text{-}2)$$

The formula (10-2-2) and the formula (10-1) have the same form, but the opposite meaning, the formula (10-1) stands for frame synchronization keeping time of the frame loss caused by the virtual missing, whereas the formula stands for the frame synchronization keeping time of the frame loss caused by the virtual alarm.

10.2.3 Scheme (3)

The following four schemes are all synthetical frame synchronization schemes. The third frame synchronization scheme is a synthetical frame synchronization scheme which loss is decided from the first intermediate state beginning at the virtual missing point (B) to the virtual missing point (F). Figure 10-6 shows the primitive state-time diagram and the simplified state-time diagram, where A is the synchronization deciding state, F is the loss deciding state. Figure 10-7 shows the state transition diagram of the transition process from state A to state F. Figure 10-8 shows the transition probability calculation diagram.

The Z transform of the transition probability $F_{AF}(t)$ at time t from state A to state F (Case of $\beta=3$, $\alpha=2$):

$$F_{AF}(t) = A(Z) \cdot B(Z) \cdot C(Z) \cdot D(Z) \cdot E(Z)$$

$$A(Z) = \frac{P_c Z^N}{1-(1-P_c)Z^N}$$

$$B(Z) = \frac{P_c Z^i}{1 - (1 - P_c)[A(1 - P_s) + P_s]Z^N}$$

$$C(Z) = \frac{P_s Z^{N-i}}{1 - AB(1 - P_s)Z^{N-i}}$$

$$D(Z) = \frac{P_c Z^i}{1 - BC(1 - P_c)[A(1 - P_c) + P_s]Z^N}$$

$$E(Z) = \frac{P_s Z^{N-i}}{1 - ABCD(1 - P_s)Z^{N-i}}$$

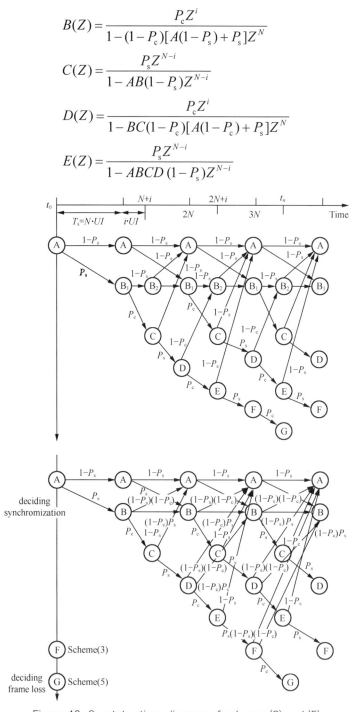

Figure 10-6 state-time diagram of scheme (3) and (5)

Chapter 10 Anti Fading Frame Synchronization

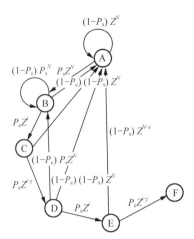

Figure 10-7 Synchronization keeping state transition of scheme (3)

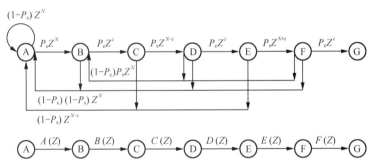

Figure 10-8 Synchronization keeping state transition
probability calculation diagram of scheme (3)

Frame synchronization average keeping time (Case of $\beta=3$, $\alpha=2$) is solved:

$$T_H = \left.\frac{\mathrm{d}F_{AF}(Z)}{\mathrm{d}Z}\right|_{Z=1}$$

$$= \left.\frac{\mathrm{d}A(Z)}{\mathrm{d}Z}\right|_{Z=1} + \left.\frac{\mathrm{d}B(Z)}{\mathrm{d}Z}\right|_{Z=1} + \left.\frac{\mathrm{d}C(Z)}{\mathrm{d}Z}\right|_{Z=1} + \left.\frac{\mathrm{d}D(Z)}{\mathrm{d}Z}\right|_{Z=1} + \left.\frac{\mathrm{d}E(Z)}{\mathrm{d}Z}\right|_{Z=1}$$

$$\left.\frac{\mathrm{d}A(Z)}{\mathrm{d}Z}\right|_{Z=1} = \frac{N}{P_s}$$

$$\left.\frac{\mathrm{d}B(Z)}{\mathrm{d}Z}\right|_{Z=1} = i + \frac{1-P_c}{P_c}\left[(1-P_s)\left.\frac{\mathrm{d}A(Z)}{\mathrm{d}Z}\right|_{Z=1} + N\right]$$

− 259 −

$$\left.\frac{dC(Z)}{dZ}\right|_{Z=1} = \frac{N-i}{P_s} + \frac{1-P_s}{P_s}\left[\left.\frac{dA(Z)}{dZ}\right|_{Z=1} + \left.\frac{dB(Z)}{dZ}\right|_{Z=1}\right]$$

$$\left.\frac{dD(Z)}{dZ}\right|_{Z=1} = i + \frac{1-P_c}{P_c}\left[(1-P_s)\left.\frac{dA(Z)}{dZ}\right|_{Z=1} + \left.\frac{dB(Z)}{dZ}\right|_{Z=1} + \left.\frac{dC(Z)}{dZ}\right|_{Z=1} + N\right]$$

$$\left.\frac{dC(Z)}{dZ}\right|_{Z=1} = \frac{N-i}{P_s} + \frac{1-P_s}{P_s}\left[\left.\frac{dA(Z)}{dZ}\right|_{Z=1} + \left.\frac{dB(Z)}{dZ}\right|_{Z=1} + \left.\frac{dC(Z)}{dZ}\right|_{Z=1} + \left.\frac{dD(Z)}{dZ}\right|_{Z=1}\right]$$

$$\therefore T_H = \frac{N}{P_c^3 P_c^2}\left(1 + P_s P_c + P_s^2 P_c^2\right) \cdot UI$$

$$\frac{T_H}{T_S} = \frac{1}{P_c^3 P_c^2}\left(1 + P_s P_c + P_s^2 P_c^2\right)$$

This is the frame synchronization average keeping time exexpression of the third frame synchronization scheme when $\beta=3$, $\alpha=2$. When $\beta \geq 1$ the general expression of the frame synchronization average keeping time of the third frame synchronization scheme:

$$\frac{T_H}{T_S} = \frac{1}{P_s^\beta P_c^{\beta-1}} \sum_{j=0}^{\beta-1}(P_s P_c)^j \tag{10-3}$$

[Case 1] $P_c = 1$

$$\frac{T_H}{T_S} = \frac{1}{P_s^\beta} \sum_{j=0}^{\beta-1} P_s^j \tag{10-3-1}$$

The formula (10-3-1) and the formula (10-1) have both the same form and the same meaning. Hence the scheme (1) is a special case of the scheme (3) when $P_c=1$ (see Figure 10-9).

[Case 2] $P_s = 1$

$$\frac{T_H}{T_S} = \frac{1}{P_c^\alpha} \sum_{j=0}^{\alpha-1} P_C^j = \frac{1}{P_c^\alpha} \sum_{j=0}^{\alpha-1} P_C^j + 1 \tag{10-3-2}$$

See Figure 10-9, Compare the simplest case of the formula (10-3-2) with that of the formula (10-2) (only compare the structures of the formula and ignore the absolute value of α): consider the definitions of the initial state and the final state:

$$\left.\frac{T_H}{T_S}\right|_{(3-2)} -1 - (1-i/N)$$

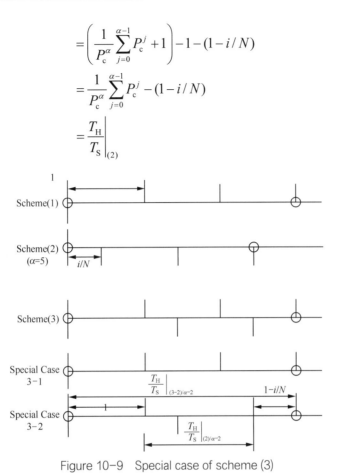

$$= \left(\frac{1}{P_c^\alpha}\sum_{j=0}^{\alpha-1}P_c^j + 1\right) - 1 - (1-i/N)$$

$$= \frac{1}{P_c^\alpha}\sum_{j=0}^{\alpha-1}P_c^j - (1-i/N)$$

$$= \frac{T_H}{T_S}\bigg|_{(2)}$$

Figure 10-9 Special case of scheme (3)

Obviously, the scheme (2) is the special case of the scheme (3) deducting the difference $1+(1-i/N)$ of the definitions of the initial state and the final state when $P_s=1$.

10.2.4 Scheme (4)

The fourth frame synchronization scheme is a synthetical frame synchronization scheme of loss deciding from the virtual missing point (B) starting the first intermediate state, to the virtual alarm point (G). Figure 10-10 shows the primitive state-time diagram and its simplified state-time diagram, where A is the synchronization deciding state, G is the loss deciding state. Figure 10-11 shows state transition diagram from state A to state G. Figure 10-12 shows the transition probability calculation diagram.

PDH for Telecommunication Network

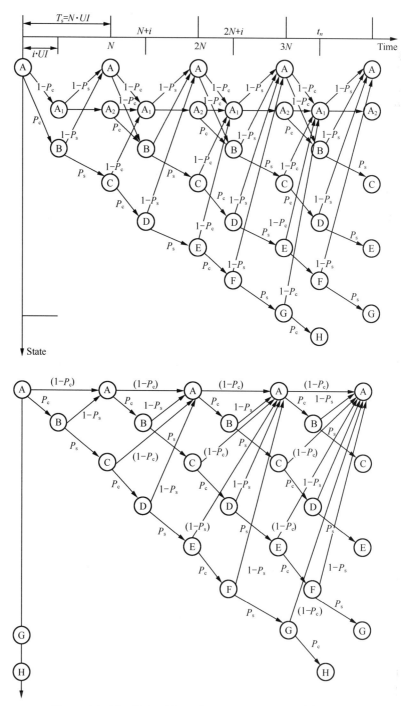

Figure 10-10 State-Time diagram of schemes (4) and (6)

Chapter 10 Anti Fading Frame Synchronization

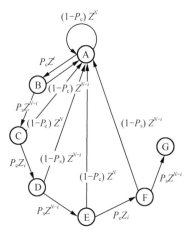

Figure 10-11 Synchronization keeping state transition diagram scheme (4)

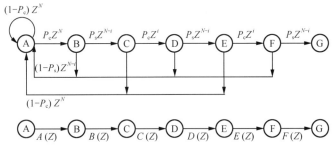

Figure 10-12 Synchronization keeping state transition
calculation diagram of scheme (4)

The Z transform of the transition probability at time t from state A to state G (Case of $\beta=3$, $\alpha=2$):

$$F_{AG}(Z) = A(Z) \cdot B(Z) \cdot C(Z) \cdot D(Z) \cdot E(Z) \cdot F(Z)$$

$$A(Z) = \frac{P_c Z^i}{1-(1-P_c)Z^N}$$

$$B(Z) = \frac{P_s Z^{N-i}}{1-A(Z)(1-P_s)Z^{N-i}}$$

$$C(Z) = \frac{P_c Z^i}{1-A(Z)B(Z)(1-P_c)Z^N}$$

$$D(Z) = \frac{P_s Z^{N-i}}{1-A(Z)B(Z)C(Z)(1-P_s)Z^{N-i}}$$

$$E(Z) = \frac{P_c Z^i}{1-A(Z)B(Z)C(Z)D(Z)(1-P_c)Z^N}$$

$$F(Z) = \frac{P_s Z^{N-i}}{1 - A(Z)B(Z)C(Z)D(Z)E(Z)(1-P_s)Z^{N-i}}$$

Frame synchronization average keeping time ($\beta=\alpha=3$) is solved:

$$T_H = \frac{dF_G(Z)}{dZ}\bigg|_{Z=1}$$

$$= \frac{dA(Z)}{dZ}\bigg|_{Z=1} + \frac{dB(Z)}{dZ}\bigg|_{Z=1} + \frac{dC(Z)}{dZ}\bigg|_{Z=1}$$

$$+ \frac{dD(Z)}{dZ}\bigg|_{Z=1} + \frac{dE(Z)}{dZ}\bigg|_{Z=1} + \frac{dF(Z)}{dZ}\bigg|_{Z=1}$$

$$\frac{dA(Z)}{dZ}\bigg|_{Z=1} = i + \frac{1-P_c}{P_c} N$$

$$\frac{dB(Z)}{dZ}\bigg|_{Z=1} = \frac{N-i}{P_s} + \frac{1-P_s}{P_s} \frac{dA(Z)}{dZ}\bigg|_{Z=1}$$

$$\frac{dC(Z)}{dZ}\bigg|_{Z=1} = \frac{dA(Z)}{dZ}\bigg|_{Z=1} + \frac{1-P_s}{P_s}\left[\frac{dA(Z)}{dZ}\bigg|_{Z=1} + \frac{dB(Z)}{dZ}\bigg|_{Z=1}\right]$$

$$\frac{dD(Z)}{dZ}\bigg|_{Z=1} = \frac{N-i}{P_s} + \frac{1-P_s}{P_s}\left[\frac{dA(Z)}{dZ}\bigg|_{Z=1} + \frac{dB(Z)}{dZ}\bigg|_{Z=1} + \frac{dC(Z)}{dZ}\bigg|_{Z=1}\right]$$

$$\frac{dE(Z)}{dZ}\bigg|_{Z=1} = \frac{dA(Z)}{dZ}\bigg|_{Z=1} + \frac{1-P_c}{P_c}\left[\frac{dA(Z)}{dZ}\bigg|_{Z=1} + \frac{dB(Z)}{dZ}\bigg|_{Z=1} + \frac{dC(Z)}{dZ}\bigg|_{Z=1} + \frac{dD(Z)}{dZ}\bigg|_{Z=1}\right]$$

$$\frac{dF(Z)}{dZ}\bigg|_{Z=1} = \frac{N-i}{P_s} + \frac{1-P_s}{P_s}\left[\frac{dA(Z)}{dZ}\bigg|_{Z=1} + \frac{dB(Z)}{dZ}\bigg|_{Z=1} + \frac{dC(Z)}{dZ}\bigg|_{Z=1} + \frac{dD(Z)}{dZ}\bigg|_{Z=1} + \frac{dE(Z)}{dZ}\bigg|_{Z=1}\right]$$

$$T_H = \frac{N}{P_s^3 P_c^3}\left(1 + P_s P_c + P_s^2 P_c^2\right) \cdot UI$$

This is the fourth frame synchronization average keeping time expression when $\beta=3$, $\alpha=2$. When $\beta=\alpha \geq 1$ the general expression of the fourth frame synchronization average keeping time is:

$$\frac{T_H}{T_S} = \frac{1}{(P_s P_c)^\beta} \sum_{j=0}^{\beta-1}(P_s P_c)^j \tag{10-4}$$

[Case 1] $P_c = 1$

$$\frac{T_H}{T_S} = \frac{1}{P_s^\beta} \sum_{j=0}^{\beta-1} P_s^j \tag{10-4-1}$$

The formula (10-4-1) and formula (10-1) have both the same form and the same meaning. Hence the scheme (1) is a special case of the scheme (4) when $P_c=1$

(see Figure 10-13).

[Case 2] $P_s = 1$

$$\frac{T_H}{T_S} = \frac{1}{P_c^\alpha} \sum_{j=0}^{\alpha-1} P_c^j \qquad (10\text{-}4\text{-}2)$$

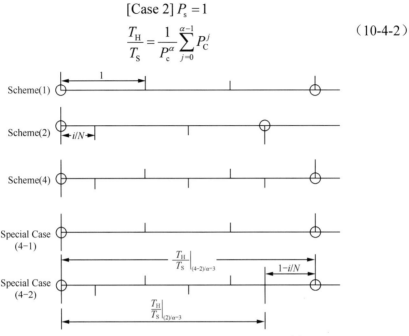

Figure 10-13 Special case diagram of scheme (4)

See Figure 10-13, consider the difference of the definitions when frame synchronization ends:

$$\left.\frac{T_H}{T_S}\right|_{(4-2)} - (1-i/N)$$
$$= \frac{1}{P_c^\alpha} \sum_{j=0}^{\alpha-1} P_c^j - (1-i/N)$$
$$= \left.\frac{T_H}{T_S}\right|_{(2)}$$

Obviously, The scheme (2) is the special case of the scheme (4) when $P_s=1$ and deducting the definition difference $(1-i/N)$.

10.2.5 Scheme (5)

The fifth frame synchronization scheme is a synthetical frame synchronization scheme whose frame loss is decided from the virtual missing point (B) starting the first intermediate state, to the virtual alarm point (G). Figure 10-6 shows the primitive state-time diagram and its simplified state-time diagram, where A is the synchronization deciding state, G is the loss deciding state. Figure 10-14 shows state transition diagram from state A to state G. Figure 10-15 shows the transition probability calculation diagram.

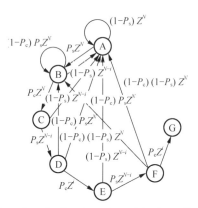

Figure 10-14 Synchronization keeping state transition of scheme (5)

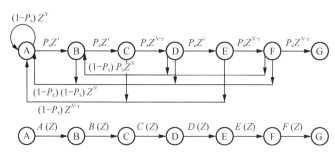

Figure 10-15 Synchronization keeping state transition calculation diagram of scheme (5)

The Z transform of the transition probability $F_{AG}(t)$ at time t from state A to state G (Case of $\beta=3$, $\alpha=3$):

$$F_{AG}(Z)=A(Z) \cdot B(Z) \cdot C(Z) \cdot D(Z) \cdot E(Z) \cdot F(Z)$$

$$A(Z)=\frac{P_s Z^N}{1-(1-P_s)Z^N}$$

$$B(Z)=\frac{P_c Z^i}{1-\left[P_s+A(Z)(1-P_s)\right](1-P_c)Z^N}$$

$$C(Z)=\frac{P_s Z^{N-i}}{1-A(Z)B(Z)(1-P_s)Z^{N-i}}$$

$$D(Z) = \frac{P_c Z^i}{1 - B(Z)C(Z)(1-P_c)Z^i}$$

$$E(Z) = \frac{P_s Z^{N-i}}{1 - A(Z)B(Z)C(Z)D(Z)(1-P_s)Z^{N-i}}$$

$$F(Z) = \frac{P_c Z^i}{1 - A(Z)B(Z)C(Z)D(Z)E(Z)(1-P_c)Z^i}$$

Frame synchronization average keeping time (Case of $\beta = \alpha = 3$) is solved:

$$T_H = \frac{dF_{AG}(Z)}{dZ}\bigg|_{Z=1}$$

$$= \frac{dA(Z)}{dZ}\bigg|_{Z=1} + \frac{dB(Z)}{dZ}\bigg|_{Z=1} + \frac{dC(Z)}{dZ}\bigg|_{Z=1}$$

$$+ \frac{dD(Z)}{dZ}\bigg|_{Z=1} + \frac{dE(Z)}{dZ}\bigg|_{Z=1} + \frac{dF(Z)}{dZ}\bigg|_{Z=1}$$

$$\frac{dA(Z)}{dZ}\bigg|_{Z=1} = \frac{N}{P_s}$$

$$\frac{dB(Z)}{dZ}\bigg|_{Z=1} = i + \frac{1-P_c}{P_c}\left[(1-P_s)\frac{dA(Z)}{dZ}\bigg|_{Z=1} + N\right]$$

$$\frac{dC(Z)}{dZ}\bigg|_{Z=1} = \frac{N-i}{P_s} + \frac{1-P_s}{P_s}\left[\frac{dA(Z)}{dZ}\bigg|_{Z=1} + \frac{dB(Z)}{dZ}\bigg|_{Z=1}\right]$$

$$\frac{dD(Z)}{dZ}\bigg|_{Z=1} = i + \frac{1-P_c}{P_c}\left[(1-P_s)\frac{dA(Z)}{dZ}\bigg|_{Z=1} + \frac{dB(Z)}{dZ}\bigg|_{Z=1} + \frac{dC(Z)}{dZ}\bigg|_{Z=1} + N\right]$$

$$\frac{dE(Z)}{dZ}\bigg|_{Z=1} = \frac{N-i}{P_s} + \frac{1-P_s}{P_s}\left[\frac{dA(Z)}{dZ}\bigg|_{Z=1} + \frac{dB(Z)}{dZ}\bigg|_{Z=1} + \frac{dC(Z)}{dZ}\bigg|_{Z=1} + \frac{dD(Z)}{dZ}\bigg|_{Z=1}\right]$$

$$\frac{dF(Z)}{dZ}\bigg|_{Z=1} = i + \frac{1-P_c}{P_c}\left[(1-P_s)\frac{dA(Z)}{dZ}\bigg|_{Z=1} + \frac{dB(Z)}{dZ}\bigg|_{Z=1} + \frac{dC(Z)}{dZ}\bigg|_{Z=1} + \frac{dD(Z)}{dZ}\bigg|_{Z=1} + \frac{dE(Z)}{dZ}\bigg|_{Z=1} + N\right]$$

$$\therefore T_H = \frac{N}{P_s^3 P_c^3}\left(1 + P_s P_c + P_s^2 P_c^2\right) + i \cdot UI$$

$$\frac{T_H}{T_S} = \frac{1}{P_s^3 P_c^3}\left(1 + P_s P_c + P_s^2 P_c^2\right) + i/N$$

This is the fifth frame synchronization average keeping time expression when $\beta=\alpha=3$. When $\alpha=\beta$, $\beta\geqslant 1$ the general expression of the fifth frame synchronization average keeping time is:

$$\frac{T_H}{T_S} = \frac{1}{(P_s P_c)^\beta} \sum_{j=0}^{\beta-1} (P_s P_c)^j + \frac{i}{N} \quad (10\text{-}5)$$

[Case 1] $P_c = 1, i = 0$

$$\frac{T_H}{T_S} = \frac{1}{P_s^\beta} \sum_{j=0}^{\beta-1} P_s^j \quad (10\text{-}5\text{-}1)$$

The formula (10-5-1) and the formula (10-1) are the same (See Figure 10-16).

[Case 2] $P_s = 1, i = N$

$$\frac{T_H}{T_S} = \frac{1}{P_s^\beta} \sum_{j=0}^{\beta-1} P_s^j + 1 \quad (10\text{-}5\text{-}2)$$

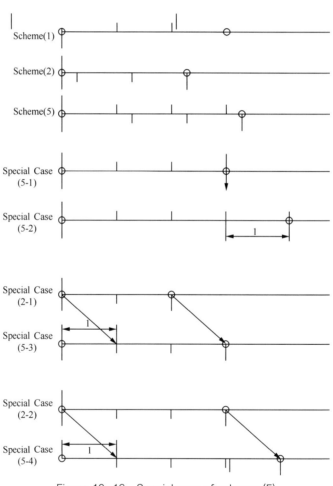

Figure 10-16 Special case of scheme (5)

The formula (10-5-2) is 1 bigger than the formula (10-1). See Figure 10-16; Because the final state of the scheme (5) is defined at the virtual alarm point, whereas the final state of the scheme (1) is defined at the virtual missing point, the difference 1 of the synchronization keeping times normalized is consistent physically.

[Case 3] $P_s = 1, i = 0$

$$\frac{T_H}{T_S} = \frac{1}{P_c^\alpha} \sum_{j=0}^{\alpha-1} P_c^j \qquad (10\text{-}5\text{-}3)$$

The formula (10-5-3) is 1 bigger than the formula (10-2). 1. See Figure 10-16, the special case (10-2-1) starts the effective transition from point A, whereas the special case (10-5-3) starts the effective transition after one frame cycle (T_s). Hence difference 1 of the synchronization keeping times normalized is consistent physically.

[Case 4] $P_s = 1, i = N$

$$\frac{T_H}{T_S} = \frac{1}{P_c^\alpha} \sum_{j=0}^{\alpha-1} P_c^j + 1 \qquad (10\text{-}5\text{-}4)$$

The formula (10-5-4) is 1 bigger than the formula (10-2-2). See Figure 10-16, for the same reason of the case 3, such a difference is consistent physically.

10.2.6 Scheme (6)

The sixth frame synchronization scheme is a synthetical frame synchronization scheme whose frame loss is decided from the virtual leak point (B) starting the first intermediate state, to the virtual alarm point (H). Figure 10-10 shows the primitive state-time diagram and its simplified state-time diagram, where A is the synchronization deciding state, H is the loss deciding state. Figure 10-17 shows state transition diagram from state A to state G. Figure 10-18 shows the transition probability calculation diagram.

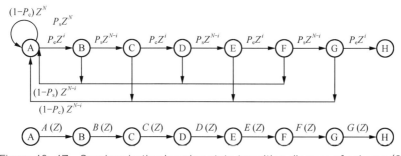

Figure 10-17　Synchronization keeping state transition diagram of scheme (6)

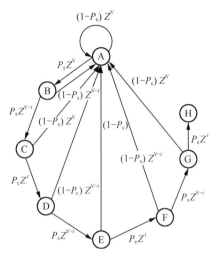

Figure 10-18 Synchronization keeping state probability diagram of scheme (5)

The Z transform of the transition probability $F_{AG}(t)$ at time t from state A to state H (Case of $\beta=3$, $\alpha=\beta+1$):

$$F_{AH}(Z) = A(z) \cdot B(z) \cdot C(z) \cdot D(z) \cdot E(z) \cdot F(z)$$

$$A(z) = \frac{P_c Z^i}{1-(1-P_c)Z^N}$$

$$B(z) = \frac{P_s Z^{N-i}}{1-A(z)(1-P_c)Z^{N-i}}$$

$$C(z) = \frac{P_c Z}{1-A(z)B(z)(1-P_c)Z}$$

$$D(z) = \frac{P_s Z^{N-i}}{1-A(z)B(z)C(z)(1-P_s)Z^{N-i}}$$

$$E(z) = \frac{P_c Z^i}{1-A(z)B(z)C(z)D(z)(1-P_c)Z^N}$$

$$F(z) = \frac{P_s Z^{N-i}}{1-A(z)B(z)C(z)D(z)E(z)(1-P_s)Z^{N-i}}$$

$$G(z) = \frac{P_c Z^i}{1-A(z)B(z)C(z)D(z)E(z)F(z)(1-P_c)Z^N}$$

Frame synchronization average keeping time (Case of $\beta=3$, $\alpha=\beta+1$) is solved:

$$T_\mathrm{H} = \frac{dF_{\mathrm{AH}}(Z)}{dz}\bigg|_{Z=1}$$

$$= \frac{dA(Z)}{dz}\bigg|_{Z=1} + \frac{dB(Z)}{dz}\bigg|_{Z=1} + \frac{dC(Z)}{dz}\bigg|_{Z=1}$$

$$+ \frac{dD(Z)}{dz}\bigg|_{Z=1} + \frac{dE(Z)}{dz}\bigg|_{Z=1} + \frac{dF(Z)}{dz}\bigg|_{Z=1}$$

$$\frac{dA(Z)}{dz}\bigg|_{Z=1} = i + \frac{1-P_\mathrm{c}}{P_\mathrm{c}}N$$

$$\frac{dB(Z)}{dz}\bigg|_{Z=1} = \frac{N-i}{P_\mathrm{s}} + \frac{1-P_\mathrm{s}}{P_\mathrm{s}}\frac{dA(Z)}{dZ}\bigg|_{Z=1}$$

$$\frac{dC(Z)}{dz}\bigg|_{Z=1} = \frac{dA(Z)}{dZ}\bigg|_{Z=1} + \frac{1-P_\mathrm{c}}{P_\mathrm{c}}\left[\frac{dA(Z)}{dz}\bigg|_{Z=1} + \frac{dB(Z)}{dz}\bigg|_{Z=1}\right]$$

$$\frac{dD(Z)}{dz}\bigg|_{Z=1} = \frac{N-i}{P_\mathrm{s}} + \frac{1-P_\mathrm{s}}{P_\mathrm{s}}\left[\frac{dA(Z)}{dz}\bigg|_{Z=1} + \frac{dB(Z)}{dz}\bigg|_{Z=1} + \frac{dC(Z)}{dz}\bigg|_{Z=1}\right]$$

$$\frac{dE(Z)}{dz}\bigg|_{Z=1} = \frac{dA(Z)}{dZ}\bigg|_{Z=1} + \frac{1-P_\mathrm{c}}{P_\mathrm{c}}\left[\frac{dA(Z)}{dZ}\bigg|_{Z=1} + \frac{dB(Z)}{dZ}\bigg|_{Z=1} + \frac{dC(Z)}{dZ}\bigg|_{Z=1} + \frac{dD(Z)}{dZ}\bigg|_{Z=1}\right]$$

$$\frac{dF(Z)}{dz}\bigg|_{Z=1} = \frac{N-i}{P_\mathrm{s}} + \frac{1-P_\mathrm{s}}{P_\mathrm{s}}\left[\frac{dA(Z)}{dZ}\bigg|_{Z=1} + \frac{dB(Z)}{dZ}\bigg|_{Z=1} + \frac{dC(Z)}{dZ}\bigg|_{Z=1} + \frac{dD(Z)}{dZ}\bigg|_{Z=1} + \frac{dE(Z)}{dZ}\bigg|_{Z=1}\right]$$

$$\frac{dG(Z)}{dZ}\bigg|_{Z=1} = \frac{dA(Z)}{dZ}\bigg|_{Z=1} + \frac{1-P_\mathrm{c}}{P_\mathrm{c}}\left[\frac{dA(Z)}{dZ}\bigg|_{Z=1} + \frac{dB(Z)}{dZ}\bigg|_{Z=1} + \frac{dC(Z)}{dZ}\bigg|_{Z=1} + \frac{dD(Z)}{dZ}\bigg|_{Z=1} + \frac{dE(Z)}{dZ}\bigg|_{Z=1} + \frac{dF(Z)}{dZ}\bigg|_{Z=1}\right]$$

$$T_\mathrm{H} = \frac{N}{P_\mathrm{s}^3 P_\mathrm{c}^3}\left(1 + P_\mathrm{s}P_\mathrm{c} + P_\mathrm{s}^2 P_\mathrm{c}^2 + P_\mathrm{s}^3 P_\mathrm{c}^3\right) - (N-i)\cdot UI$$

$$\frac{T_\mathrm{H}}{T_\mathrm{S}} = \frac{1}{P_\mathrm{s}^3 P_\mathrm{c}^4}\left(1 + P_\mathrm{s}P_\mathrm{c} + P_\mathrm{s}^2 P_\mathrm{c}^2 + P_\mathrm{s}^3 P_\mathrm{c}^3\right) - \left(1 - \frac{i}{N}\right)$$

This is the sixth frame synchronization average keeping time expression when $\beta=3$, $\alpha=4$. When $\beta \geqslant 1$, $\alpha=\beta+1$ the general expression of the sixth frame synchronization average keeping time is:

$$\frac{T_\mathrm{H}}{T_\mathrm{S}} = \frac{1}{P_\mathrm{s}^\beta P_\mathrm{c}^\beta}\sum_{j=0}^{\beta}(P_\mathrm{s}P_\mathrm{c})^j - \left(1 - \frac{i}{N}\right) \quad (10\text{-}6)$$

[Case 1] $P_\mathrm{c} = 1, i = 0$

$$\frac{T_H}{T_S} = \frac{1}{P_s^\beta}\sum_{j=0}^{\beta} P_s^j - 1$$

$$= \frac{1}{P_s^\beta}\sum_{j=0}^{\beta-1} P_s^j \quad (10\text{-}6\text{-}1)$$

The formula (10-5-1) and formula (10-1) are the same (See Figure 10-19).

[Case 2] $P_s = 1, i = 1$

$$\frac{T_H}{T_S} = \frac{1}{P_s^\beta}\sum_{j=0}^{\beta} P_s^j \quad (10\text{-}6\text{-}2)$$

$$= \frac{1}{P_s^\beta}\sum_{j=0}^{\beta-1} P_s^j + 1$$

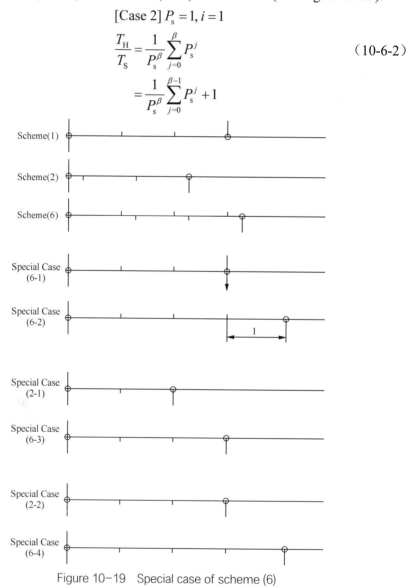

Figure 10−19 Special case of scheme (6)

The formula (10-6-2) is 1 bigger than the formula (10-1). See Figure 10-19;

Because the final state of the scheme (6) is defined at the virtual alarm point, whereas that of the scheme (1) is defined at the virtual missing point, the difference I of the two synchronization keeping times normalized is consistent physically.

[Case 3] $P_s = 1, i = 0$

$$\frac{T_H}{T_S} = \frac{1}{P_c^\alpha} \sum_{j=0}^{\alpha-1} P_c^j - 1 \qquad (10\text{-}6\text{-}3)$$

The formula (10-6-3) and the formula (10-2-1) are the same in the form and consistent physically. See Figure 10-19, the only difference of two special cases is that α takes different values.

[Case 4] $P_s = 1, i = N$

$$\frac{T_H}{T_S} = \frac{1}{P_c^\alpha} \sum_{j=0}^{\alpha-1} P_c^j \qquad (10\text{-}6\text{-}4)$$

The formula (10-6-4) and the formula (10-2-2) are in the same form and agreed physically. See Figure 10-19, the only difference of two special cases is that α takes different values.

10.3 FRAME LOSS AVERAGE KEEPING TIME

10.3.1 Frame Loss deciding Time

As described already, when the frame loss occurs, only the first frame synchronization scheme exists the concept of frame loss deciding time. When a frame loss occurs, the input group of code stream already slips, but the frame synchronization mechanism is still in the frame synchronization deciding state. Obviously, if there is no virtual alarm (i.e. $P_e=0$) at the original frame synchronization position, the frame loss will be decided after βT_s later; If the virtual alarm ($P_E \neq 0$) occurs, the frame loss deciding time will be extended. The state transition diagram and its simplified example of the frame loss deciding process are shown in Figure 10-20.

The Z transform ($\beta=3$) of the transition probability $F_{AD}(t)$ at time t from the state A

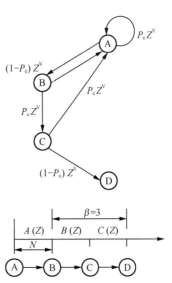

Figure 10-20 Frame loss deciding process state transition diagram

frame loss occurring to state D frame loss decided:

$$F_{AD}(Z) = A(Z) \cdot B(Z) \cdot C(Z)$$

$$A(Z) = \frac{(1-P_c)Z^N}{1-P_c Z^N}$$

$$B(Z) = \frac{(1-P_c)Z^N}{1-A(Z)P_c Z^N}$$

$$C(Z) = \frac{(1-P_c)Z^N}{1-A(Z) \cdot B(Z)P_c Z^N}$$

The frame loss average deciding time:

$$T_H = \frac{dF_{AD}(Z)}{dZ}\bigg|_{Z=1}$$

$$= \frac{dA(Z)}{dZ}\bigg|_{Z=1} + \frac{dB(Z)}{dZ}\bigg|_{Z=1} + \frac{dC(Z)}{dZ}\bigg|_{Z=1}$$

$$\frac{dA(Z)}{dZ}\bigg|_{Z=1} = \frac{N}{1-P_c}$$

$$\frac{dB(Z)}{dZ}\bigg|_{Z=1} = \frac{N}{1-P_c} + \frac{P_c}{1-P_c}\frac{dA(Z)}{dZ}\bigg|_{Z=1}$$

$$\frac{dC(Z)}{dZ}\bigg|_{Z=1} = \frac{N}{1-P_c} + \frac{P_c}{1-P_c}\left[\frac{dA(Z)}{dZ}\bigg|_{Z=1} + \frac{dB(Z)}{dZ}\bigg|_{Z=1}\right]$$

$$\therefore T_D = \frac{N}{(1-P_c)^3}\left[1+(1-P_c)+(1-P_c)^2\right] \cdot UI$$

The above formula is the frame loss average deciding time when $\beta=3$. When $\beta \geqslant 1$, the general formula of frame loss average deciding time:

$$\frac{T_D}{T_S} = \frac{1}{(1-P_c)^\beta}\sum_{j=0}^{j-1}(1-P_c)^j$$

$$= \frac{1}{(1-P_c)^\beta}\left[\frac{1-(1-P_c)^\beta}{P_c}\right]$$

$$= \frac{1}{P_c}\left[\frac{1}{(1-P_c)^\beta}-1\right]$$

$$= \frac{1}{P_c}\left[1+\beta P_c \frac{\beta(\beta+1)}{2}P_c^2 + L - 1\right]$$

$$\approx \beta + \frac{\beta(\beta+1)}{2} P_c$$
$$\frac{T_D}{T_S} \approx \beta + \frac{\beta(\beta+1)}{2} P_c \qquad (10\text{-}7)$$

10.3.2 Frame Synchronization Search Time

See Figure 10-1.

After frame loss deciding, the frame synchronization scheme recommended by CCITT starts frame synchronization search immediately which repeats at several non-synchronization position until a new synchronization position. In the case of real frame loss, if the slipped loss frame is delayed by $j \cdot$ UI, j position will be searched; whereas in the case of false frame loss. because the original synchronization position is misunderstood as a non-synchronization position, hence N position will be searched to reach the new synchronization position.

After each new reset, the anti fading synchronization scheme starts immediately the search for new frame synchronization. The schemes (2), (4) and (5) search bit by bit as the same as the scheme (1). In the case of real frame loss, j positions will be searched; whereas in the case of false frame loss, if the frame loss occurs at j. UI, $N-i$ positions will be searched. The schemes (3) and (5) search frame by frame, and only one frame is need to search.

The search time bit by bit and the search time frame by frame are calculated respectively below.

(1) Bit by bit search time

The non-synchronization positions are searched because the real frame alignment signal doesn't exist. If the virtual alarm doesn't occur, the search time is only $i \cdot$ UI: whereas if the virtual alarm occurs, the bit by bit search time will extend. The state transition diagram of bit by bit search process (i.e. from state b_i to b_{i+1}) is shown in Figure 10-21 and Figure 10-22.

Figure 10-21 is the state transition diagram of bit by bit search process of the scheme (1). If the virtual alarm $(1-P_c)$ doesn't occur, after 1.UI it can transit from b_i to b_{i+1}; if a virtual alarm occurs and no another virtual alarm $(1-P_c)$ occurs after the next frame ($N \cdot$ UI), it will transit from b_i to b_{i+1} after $(N+1)$UI; if virtual alarms occur α times consecutively. the system enters false synchronization and transits into the frame synchronization keeping state, in this case, the virtual alarms have not occurred in the consecutive β frames in order to transit from b_i to b_{i+1}.

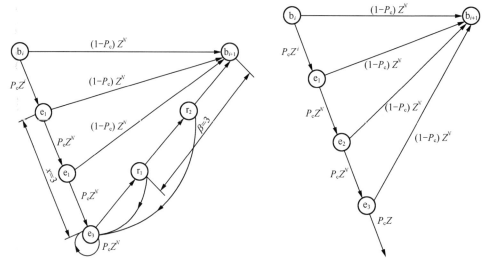

Figure 10-21　State transition diagram of bit by bit search process of scheme (1)

Figure 10-22　State transition diagram of bit by bit search process of scheme (2)-(6)

Figure 10-22 is the state transition diagram of bit by bit search process of the schemes (2), (4) and (6). For the scheme (2), if α virtual alarms occur consecutively, the system resets false frame synchronization once, then searches continuously. Because the false frame synchronization reset doesn't affect the later searches essentially, the later search processes will be carried out in the same way as long as the virtual alarms occur consecutively. For the schemes (4) and (6), if a real frame alignment signal is found between two consecutive virtual alarms, the previous search will not be in effect at all, hence the search continues; if no new frame alignment signal is found between two consecutive virtual alarms, the false frame synchronization resets only once for α consecutive virtual alarms, it will not affect the later search essentially. Hence, as long as the virtual alarms occur consecutively, the later search processes will be carried out in the same way. This is just what is described in Figure 10-22.

Compare Figure 10-21 and Figure 10-22, The former is more complicated to solve, whereas the later is simpler to solve. For approximate calculation, the precision is enough using Figure 10-22 instead of Figure 10-21. Hence, Figure 10-22 is used to solve the bit by bit search time for schemes (1), (2), (4) and (5).

The Z transform of the transition probability $F_b(t)$ from state b_i to b_{i+1} is gotten from Figure 10-22:

Chapter 10 Anti Fading Frame Synchronization

$$F_b(z) = (1-P_c)Z + P_c(1-P_c)Z^{N+1} + P_c^2(1-P_c)Z^{2N+1} + \cdots$$
$$= (1-P_c)Z\left[1 + P_cZ^N + (P_cZ^N)^2 + \cdots\right]$$
$$= (1-P_c)Z \cdot \frac{1}{1-P_cZ^N}$$

The average transition time from state b_i to b_{i+1}:

$$t_a = \frac{dF_b(Z)}{dZ}\bigg|_{Z=1}$$
$$= 1 + \frac{P_c}{1-P_c}N \cdot UI \qquad (10\text{-}8)$$
$$\frac{t_a}{T_s} = \frac{1}{N} + \frac{P_c}{1-P_c}$$

(2) Frame by frame search time

Figure 10-23 shows the state transition diagram of frame by frame search process of schemes (3) and (5). In the case of loss frame, there is no real frame synchronization signal certainly which is one frame ($N \cdot UI$) distant from the original frame position. If no virtual alarm occurs at this position, only one frame cycle (T_s) is need to complete frame by frame search process; whereas if a virtual alarm occurs at this position, the frame by frame search will extend.

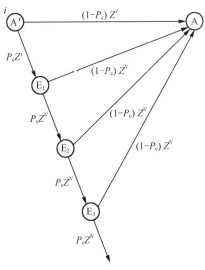

Figure 10-23 The state transition diagram of frame by frame search process of schemes (3) and (5).

The Z transform of the transition probability $F_s(t)$ from state A to state A is gotten from Figure 10-23:

$$F_s(t) = (1-P_c)Z^N + P_c(1-P_c)Z^{2N} + P_c^2(1-P_c)Z^{3N} + \cdots$$
$$= (1-P_c)Z^N \left[1 + P_c Z^N + (P_c Z^N)^2 + \cdots\right]$$
$$= (1-P_c)Z^N \cdot \frac{1}{1-P_c Z^N}$$

The average transition time from state A to state A:

$$t_a = \left.\frac{dF_s(Z)}{dZ}\right|_{Z=1}$$
$$= N + \frac{P_c}{1-P_c}N$$

$$\frac{t_a}{T_s} = 1 + \frac{P_c}{1-P_c} \tag{10-9}$$

(3) Frame synchronization average search time

According to Formula (10-8) and (10-9), get the frame synchronization average search times of the various schemes:

Assume: $j=N/2$, $i=N/2$,

Schemes				Frame synchronization average search time (T_A/T_S)
(1)		Real frame loss		$\frac{P_c}{1+P_c}\frac{N}{2} + 1/2$
(1)		False frame loss		$\frac{P_c}{1+P_c}N + 1$
(2)	(4)		(6)	$1/2 + \frac{P_c}{1+P_c}N$
	(3)		(5)	$1 + \frac{P_c}{1+P_c}$

10.3.3 Frame Synchronization Deciding Time

According to Figure 10-1, the comparison diagram between the frame synchronization keeping process and frame synchronization deciding process may be abstracted. Obviously, the two processes are similar; the difference is actually the probability difference and the initial time difference.

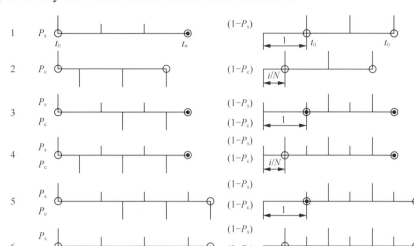

Figure 10-24 Frame Synchronization Keeping Deciding Processes Comparison

According to the similarity, the probability corresponding relationship and the process initial time difference of the two processes, the corresponding frame synchronization deciding time $\left(\dfrac{T_H}{T_S}\right)$ calculation formula can be written directly from the frame synchronization keeping time formula $\left(\dfrac{T_W}{T_n}\right)$.

(1) Scheme (1)

The state transition diagram and its simplest example are shown in Figure 10-25.

$$\left.\dfrac{T_W}{T_S}\right|_{(1)} = \dfrac{1}{(1-P_s)^\beta} \sum_{j=0}^{\beta-1}(1-P_s)^j - 1$$

$$= \dfrac{1}{(1-P_s)^\beta} \left[\dfrac{1-(1-P_s)^\beta}{P_s}\right] - 1$$

$$= \dfrac{1}{P_s}\left[\dfrac{1}{(1-P_s)^\beta} - 1\right] - 1$$

$$= \dfrac{1}{P_s}\left[1 + \beta P_s \dfrac{\beta(\beta+1)}{2}P_s^2 + LL - 1\right] - 1$$

$$\approx (\beta-1)+\frac{\beta(\beta+1)}{2}P_s$$

$$\left.\frac{T_W}{T_S}\right|_{(1)} \approx (\beta-1)+\frac{\beta(\beta+1)}{2}P_s \qquad (10\text{-}10)$$

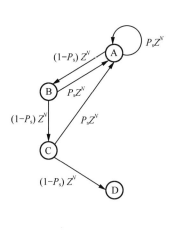

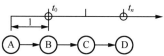

Figure 10-25 The state transition diagram its simplest example of scheme (1)

Scheme	Probability difference		Start time difference	
	Keep process	Deciding process	Keep process	Deciding process
1	P_s	$1-P_s$	0	1
2	P_c	$1-P_c$	0	i/N
3	P_s	$1-P_s$	0	1
	P_c	$1-P_c$		
4	P_s	$1-P_s$	0	i/N
	P_c	$1-P_c$		
5	P_s	$1-P_s$	0	1
	P_c	$1-P_c$		
6	P_s	$1-P_s$	0	i/N
	P_c	$1-P_c$		

(2) Scheme (2)

The state transition diagram and its simplest example are shown in Figure 10-26.

$$\left.\frac{T_W}{T_S}\right|_{(2)} = \frac{1}{(1-P_s)^\alpha}\sum_{j=0}^{\alpha-1}(1-P_s)^j - (1-i/N) - i/N$$

$$= \frac{1}{(1-P_s)^\alpha}\left[\frac{1-(1-P_s)^\alpha}{P_s}\right] - 1$$

$$= \frac{1}{P_s}\left[\frac{1}{(1-P_s)^\alpha} - 1\right] - 1 \quad (10\text{-}11)$$

$$= \frac{1}{P_s}\left[1 + \alpha P_s \frac{\alpha(\alpha+1)}{2}P_s^2 + LL - 1\right] - 1$$

$$\approx (\alpha-1) + \frac{\alpha(\alpha+1)}{2}P_s$$

$$\left.\frac{T_W}{T_S}\right|_{(2)} \approx (\alpha-1) + \frac{\alpha(\alpha+1)}{2}P_s$$

(3) Scheme (3)

The state transition diagram and its simplest example are shown in Figure 10-27.

$$\left.\frac{T_W}{T_S}\right|_{(3)} = \frac{1}{(1-P_s)^\beta(1-P_s)^{\beta-1}}\sum_{j=0}^{\beta-1}[(1-P_s)(1-P_c)]^j - 1$$

$$= \frac{(1-P_s)}{[(1-P_s)(1-P_c)]^\beta}\left\{\frac{1-[(1-P_s)(1-P_c)]^\beta}{1-(1-P_s)(1-P_c)}\right\} - 1$$

$$= \frac{(1-P_s)}{1-(1-P_s)(1-P_c)}\left\{\frac{1}{[(1-P_s)(1-P_c)]^\beta} - 1\right\} - 1$$

$$\approx \frac{(1-P_s)}{(P_s+P_c)}\left\{\frac{1}{[1-(P_s+P_c)]^\beta} - 1\right\} - 1 \quad (10\text{-}12)$$

$$= \frac{(1-P_s)}{(P_s+P_c)}\left\{1 + \beta(P_s+P_c) + \frac{\beta(\beta+1)}{2}(P_s+P_c)^2 + \cdots - 1\right\} - 1$$

$$\approx (\beta-1) + \frac{\beta(\beta-1)}{2}P_s + \frac{\beta(\beta+1)}{2}P_c$$

$$\left.\frac{T_W}{T_S}\right|_{(3)} \approx (\beta-1) + \frac{\beta(\beta-1)}{2}P_s + \frac{\beta(\beta+1)}{2}P_c$$

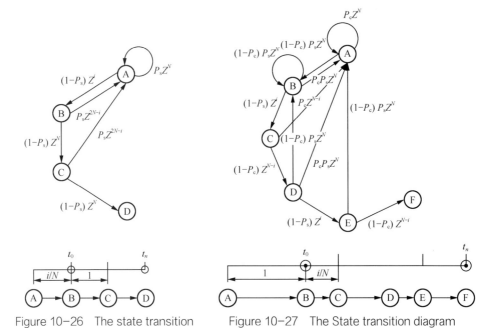

Figure 10-26 The state transition diagram and its simplest example of scheme (2)

Figure 10-27 The State transition diagram and its simplest example of scheme (3)

(4) Scheme (4)

The state transition diagram and its simplest example are shown in Figure 10-28.

$$\left.\frac{T_W}{T_S}\right|_{(4)} = \frac{1}{(1-P_s)^\beta (1-P_s)^{\beta-1}} \sum_{j=0}^{\beta-1} [(1-P_s)(1-P_c)]^j - i/N$$

$$= \frac{(1-P_s)}{[(1-P_s)(1-P_c)]^\beta} \left\{ \frac{1-[(1-P_s)(1-P_c)]^\beta}{1-(1-P_s)(1-P_c)} \right\} - i/N$$

$$= \frac{(1-P_s)}{1-(1-P_s)(1-P_c)} \left\{ \frac{1}{[(1-P_s)(1-P_c)]^\beta} - 1 \right\} - i/N$$

$$\approx \frac{1}{(P_s+P_c)} \left\{ \frac{1}{[1-(P_s+P_c)]^\beta} - 1 \right\} - i/N$$

$$= \frac{1}{(P_s+P_c)} \left\{ 1 + \beta(P_s+P_c) + \frac{\beta(\beta+1)}{2}(P_s+P_c)^2 + \cdots - 1 \right\} - i/N$$

$$\approx \left(\beta - \frac{i}{N} \right) + \frac{\beta(\beta+1)}{2}(P_s+P_c)$$

$$\left.\frac{T_W}{T_S}\right|_{(4)} \approx \left(\beta - \frac{i}{N}\right) + \frac{\beta(\beta+1)}{2}(P_s + P_c) \qquad (10\text{-}13)$$

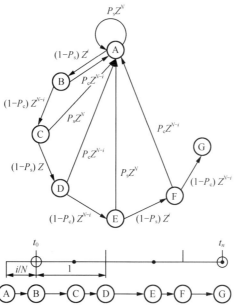

Figure 10-28 The State Transition Diagram and its Simplest Example of Scheme (4)

(5) Scheme (5)

The state transition diagram and its simplest example are shown in Figure 10-29.

$$\left.\frac{T_W}{T_S}\right|_{(5)} = \frac{1}{(1-P_s)^\beta(1-P_s)^\beta} \sum_{j=0}^{\beta-1}[(1-P_s)(1-P_c)]^j + \frac{i}{N} - 1$$

$$= \frac{(1-P_s)}{[(1-P_s)(1-P_c)]^\beta}\left\{\frac{1-[(1-P_s)(1-P_c)]^\beta}{1-(1-P_s)(1-P_c)}\right\} + \frac{i}{N} - 1$$

$$= \frac{(1-P_s)}{1-(1-P_s)(1-P_c)}\left\{\frac{1}{[(1-P_s)(1-P_c)]^\beta} - 1\right\} + \frac{i}{N} - 1$$

$$\approx \frac{1}{(P_s+P_c)}\left\{\frac{1}{[1-(P_s+P_c)]^\beta} - 1\right\} + \frac{i}{N} - 1$$

$$= \frac{1}{(P_s+P_c)}\left\{1 + \beta(P_s+P_c) + \frac{\beta(\beta+1)}{2}(P_s+P_c)^2 + \cdots - 1\right\} + \frac{i}{N} - 1$$

$$\approx \left(\beta - \frac{N-i}{N}\right) + \frac{\beta(\beta+1)}{2}(P_s + P_c)$$

$$\left.\frac{T_W}{T_S}\right|_{(5)} \approx \left(\beta - \frac{N-i}{N}\right) + \frac{\beta(\beta+1)}{2}(P_s + P_c) \tag{10-14}$$

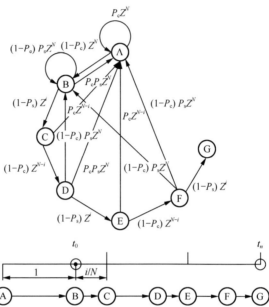

Figure 10-29 The state transition diagram and its simplest example of scheme (5)

(6) Scheme (6)

The state transition diagram and its simplest example are shown in Figure 10-30.

$$\left.\frac{T_W}{T_S}\right|_{(6)} = \frac{1}{(1-P_s)^\beta (1-P_s)^{\beta+1}} \sum_{j=0}^{\beta} [(1-P_s)(1-P_c)]^j - \left(1-\frac{i}{N}\right) - \frac{i}{N}$$

$$= \frac{(1-P_c)}{[(1-P_s)(1-P_c)]^{\beta+1}} \left\{ \frac{1-[(1-P_s)(1-P_c)]^{\beta+1}}{1-(1-P_s)(1-P_c)} \right\} - 1$$

$$= \frac{(1-P_c)}{1-(1-P_s)(1-P_c)} \left\{ \frac{1}{[(1-P_s)(1-P_c)]^{\beta+1}} - 1 \right\} - 1$$

$$\approx \frac{(1-P_c)}{(P_s+P_c)} \left\{ \frac{1}{[1-(P_s+P_c)]^{\beta+1}} - 1 \right\} - 1$$

$$= \frac{(1-P_c)}{(P_s+P_c)} \left\{ 1 + \beta(P_s+P_c) \frac{(\beta+1)(\beta+2)}{2}(P_s+P_c)^2 + L - 1 \right\} - 1$$

$$\approx (1-P_c) \left\{ (\beta+1) + \frac{(\beta+1)(\beta+2)}{2}(P_s+P_c) \right\} - 1$$

$$\approx \beta + \frac{(\beta+1)(\beta+2)}{2} P_s + \frac{\beta(\beta+1)}{2} P_c \qquad (10\text{-}15)$$

$$\left. \frac{T_W}{T_S} \right|_{(6)} \approx \beta + \frac{(\beta+1)(\beta+2)}{2} P_s + \frac{\beta(\beta+1)}{2} P_c$$

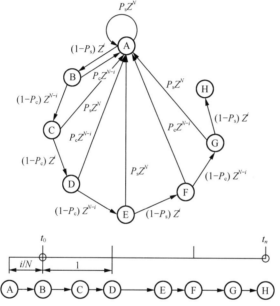

Figure 10-30 The State transition diagram and its simplest example of scheme (6)

10.3.4 Frame Synchronization Reset Delay

There is no frame synchronization reset problem in the scheme recommended by CCITT (i.e. scheme (1)). If frame synchronization is decided, in fact, when the frame synchronization deciding process begins, the system enters frame synchronous state already.

There are frame synchronization reset problems in all the anti fading frame synchronization schemes (i.e. schemes (2)-(6). Because during the frame synchronization search and frame synchronization deciding processes, the system is keeping the original frame state, only after the new frame synchronization reset is carried out, then the system enters a new frame state.

See Figure 10-1, the frame synchronization deciding process is completed, the schemes (2), (5) and (6) take the new frame synchronization reset immediately, hence the frame synchronization reset delay is 0; whereas at the end of the frame synchronization deciding process, schemes (3) and (4) is being conservative (virtual missing) followed, the new frame synchronization reset has to postponed to the refresh point (virtual alarm). Hence frame synchronization reset delay exists. In the case of real frame loss, when the slip time interval is equal to $\frac{j}{N}T_S$, the reset delay is $\frac{i}{N}T_S$; whereas in the case of false frame loss, when the virtual alarm time is $\frac{j}{N}T_S$, the reset delay is $\frac{N-i}{N}T_S$. In average, reset delay is $T_S/2$.

10.3.5 Frame Loss Keeping Time

The frame loss keeping time calculation formula:

$$\frac{T_L}{T_S} = \frac{T_D}{T_S} + \frac{T_A}{T_S} + \frac{T_W}{T_S} + \frac{T_Z}{T_S}$$

where, $\frac{T_D}{T_S}$ —frame loss deciding time;

$\frac{T_A}{T_S}$ —frame synchronization search time;

$\frac{T_W}{T_S}$ —frame synchronization deciding time;

$\frac{T_Z}{T_S}$ —frame synchronization reset delay.

According to the formula derived above, we may get the frame loss keeping time formula (See Table).

10.4 COMPARISON OF FRAME SYNCHRONIZATION SCHEMES

10.4.1 Frame Synchronization Average Keeping Time

The theoretical formula of the frame synchronization average keeping times of the six typical schemes are concluded below:

Chapter 10 Anti Fading Frame Synchronization

Where
$$P_S = 1 - (1 - P_e)^n$$
$$P_e = \left(\frac{1}{2}\right)^n$$

Under the condition of $n=10$, $\beta=3$, the special calculation result is shown in Figure 10-31.

Schemes	T_H/T_S
(1)	$\dfrac{1}{P_S^\beta} \sum_{j=0}^{\beta-1} P_S^j$
(2)	$\dfrac{1}{P_c^\alpha} \sum_{j=0}^{\alpha-1} P_e^j \cdot \left(1 - \dfrac{i}{N}\right)$
(3)	$\dfrac{1}{P_S^\beta P_c^{\beta-1}} \sum_{j=0}^{\beta-1} (P_S P_c)^j$
(4)	$\dfrac{1}{(P_S P_c)^\beta} \sum_{j=0}^{\beta-1} (P_S P_c)^j$
(5)	$\dfrac{1}{(P_S P_c)^\beta} \sum_{j=0}^{\beta-1} (P_S P_c)^j + \dfrac{i}{N}$
(6)	$\dfrac{1}{P_S^\beta P_c^{\beta-1}} \sum_{j=0}^{\beta-1} (P_S P_c)^j \cdot \left(1 - \dfrac{i}{N}\right)$

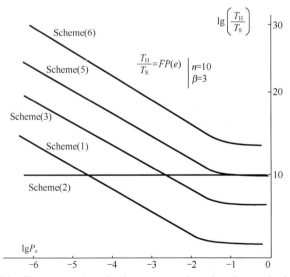

Figure 10-31 Frame synchronization average keeping time calculation curve

Table Frame Loss Keeping Time Calculation Formula:

Schemes	$T_H T_S$	$T_A T_S$	$T_H T_S$		
1 (real frame loss)	$\frac{\beta(\beta+1)}{2}P_c + \beta$	$\frac{1}{2}+\frac{N}{2}P_c$	$(\beta-1)+\frac{\beta(\beta+2)}{2}P_s$	0	$\left(2\beta-\frac{1}{2}\right)+\frac{\beta(\beta+1)}{2}P_s+$ $\frac{\beta(\beta+1)+N}{2}$
1 (real frame loss)	/	$1+NP$	$(\beta-1)+\frac{\beta(\beta+2)}{2}P_s$	0	$\beta+\frac{\beta(\beta+1)}{2}P_s+NP_c$
2	/	$\frac{1}{2}+\frac{N}{2}P_c$	$(\alpha-1)+\frac{\alpha(\alpha+2)}{2}P_s$	0	$\left(\alpha-\frac{1}{2}\right)+\frac{\alpha(\alpha+2)}{2}P_s+$ $\frac{N}{2}P_c$
3	/	$1+P_c$	$(\beta-1)+\frac{\beta(\beta-1)}{2}P_s+$ $\frac{\beta(\beta+1)}{2}P_c$	1/2	$\left(\beta+\frac{1}{2}\right)+\frac{\beta(\beta-1)}{2}P_s+$ $\frac{\beta(\beta+1)+2}{2}P_c$
4	/	$\frac{1}{2}+\frac{N}{2}P_c$	$\left(\beta-\frac{1}{2}\right)+\frac{\beta(\beta+1)}{2}(P_s+P_c)$	1/2	$\left(\beta+\frac{1}{2}\right)+\frac{\beta(\beta+1)}{2}P_s+$ $\frac{\beta(\beta+1)+N}{2}P_c$
5	/	$1+P$	$\left(\beta-\frac{1}{2}\right)+\frac{\beta(\beta+1)}{2}(P_s+P_c)$	0	$\left(\beta-\frac{1}{2}\right)+\frac{\beta(\beta+1)}{2}P_s+$ $\frac{\beta(\beta+1)+2}{2}P_c$
6	/	$\frac{1}{2}+\frac{N}{2}P_c$	$\beta+\frac{(\beta+1)(\beta+2)}{2}P_s+$ $\frac{\beta(\beta+1)}{2}P_c$	0	$\left(\beta+\frac{1}{2}\right)+\frac{\beta(\beta+3)}{2}P_s+$ $\frac{\beta(\beta+1)+N}{2}P_c$

See from Figure 10-31, except the scheme (2), the frame synchronization average keeping times increase as the average error code rates decrease. The change slopes are equal and depend on P_s^β; the absolute values are not equal, the times of the differences depend on P_e and the corresponding exponents:

Scheme	(1)	(3)	(4)	(5)	(6)
	1	$1/P_c^{\beta-1}$	$1/P_c^\beta$	$1/P_c^\beta$	$1/P_c^{\beta+1}$
	1	$2^{n(\beta-1)}$	$2^{n\beta}$	$2^{n\beta}$	$2^{n(\beta+1)}$

In the case of $P_e = 1 \times 10^{-1}$, the relationship calculation curve of frame

Chapter 10 Anti Fading Frame Synchronization

synchronization average keeping time and the elementary parameters (β,n) is shown in Figure 10-32.

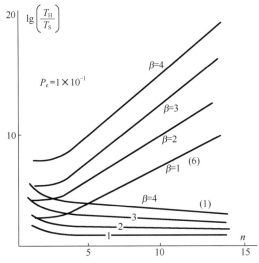

Figure 10-32 Frame synchronization parameter design of scheme (6)

From Figure 10-32, in the condition of $P_e=1\times10^{-1}$ (the worst case), the frame synchronization average keeping time of the scheme (6) increases monotonously as the elementary parameters increase and the ratio ($2^{n(\beta+1)}$) of the frame synchronization average keeping times of the scheme (6) and (1) also increases remarkably.

Consider the frame synchronization keeping time, the elementary parameters of the scheme (6) call be designed in two ways: Keeping the original (using scheme (1)) average frame synchronization keeping time unchanged, decrease the elementary parameters to enhance the transmission efficiency and simplify the circuits; or Keeping the original (using scheme(1)) elementary parameters unchanged to improve the frame synchronization average keeping time remarkably.

10.4.2 Frame Loss Average Keeping Time

The theoretical formula of frame loss average keeping times of the six typical schemes are concluded below:

Schemes		T_H/T_S
(1)	Real frame loss	$2\beta - \dfrac{1}{2} + \dfrac{\beta(\beta+1)}{2}P_s + \dfrac{\beta(\beta+1)+N}{2}P_c$
	False frame loss	$\beta + \dfrac{\beta(\beta+2)}{2}P_c + \dfrac{N}{2}P_c$

续表

Schemes	T_H/T_S
(2)	$\alpha - \dfrac{1}{2} + \dfrac{\alpha(\alpha+1)}{2}P_c + \dfrac{N}{2}P_c$
(3)	$\beta + \dfrac{1}{2} + \dfrac{\beta(\beta-1)}{2}P_s + \dfrac{\beta(\beta+1)+2}{2}P_c$
(4)	$\beta + \dfrac{1}{2} + \dfrac{\beta(\beta-1)}{2}P_s + \dfrac{\beta(\beta+1)+N}{2}P_c$
(5)	$\beta + \dfrac{1}{2} + \dfrac{\beta(\beta-1)}{2}P_s + \dfrac{\beta(\beta+1)+2}{2}P_c$
(6)	$\beta + \dfrac{1}{2} + \dfrac{(\beta-1)(\beta+2)}{2}P_s + \dfrac{\beta(\beta+1)+N}{2}P_c$

In the condition of $n=10$, $\beta=3$, $N=848$, the special calculation result is shown in Figure 10-33.

Schemes		T_D/T_S	T_L/T_S		T_Z/T_S
			T_A/T_S	T_W/T_S	
(1)	Real frame loss	$\dfrac{1}{P_c}\left[\dfrac{1}{(1-P_c)^\beta} - 1\right]$	$\dfrac{1}{2} + \dfrac{P_c}{1+P_c}N/2$	$\dfrac{1}{P_s}\left[\dfrac{1}{(1-P_s)^\beta} - i\right] - 1$	0
	False frame loss	/	$1 + \dfrac{P_c}{1+P_c}N$	$\dfrac{1}{P_s}\left[\dfrac{1}{(1-P_s)^\beta} - 1\right] - 1$	0
(2)		/	$\dfrac{1}{2} + \dfrac{P_c}{1+P_c}N/2$	$\dfrac{1}{P_s}\left[\dfrac{1}{(1-P_s)^\alpha} - 1\right] - 1$	0
(3)		/	$1 + \dfrac{P_c}{1+P_c}$	$\dfrac{(1-P_s)}{1-(1-P_s)(1-P_c)}\left\{\dfrac{1}{[(1-P_s)(1-P_c)]^\beta} - 1\right\} - 1$	1/2
(4)		/	$\dfrac{1}{2} + \dfrac{P_c}{1+P_c}N/2$	$\dfrac{1}{1-(1-P_s)(1-P_c)}\left\{\dfrac{1}{[(1-P_s)(1-P_c)]^\beta} - 1\right\} - \dfrac{i}{N}$	1/2

Chapter 10 Anti Fading Frame Synchronization

续表

Schemes	T_L/T_S			T_Z/T_S
	T_D/T_S	T_A/T_S	T_W/T_S	
(5)	/	$1+\dfrac{P_c}{1+P_c}$	$\dfrac{1}{1-(1-P_s)(1-P_c)}\left\{\dfrac{1}{[(1-P_s)(1-P_c)]^{\beta}}-1\right\}+\dfrac{i}{N}-1$	0
(6)	/	$\dfrac{1}{2}+\dfrac{P_c}{1+P_c}N/2$	$\dfrac{(1-P_c)}{1-(1-P_s)(1-P_c)}\left\{\dfrac{1}{[(1-P_s)(1-P_c)]^{\beta+1}}-1\right\}-1$	0

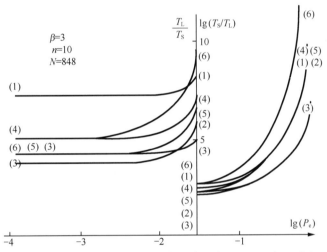

Figure 10-33 Frame loss average keeping time curve (special case)

We can see from Figure 10-33, When $P_e \geqslant 1\times 10^{-1}$, T_L increases seriously as P_e increases; When $P_e \leqslant 1\times 10^{-2}$, T_L decreases slowly as P_e decreases; When P_e is between 1×10^{-1}-1×10^{-2}, the values of T_L in various schemes cross, for example: in the range of $P_e \geqslant 1\times 10^{-1}$, $\left.\dfrac{T_L}{T_S}\right|_{(6)}$ is the biggest, $\left.\dfrac{T_L}{T_S}\right|_{(3)}$ is the smallest; in the range of $P_e \leqslant 1\times 10^{-2}$, $\left.\dfrac{T_L}{T_S}\right|_{(1)}$ is the biggest, and $\left.\dfrac{T_L}{T_S}\right|_{(2)}$ is the smaleest.

The value of $\left.\dfrac{T_L}{T_S}\right|_{max}$ when $P_e=1/2$ and the value of $\left.\dfrac{T_L}{T_S}\right|_{1\times 10^{-6}}$ when $P_e=1\times 10^{-6}$

are calculated below:

Scheme	(1)	(2)	(3)	(4)	(5)	(6)
$\left.\dfrac{T_L}{T_S}\right\|_{max}$	$1.074\ 91\times 10^9$	$1.074\ 91\times 10^9$	$1.052\ 628\times 10^9$	$1.077\ 945\times 10^9$	$1.077\ 945\times 10^9$	$1.103\ 816\times 10^9$
$\left.\dfrac{T_L}{T_S}\right\|_{max}$	5.910 527	2.913 659	3.506 865	3.915 979	3.506 896	3.919 560

See Figure 10-34, consider frame loss keeping time, the elementary parameter (β, n) can be designed as bellow: if the domain of the elementary parameters are $\beta=[1, 4]$; $n=[2, 7]$, in the worst case $(P_e=1\times 10^{-1})$, the average frame loss time is better than the scheme (1) $(\beta=4, n=10, \lg\left(\dfrac{T_L}{T_S}\right)=2.02)$; if the elementary parameters $\beta=3$, $n=5$, in the case of $P_e=1\times 10^{-1}$, the average frame loss time is the smallest when $\beta=3$ and is better than the design of scheme (1) $\beta=3$, $n=10$.

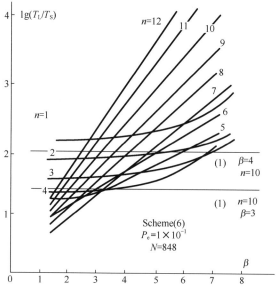

Figure 10-34 Frame loss parameters design of scheme (6)

If only consider the frame loss average keeping time, in the worse environment $(P_e \leqslant 1\times 10^{-1})$, the frame loss keeping time of the scheme (1) is the longest; in the extreme case $(P_e=1/2)$, the average frame loss time of the scheme (6) will be 10^3 times higher than that of the other schemes (1), (2), (4), (5), whereas the schemes (5) and (4) are fairly close to the scheme (1). Hence under worse environment the system should use the scheme (5) and (4).

10.4.3 The Synthetical Criterion of the Frame Synchronization System

The purpose of giving the Synthetical criterion is to give a quantity which is used to describe the overall performance of the frame synchronization system. We already introduce two elementary performance indices of the frame synchronization System, i.e. frame synchronization average keeping time (T_H) and frame loss average keeping time (T_L). The purpose of this section is to derive a synthetical criterion from these two elementary performance indices, which will be used as the elementary base to evaluate the frame synchronization schemes.

From the engineering practice, a frame synchronization system always want: the average frame synchronization keeping time (T_H) will be as long as possible, whereas the average frame loss frequency (F_L) will be as low as possible. Hence we can use the ratio of the frame synchronization keeping time to the loss frame frequency as the synthetical criterion. For convenient use, the synthetical criterion should be represented by dimensionless values as much as possible. Such criterion can be defined as the quality factor of the frame synchronization system and is described below:

$$K = \frac{T_H}{T_S} \bigg/ \frac{1}{\dfrac{T_H}{T_S} + \dfrac{T_L}{T_S}}$$

$$= \frac{T_H}{T_S}\left(\frac{T_H}{T_S} + \frac{T_L}{T_S}\right)$$

Figure 10-35 gives the quality factor calculation curves of the six typical frame synchronization systems in the condition of $\beta=3$, $n=10$ and $N=848$. It shows that starting from $P_e=1/2$, as the error code rate decreases, the quality factors of all the schemes decrease at the beginning, after the lowest point, all increase monotonously. Such a phenomenon is because of the result of synthetical function of that the frame synchronization keeping time lengthens and the frame loss keeping time shorten. In the six typical schemes, the quality factor of the scheme (6) is highest, even though in condition of the high error code rate $P_e=1/2$, its frame loss average keeping time is the longest, but because of its frame synchronization average keeping time is also the longest, its quality factor is the highest; The schemes (5) and (4) have the same good quality factor; in the case of high error code rate, the quality factor of the scheme (2) is the lowest and worse than the scheme (1), in the case of low error code rate, the quality factors of the schemes (2) and (1) are the equal and all worse than the other schemes.

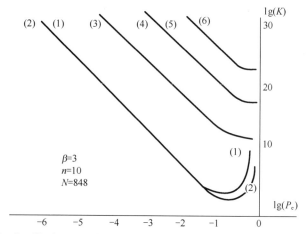

Figure 10-35 Quality factor calculation curves of the frame synchronization systems

Figure 10-36 gives the relationship curve of the quality factor and the elementary parameters. We can see, when the schemes (6) and (1) use the same elementary parameter ($\beta=3$, $n=10$), their quality factors will be improved 10^{14} times; when the elementary parameters are $\beta=3$, $n=4$, in any error code rate, its quality factor is better than the scheme (1): when the elementary parameters are $\beta=2$, $n=10$, in the range of $P_e=1\times10^{-8}$ to $P_e=1/2$, the quality factor of the scheme (6) is better than that of scheme (1).

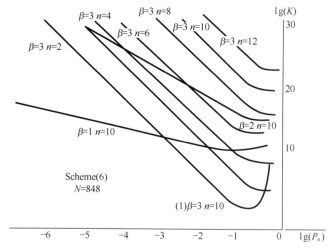

Figure 10-36 Synthentical parameter design of scheme (6)

Chapter 10　Anti Fading Frame Synchronization

Some descriptions about anti fading frame synchronization:

(1) In the case of error codes, the frame synchronization keeping time of the first frame synchronization scheme is generally shorter, and it doesn't have ability to protect from burst error codes mechanically. Hence, although this scheme is used all over the world, it is not a better scheme.

(2) Mechanically the second scheme (i.e. the basic anti fading frame synchronization scheme) has ability anti burst error codes, but the ability of recognizability of anti errors. Hence although it has anti fading ability, but its frame synchronization average keeping time is shorter.

(3) The rest four schemes (i.e. the synthetical schemes) have both the ability of anti burst error codes and the recognizably of anti errors, hence they have both the anti fading ability and the advantage that the frame synchronization average keeping time is longer.

(4) Tne scheme (3) can not use the first recognition point and the $\beta+1$ th recognition point (i.e. reset point); the scheme (4) can not use the $\beta+1$ th recognition point; the scheme (5) can not use the first recognition point; whereas the scheme (6) uses all the time resource usable. Hence the scheme has the highest quality factor; the schemes (5) and (4) have the same quality factor; the scheme (3) has a lower quality factor.

(5) There is a mechanical difference between a anti fading frame synchronization and the classic frame synchronization (CCITT recommended). The classic frame synchronization is based on monitoring if the original frame synchronization is lost, as long as the loss is decided the search is repeated, hence the virtual missing event caused by error code will affect, so it is related closely to error code phenomenon. The basic anti fading frame synchronization is based on detecting if a new frame synchronization pattern occurs, as long as it is decided the reset is carried out, hence the virtual alarm event caused by information code random combination will affect, so it is related closely to information code structure. A synthetical frame synchronization is based on detecting a new frame synchronization and monitoring the original frame synchronization, the new frame synchronization is carried out only when the new frame synchronization is decided and loses the original synchronization simultaneously, so it has not only the anti fading function but also better quality factor.

(6) The scheme (5) has both longer frame synchronization keeping time and shorter frame loss keeping time; hence the devices are made, which are used successfully.

Chapter 11　Engineering Application Design

11.1　GENERAL PLESIOCIIRONOUS GROUP MULTIPLEX DESIGN

The general plesiochronous digital multiplexes are the multiplex from 2048kbit/s to 8448kbit/s, the multiplex from 8448kbit/s to 34 368kbit/s and the multiplex from 34 368kbit/s to 139 26kbit/s. They all belong to n-$(n+1)$ digital multiplex series, and all use positive justification plesiochronous multiplex techniques.

11.1.1　Elementary Parameters Design

When using positive justification plesiochronous multiplex, the elementary parameters design methods are summarized in the following:

$$\left(\frac{1}{m} - \frac{f_1}{f_h}\right) L_s = K + S \qquad (11\text{-}1)$$

$$Q = \frac{L_s}{m} - K \qquad (11\text{-}2)$$

$$\Delta S = \left(\left|\frac{\Delta f_1}{f_1}\right| + \left|\frac{\Delta f_h}{f_h}\right|\right) Q \qquad (11\text{-}3)$$

$$A_{\text{jpp}} = F(S \pm \Delta S) \qquad (11\text{-}4)$$

Where, m—number of multiplex tributaries;

f_1, Δf_1—tributary normal rate and its tolerance;

f_h, Δf_h—multiplex normal rate and its tolerance;

L_s—frame length;

Q—number of information bits per tributary in a frame;

K—average number of non-information bits per tributary in a frame;

Chapter 11 Engineering Application Design

S, ΔS—normal justification rate and its range;

A_{jpp}—maximum peak-peak of stuffing jitter; the calculating value of $A_{jpp}=F(S)$ is shown in the table below and the corresponding application curves are shown in Figure 11-1.

A_{jpp}(UI%)	P	S
100.0	1	1
50.0	2	1/2
33.3	3	1/3, 2/3
25.0	4	1/4, 3/4
20.0	5	1/5, 2/5, 3/5, 4/5
16.7	6	1/6 5/6
14.3	7	1/7, 2/7, 3/7, 4/7, 5/7, 6/7
12.5	8	1/8, 3/8, 5/8, 7/8
11.1	9	1/9, 2/9, 4/9, 5/9, 7/9, 8/9
10.0	10	1/10, 3/10, 7/10, 9/10
9.1	11	1/11, 2/11, 3/11, 4/11, 5/11, 6/11, 7/11, 8/11, 9/11, 10/11
8.4	12	1/12, 5/12, 7/12, 11/12
7.7	13	1/13, 2/13, 3/13, 4/13, 5/13, 7/13, 8/13, 9/13, 10/13, 11/13, 12/13
7.1	14	1/14, 3/14, 5/14, 9/14, 11/14, 13/14
6.7	15	1/15, 2/15, 4/15, 7/15, 8/15, 11/15, 13/15, 14/15
6.2	16	1/16, 3/16, 5/16, 7/16, 9/16, 11/16, 13/16, 15/16
5.9	17	1/17, 2/17, 3/17, 4/17, 5/17, 6/17, 7/17, 8/17, 9/17, 10/17, 11/17, 12/17, 13/17, 14/17, 15/17, 16/17
5.5	18	1/18, 5/18, 7/18, 11/18, 13/18, 17/18
5.3	19	1/19, 2/19, 3/19, 4/19, 5/19, 6/19, 7/19, 8/19, 9/19, 10/19, 11/19, 12/19, 13/19, 14/19, 15/19, 16/19, 17/19, 18/19
5.0	20	1/20, 3/20, 7/20, 9/20, 11/20, 13/20, 17/20, 19/20

PDH for Telecommunication Network

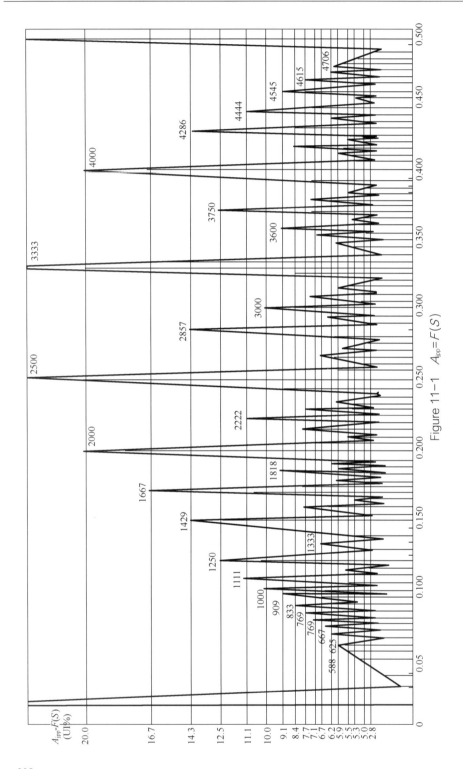

Figure 11−1 $A_{jpp}=F(S)$

Chapter 11 Engineering Application Design

Where, $m, f_l, \Delta f, f_h, \Delta f_h$ are given; $L_s, K, Q, S, \Delta S$ and A_{jpp} are designed; among the designed values, the frame length L_s is a variable, its range is discussed as following:

(1) L_{smin}

In addition to information code, the shortest frame consists of at least one frame alignment signal bit and three stuffing indication bits, i.e. $K_{min}=4$. Then

$$L_{min} = 4 / \left(\frac{1}{m} - \frac{f_l}{f_h} \right) \tag{11-5}$$

(2) L_{smax}

When the frame length is big enough, even satisfying the condition $L_s \gg mk$, the following approximate formula is established:

$$\Delta S \approx \left(\left| \frac{\Delta f_l}{f_l} \right| + \left| \frac{\Delta f_h}{f_h} \right| \right) \frac{L_s}{m} \tag{11-6}$$

We can see that ΔS is approximately proportional to L_s. When L_s increases to a certain extent, ΔS increases an extent correspondingly such that it is difficult to find A_{jpp} satisfying the requirement in the curve of $A_{jpp}=F(S)$. We can see from the curve of $A_{jpp}=F(S)$ that when $\Delta S \leqslant 0.02$, A_{jpp} may avoid the peak to find a proper valley range. Hence:

$$L_{smax} = 0.02m / \left(\left| \frac{\Delta f_l}{f_l} \right| + \left| \frac{\Delta f_h}{f_h} \right| \right) \tag{11-7}$$

For each proper L_s in the range of $[L_{smin}, L_{smax}]$, we can get a group designed values of $K, Q, S, \Delta S$ and A_{jpp}. If A_{jpp} is over the predictive or given value, L_s must be selected again until A_{jpp} satisfies the design requirement. Besides, the other factors should be considered in the design of the elementary parameters especially the channel utilization and the equipment economy should be considered. When L_s closes to its maximum L_{smax}, because the order-wire digits are few, the channel utilization is higher, which is normally called a high efficient scheme; whereas when L_s closes to its minimum L_{smin} and the timing sequence is easily generated, the number of elements is small, which is normally called an economical scheme. Obviously, the high efficient scheme needs more equipment, whereas the economical scheme has lower channel utilization. There is a problem to exchange the channel capacity and the number of equipments.

11.1.2 Transition Process Design

The general formula of the justification transition process is as below:

$$\Delta t_s = \Delta t_0 + \left[\frac{mf_1}{f_h}(x+g) - x \right] \cdot UI \qquad (11\text{-}8)$$

where, Δt_x—The read/write time difference when read out the xth information code from the justification buffer, its unit is the tributary time slot width (UI);

Δt_0—Δt_x when $x=0$, i.e. the initial read/write time difference;

m—The number of tributaries;

f_l—The nominal of the tributary clock frequency;

f_h—The nominal of the multiplex clock frequency;

g—The number of the entire beats stoped before read out the xth information code.

After the frame structure is determined, we can calculate the transition process curve. From the curve, we can get the maximum (Δt_{xmax}), the minimum (Δt_{xmin}) and the initial (Δt_0) of the read/write time difference as well as the width (d) of the steady justification decision area. Then we can determine the buffer size, the initial time difference setting and the control decision time wave design. The practices have proved that the design of the transition process is easily ignored, however it is very important.

11.1.3 Frame Loss Probability and Search Characteristics Design

When the channel average error rate $P_e \leqslant 1 \times 10^{-3}$, the following formulas may be used in the engineering design:

$$t_f = T_s / (nP_e)^\beta \qquad (11\text{-}9)$$

$$t_s = T_s / C_\eta^{d+1} \cdot P_e^{d+1} \qquad (11\text{-}10)$$

$$t_r = \left(\alpha + \beta - \frac{1}{2} \right) T_s \qquad (11\text{-}11)$$

where, t_f—average time interval between two consecutive synchronization frame losses;

T_s—frame period;

n—length of the frame alignment signal;

β—number of synchronization protection frames;

t_s—average time interval between two consecutive stuffing frame losses;

η—code length of the stuffing indication;

d—tolerance bit number of the stuffing indication code;

t_r—average frame loss time;

α—number of the search protection frames.

where the frame period (T_s) and the frame alignment signal length (n) are determined by the synchronization multiplex design; α and β normally are:

$$\alpha = \frac{8}{0.3n} \quad (11\text{-}12)$$

$$\beta = \frac{8}{3 - \lg n} \quad (11\text{-}13)$$

The justification indication normally uses simple maximum likelihood decision protection, the indication code length (η) normally takes 3-5 bits, its tolerance bit number (d) normally takes 1-2 bits.

11.1.4 Partition of the Basic Units

How an entire multiplexer is partitioned into basic units is also an important design problem. While partitioning basic units, usually the following aspects should be considered: (1) the wires should be as few as possible to connect conveniently; (2) the basic units should be justified independently and tested individually; (3) the basic units are easily standardized and generalized.

Now, a basic unit partitioning method used usually internationally is shown in Figure 11-2. Where, G is a justification plug-in unit, D is a recovery plug-in unit, T_1/T_2 is a synchronization multiplex plug-in unit, R_1/R_2 is a synchronization demultiplex plug-in unit; CL is a clock source unit; in addition, there are also a monitor and alarm plug-in unit and a multiplex clock and information code interface plug-in unit. For example, the products of France in middle of seventies were partitioned into 15 basic units; the products of Japan in the later of seventies were partitioned into 12 basic units. The difference is that T_1/T_2 and R_1/R_2 are in one plug-in unit or partitioned into two plug-in units respectively, which depend on the integration scale of the packages and the area of the plug-in boards. The advantages of such a basic unit partitioning method are obvious, the link wires between, plug-in units are few, and the signals SZ, SV, f_m are used commonly by all the tributaries; and all the plug-in units can be justified independently and tested individually; Especially the plug-in units G and D are easily standardized and used generally in many kinds of multiplexers. Such partitions are helpful in ASIC design.

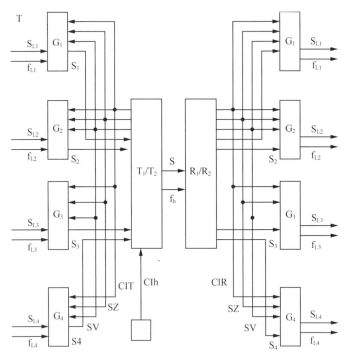

Figure 11-2 Chart of Standard multiplexers

11.1.5 Design Examples

The Figure 11-3 to Figure 11-6 are the principles of 2048/8448kbits/s multiplexers, the main time waves are shown in Figure 11-7 and 11-8. Such multiplexers accord with the specifications of the standard G.742 of CCITT; all of shown in the Figures are the practical, circuits at that time. It should be mentioned that some important improvements are made on number of parts of these circuits now, however the principles shown here still are valuable for reference in certain extent. Now, these prototype circuits are made into ASIC.

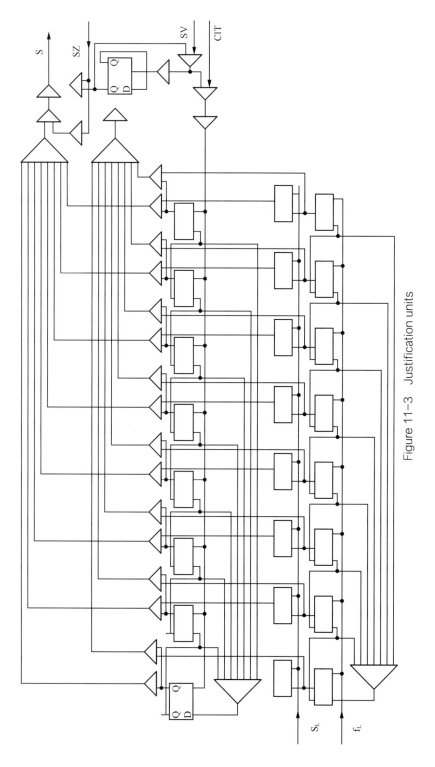

Figure 11-3　Justification units

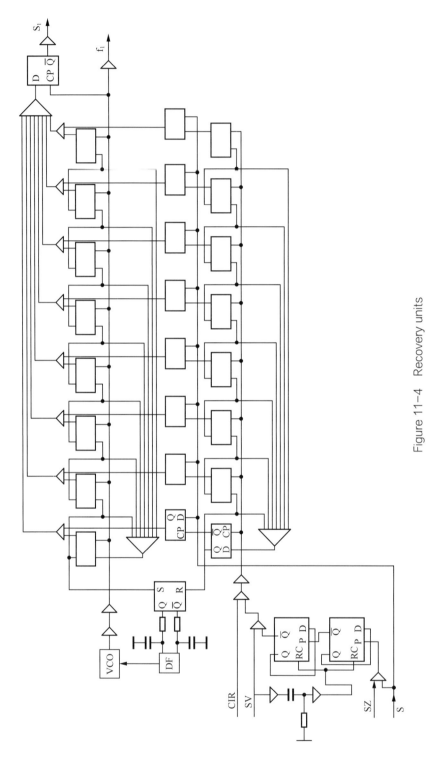

Figure 11-4　Recovery units

Chapter 11　Engineering Application Design

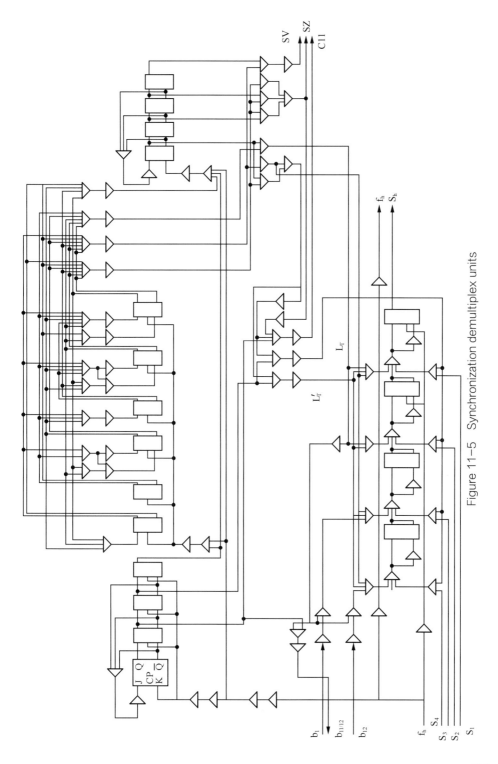

Figure 11–5　Synchronization demultiplex units

PDH for Telecommunication Network

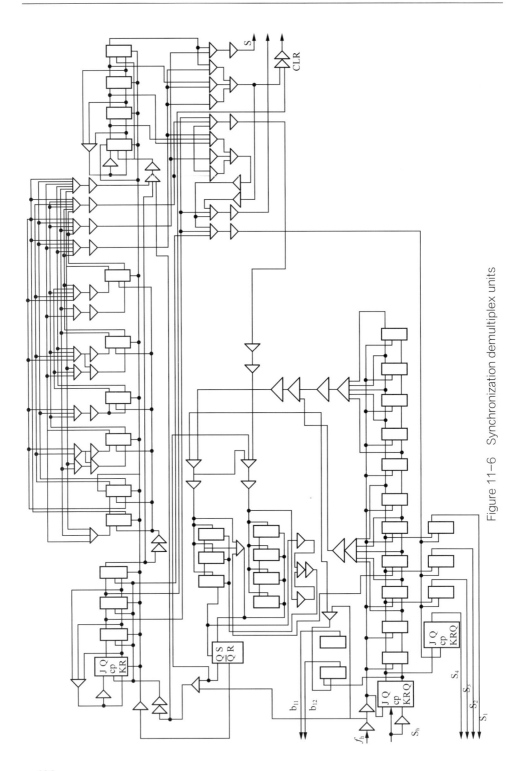

Figure 11-6 Synchronization demultiplex units

- 306 -

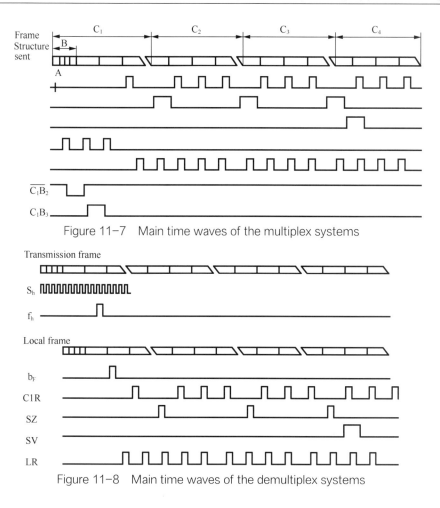

Figure 11-7 Main time waves of the multiplex systems

Figure 11-8 Main time waves of the demultiplex systems

11.2 STANDARD/NON-STANDARD RATES TOLERANCE DESIGN

11.2.1 Introduction

Recently, some engineering projects raise a series of problems about how the non-standard rate binary digits are transmitted through the standard rate channels. For example, how the digital television binary digits (32 063.989kbit/s) are transmitted through the standard 3rd group (34 368kbit/s); how the digital video phone binary digits (8012.8kbit/s) are transmitted through the standard 2nd group (8448kbit/s); how the digital cloud chart binary digits (1330.8kbit/s) are transmitted through the standard primary group (2048kbit/s) and etc. Documents prove that the plesiochronous multiplex techniques using positive justification may solve well such problem that the non-standard streams are transmitted through the standard channels; especially it is more suitable if the non-standard binary digits have an independent clock and the rates are relative high.

11.2.2 Transmission of Non-Standard Binary Digits through Standard Channels

Non-standard binary digits may be considered as a tributary multiplexed, whereas the standard channel is considered as the multiplex path. Then the transmission of the non-standard binary digits through the standard channels may be considered as the plesiochronous multiplex when the multiplex tributary number $m=1$. That is a single channel plesiochronous multiplex. Replace $m=1$ into the basic formula of the justification, the following formula is derived:

$$\left(1-\frac{f_1}{f_h}\right)L_s = K + S \tag{11-14}$$

where, f_1—Non-standard binary digit rate, i.e. the tributary rate;

f_h—Standard channel capacity, i.e. multiplex rate;

L_s—Frame length (see Figure 11-9);

K—The number of non information bits in a frame;

S—Standard justification ratio.

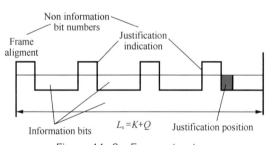

Figure 11-9　Frame structure

$$\Delta S \approx \left(\left|\frac{\Delta f_h}{f_h}\right| + \left|\frac{\Delta f_1}{f_1}\right|\right)Q \tag{11-15}$$

where, Q—The number of information bits in a frame;

$\dfrac{\Delta f_h}{f_h}$—Multiplex rate relative tolerance;

Chapter 11 Engineering Application Design

$\dfrac{\Delta f_1}{f_1}$ —Non-standard binary digit rate relative tolerance;

ΔS—The range of justification ratio:

$$\dfrac{4}{1-\dfrac{f_1}{f_h}} \leqslant L_s \leqslant \dfrac{0.02}{\left|\dfrac{\Delta f_1}{f_1}\right|+\left|\dfrac{\Delta f_h}{f_h}\right|} \qquad (11\text{-}16)$$

In the above formulas, f_l, f_h, $\dfrac{\Delta f_1}{f_1}$, $\dfrac{\Delta f_h}{f_h}$ are given; L_s is a variable in the design. Whenever L_s takes a value from its range, a group of $K+S$ is calculated, where the integer part is K and the fraction part is S. Then ΔS is gotten. According to $S\pm\Delta S$, the stuffing jitter A_{jpp} can be gotten from $A_{jpp}=F(S)$ curve. If A_{jpp} satisfies the requirement, then a group of parameters L_s, K, Q, is gotten, or else L_s takes a new value again until find a satisfied design.

11.2.3 Design Examples

The binary digit rate of the cloud chart of the meteorological satellite is 1330.8kbit/s $\pm 1\times 10^{-4}$ which is transmitted through the quarter of the primary group 2112kbit/s$\pm 1\times 10^{-4}$:

$$f_l = 1330.8\text{kbit/s} \pm 1\times 10^{-4}$$
$$f_h = 2112\text{kbie/s} \pm 1\times 10^{-4}$$
$$\begin{cases} 0.369886 L_s = K + S \\ Q = L_s - K \\ \Delta S = 2\times 10^{-4} \cdot Q \end{cases}$$
$$2.5 < L_s \leqslant 100$$

The design results are as below:

	The simplest Scheme	Common sequence with multiframe	The high efficient Scheme
L_s	12	16	99
K	4	5	36
Q	8	11	63
S	0.438 636	0.918 182	0.618 714
ΔS	0.001 600	0.002 200	0.012 600
A_{jpp}	6.2UI%	8.4UI%	12.5UI%

The simplest scheme uses the smallest amount of equipment but has the lowest efficiency; the high efficient scheme has the highest channel utilization but uses the largest amount of equipment. the other factors should be also considered to decide a scheme. Using the scheme of common sequence with multiframe, the frame structure and the read/write time difference (Δt_x) are calculated as following (Figure 10-10):

$$\Delta t_x = [0.63(x+g) - x] \cdot UI$$

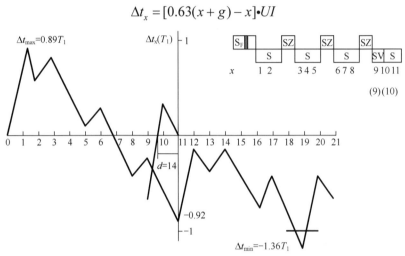

Figure 11-10 Justification transition process of meteorological data transmission

The design results are:

$$L_s = 16$$
$$Q = 11$$
$$S = 0.9182$$
$$A_{jpp} = 8.4\% UI$$
$$\Delta t_{min} = -1.36 UI$$
$$\Delta t_{max} = +0.89 UI$$
$$d = 1.4$$

11.2.4 Compatible Multiplex of Different Tributary Rates

In the previous section the tributary rate is non-standard but the multiplex rate is standard, which is solved completely by using the general design method of positive justification. In this section, another compatible problem of standard/non-standard rates will be discussed. In a general or a standard multiplexer, the nominal of all the multiplex tributaries are the same. Here the problem is that the multiplex tributary rates are different such that their rates may be either standard or non-standard. For example, require several 2048kbit/s tributaries and several 8448kbit/s tributaries are multiplexed simultaneously into a 34 368kbit/s channel (see Figure 11-11). Such a problem may also be solved by using the general design method of positive justification. Now it is described using the particular example in Figure 11-11.

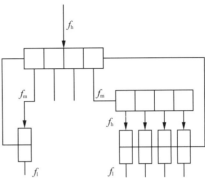

Figure 11-11 2048/8448/34 368kbit/s compatible design

According to the specifications of CCITT recommendation G.751, the particular parameters of the 8448/34 368kbit/s multiplexer are specified as bellow:

$$f_h = 34\ 368\text{kHz} \pm 20\text{ppm}$$
$$f_l = 8448\text{kHz} \pm 30\text{ppm}$$
$$m = 4$$
$$L_s = 1536$$
$$Q = 378$$
$$f_m = Qf_h/L_s = 845\ 770\text{Hz}$$

Let $f_h' = f_m/4$ be multiplex channel capacity and $f_1' = 2048\text{kbit/s}$ be tributary rate, calculate similar with the previous section according to the general formulas:

$$f_h' = 2\ 114\ 437.5\text{Hz}$$
$$f_1' = 2\ 048\ 000\text{Hz} \pm 50\text{ppm}$$
$$m' = 1$$
$$\left(1 - \frac{f_1'}{f_h'}\right)L_s' = K' + S'$$

To simplify the equipment, the f_h' frame and the f_h frame should have a simple multiple relationship and also satisfy the requirement of the lowest number of service digits. Here are:

$$T_s' = 2Tf_s$$
$$L_s' = 2 \cdot \frac{Q}{m'} = 189$$
$$K' + S' = 5.040\ 08$$
$$K' = 5$$
$$S' = 0.940\ 08$$

The f_h' frame structure is arranged particularly as shown in Figure 11-12. In the f_h frame. SZ of the fourth 8Mbit/s tributary is used to transmit the frame alignment signal of the f_1' tributary, then f_m may be used as

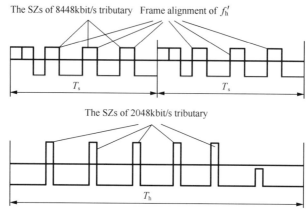

Figure 11-12 f_h' frame structure design

f_h' after demultiplex without any additional control. The f_h' from which SZ' is deducted is just the information code of the f_1' tributary. Hence the plug-in units of the general justification and recovery may be used, and all the control signals SZ', SV' and f_m may be generated from the f_h sequence source and the equipments are quite simple.

Get the stuffing jitter of f_1' in the following:

$$Q' = \frac{L_s'}{m'} - K' = 184$$

$$\Delta S' = \left(\left| \frac{\Delta f_1'}{f_1'} \right| + \left| \frac{\Delta f_h'}{f_h'} \right| \right) Q' = 0.013$$

$$\therefore S_{max}' = 0.953\ 08$$
$$S_{min}' = 0.927\ 08$$

From the curve $A_{jpp} = F(S \pm \Delta S)$ (See Figure 11-1) get:

$$A_{jpp}' = 7.1\% \text{UI}$$

Obviously, Using general design method of positive justification plesiochronous multiplex not only solves the problem of the non-standard rates of binary digits transmitted through the standard rate channels but also solves the problem of plesiochronous multiplex of different tributary rates in a digital channel simultaneously. Similarly for the later, it is feasible that the tributary rate is either standard or non-standard.

11.3 PLESIOCHRONOUS/SYNCHRONOUS COMPATIBLE DESIGN

11.3.1 Introduction

In the world wide engineering practice, the generality of the plesiochronous multiplex is blameless, but in some specific cases, the synchronous multiplex really has some benefits. From the point of view of engineering, we hope the selected equipment suits its particular conditions as much as possible; but from the point of view of manufacturing, we certainly hope our products could suit various possible conditions. Hence from the point of view of research, it is meaningful project that let the plesiochronous multiplex and the synchronous multiplex compatible whose cost is accepted by engineering.

From the relationship of the tributary binary digits and multiplex binary digits, no matter they are synchronous or not, all can be multiplexed and demultiplexed in a plesiochronous multiplexer, and correspondingly all generate stuffing jitters and stuffing errors. Here "plesiochronous/synchronous multiplex compatible" means that such a multiplexer either implements plesiochronous multiplex for the asynchronous binary digits or implements synchronous multiplex for the synchronous binary digits. Particularly speaking, when implementing such synchronous multiplex/demultiplex, the tributary binary digits will not generate the stuffing jitters, or neither stuffing jitters nor stuffing errors.

In this Chapter, we will modify a little bit of CCITT plesiochronous multiplexer to realize synchronous multiplex compatible. two particular compatible methods will be discussed.

11.3.2 Instruction Justification Compatible method

Known from the Chapter of the positive justification principles, if the justification ratio (S) strictly obeys the formula:

$$S=\frac{q}{p}, (p,q)=1 \qquad (11\text{-}17)$$

where p,q are non-zero positive real numbers of relative prime with each other. Then, after multiplex/demultiplex the tributary may be believed no stuffing jitters, i.e.

$$A_{\text{jpp}} \approx 0 \qquad (11\text{-}18)$$

One of the methods to insure the above formula established is that the tributary clock (f_1) is generated from the multiplex clock (f_h), and it accords with the following relationship:

$$f_1 = \frac{Q-S}{L_s} \cdot f_h \quad (11\text{-}19)$$

where, the frame length (L_s), the tributary information code number (Q), the tributary clock and its tolerance ($f_1 \pm \Delta f_1$), multiplex clock and its tolerance ($f_h \pm \Delta f_h$) are all particularly specified in the corresponding CCITT recommendations. Here these constraints must be satisfied when f_h generates f_1. Particularly speaking, S should satisfy the following formula:

$$\Delta f_1' = \frac{Q-S}{L_s} \cdot \Delta f_h \leq \Delta f_1 \quad (11\text{-}20)$$

The corresponding values of S is shown in the Table 11-1.

Table 11-1 Table of S values

Recommendation	S	f'_{max}(Hz)	f'_{min}(Hz)
G.742(2/8Mbit/s)	8/19	2 048 093	2 047 970
	3/7	2 048 018	2 017 895
G.751(8/34 Mbit/s)	7/16	8 448 129	8 447 788
	3/7	8 448 328	8 447 987
G.751(34/140 Mbit/s)	5/12	34 368 628	34 367 597
	8/19	34 368 419	34 367 388

The logic diagram of generating tributary clock from the multiplex clock (f_h) is shown in Figure 11-13. In consecutive p frames, deduct the clock beats at the stuffing position (SV) of q frames; then smooth it through the phase-lock loop. Pay attention that S should select p as small as possible to simplify equipment.

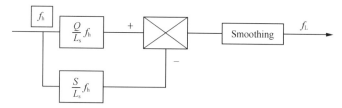

Figure 11-13 f_1 generation logic diagram

The principle of generating particular tributary clock is shown in Figure 11-14. The original synchronous multiplex sequence already has the corresponding frequency: the tributary synchronous multiplex clock $f_m = \dfrac{Q}{L_s} \cdot f_h$; the stuffing position pulse frequency is f_h/L_s. As long as it is specified that q frames are justified regularly in p frames, the tributary clock (f_1) can be generated strictly.

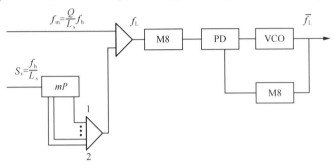

Figure 11-14 f_1 generation principle diagram

A complete plesiochronous/synchronous compatible multiplex diagram is shown in Figure 11-15. The tributary which generates tributary clock (f_1) independently is in plesiochronous state. The tributary clock (f_1') generated from this multiplexer is in synchronous state. Compare this synchronous state with the normal synchronous state, it has the ability to justify the read write time difference automatically. which is the advantage that the normal synchronous state do not have. However. a relevant shortage is that there exist still stuffing errors in this kind of synchronous state because it is the same as plesiochronous multiplex that the receive end implements the recovery operation according to the instructions. Besides, we can see from Figure 11-15 that such as compatible scheme only adds a simple additional plug-in unit which generates f_1' from f_h, however the original plesiochronous multiplex equipment has no change.

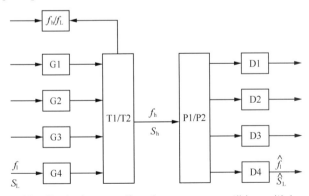

Figure 11-15 Plesiochronous/Synchronous compatible multiplex diagram

11.3.3 Multiframe Fixed Control Compatible Method

The tributary clock is generated from the multiplex clock in a fixed control method that deducts the clock beats at q SVs from the consecutive p frames, in this way, the synchronous multiplex may also be implemented. Particularly, p frames constitute a multiframe, using SZ of the corresponding synchronous multiplex tributary to transmit the multiframe alignment signal, whereas the corresponding SZ of the plesiochronous tributary transmits the justification instruction as normal. Then the plesiochronous multiplex and the synchronous multiplex may be compatible in the same multiplexer.

There are normally three stuffing slots corresponding to a multiplex tributary in a primary frame. Hence if the instructions are transmitted as the same as the plesiochronous multiplex using three slots to transmit one bit of multiframe alignment signal, the error probability of the signal is:

$$P_s = \sum_{x=2}^{n} C_n^x (1-P_e)^{n-x} \cdot P_e^x \approx 3P_e^2 \quad (11\text{-}21)$$

A multiframe consists of p primary frames. If the length of the multiframe alignment signal is p where q bits are "1", then the pattern of the multiframe alignment signal and the control signals deducting from the SV beats in the primary frames are unified. Thus at the sending end, the multiframe alignment signals are used to forcedly control the SV whether the information codes are transmitted, whereas at the receiving end the refined multiframe alignment signals are correspondingly used to forcedly control the SVs to recover the original tributary binary digits. If the multiframe alignment signal pattern is designed properly, the code distance between various shift patterns can be increased. Suppose the tolerance bit number is D, the multiframe loss probability is:

$$P_1 = \sum_{y=D+1}^{p} C_p^y (1-P_s)^{p-y} \cdot P_s \quad (11\text{-}22)$$

where, $P_s \approx 3P_e^2$. The multiframe loss probability is obviously far lower than the primary frame loss probability.

For example, In the special cases of G.742 (2/8Mbit/s) and G.751 (8/34Mbit/s), let $p=7$, $q=3$, i.e. $s=3/7$, and the frame alignment pattern is a pseudo-random pattern

(0001101) of the length p, then the search and refinement circuits of the multiframe alignment signal can be quite simple (See Figure 11-16).

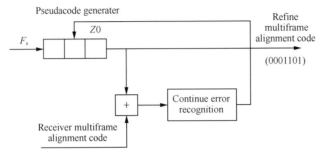

Figure 11-16   Multiframe alignment search and refine diagram

The diagram of plesiochronous/synchronous compatible multiplexer fixed controlled by multiframe is shown in Figure 11-17. Compare with normal plesiochronous multiplex equipment, such a compatible equipment has two more plug-in units (T and R), and the justification and the recovery plug-in units have one more control signal line (KZ) respectively. The plug-in units T, R and the corresponding KZ control relationship, which are used in the compatible design of the CCITT recommendation G.742 (2/8Mbit/s), are shown in Figure 11-18 and Figure 11-19.

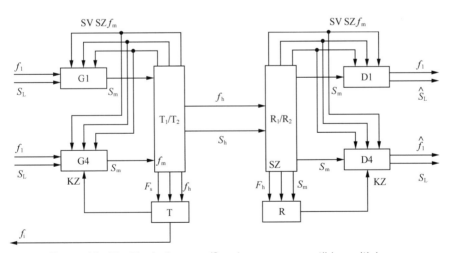

Figure 11-17   Plesiochronous/Synchronous compatible multiplexer

PDH for Telecommunication Network

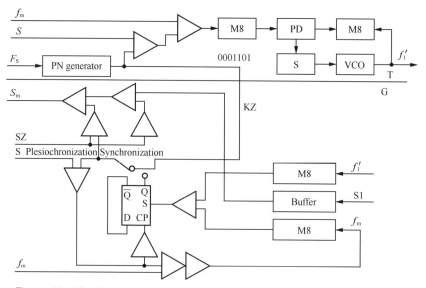

Figure 11-18　Plug-in unit T and its control relationship with plug-in unit G

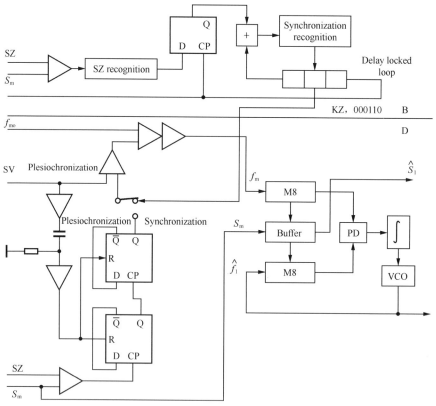

Figure 11-19　Plug-in unit R and its control relationship with plug-in unit D

Chapter 11 Engineering Application Design

Figure 11-18 gives the principle diagram of the plug-in unit T and the internal control relationship of the justification plug-in units. The frame frequency signal Fs drives the pseudo-random generator whose output is also used as multiframe alignment signal of the forcedly control signal (KZ). The combination of the signals KZ, SV and f_m generates synchronous tributary clock (f_1'). When a tributary is in synchronous multiplex state, the working state selection switch of the justification plug-in unit (G) is set in synchronous state, then KZ directly controls the read clock (f_m) of the buffer, the content of the corresponding KZ is added into the tributary binary digits simultaneously and is transmitted to the corresponding tributary at the receiving end.

Figure 11-19 gives the principle diagram of the plug-in unit R and the internal control relationship of the recovery plug-in units. The information code (S_m) of the synchronous multiplex tributary and SZ signal are ANDed to extract the content of stuffing position, which passes through the SZ recognition circuit that is the same with the plug-in unit D to get corrected multiframe alignment signal, then after sampled by the frame frequency signals, it enters the delay locked loop to refine further. Through frame frequency signal sample, the multiframe alignment signal pattern will be delayed one frame time. Hence the output signal (KZ) should be taken out before the delay lock loop to compensate the one frame delay. When the tributary synchronous multiplex is working, the working state selection switch of the recovery plug-in unit (D) should be set in synchronous working state. At this time, the signal KZ directly selects signal SV according to the pattern 0001101, thereby the recovery control to tributary clock f_m is completed.

We see from the above description that such a compatible scheme need two more plug-in units (T and D), but these two plug-in units are relative simple and all are digital circuits. Besides the justification plug-in unit and recovery plug-in unit need only an additional switch to be compatible for two kinds of working states. Because such a compatible scheme uses multiframe structure and fixed control, even in the state of multiframe loss at the receiving end the pseudo-code generator is still operating, the recovery implements the recovery control according to the ratio of $S = \dfrac{q}{p}$. Hence, there is no stuffing frame loss in this kind of scheme. It is the same with the previous scheme that because the tributary clock is extracted from the clock (f_1') internally generated in the multiplexer, hence the forcedly control ratio always keeps $S = \dfrac{q}{p}$, thus no stuffing jitter will occur.

11.4 2/34Mbit/s MULTIPLEX DESIGN

This section will take 2/34Mbit/s multiplex as example to discuss particularly the design method of the $n-(n+2)$ digital multiplex equipment.

11.4.1 2/34Mbit/s Multiplex Scheme Comparison

Several typical schemes about the 2/34Mbit/s multiplex are compared bellow:

(1) Standard Multiplexer Combinational Scheme(Figure 11-20)

This kind of multiplex scheme consists of one 8448/34368kbit/s multiplexer according with the recommendation G.751 and four 2048/8448kbit/s multiplexers according with the recommendation G.742. Each multiplexer has 12 standard plug-in units, the entire multiplex system has 16 standard plug-in units.

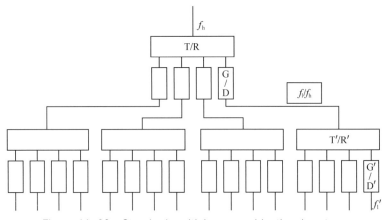

Figure 11-20 Standard multiplexer combinational system

The overall stuffing frame loss probability ($P_{s\Sigma}$) and the tributary overall stuffing jitter ($A_{jpp\Sigma}$) of this kind of combinational scheme are respectively:

$$P_{s\Sigma} \approx P_s + P_s'$$

$$A_{jpp\Sigma} = \dfrac{1}{m} A_{jpp} + A_{jpp}$$

where, P_s, A_{jpp} — Frame loss probability and stuffing jitter of 8448/34 368kbit/s multiplexer;

P'_s, A'_{jpp} —Frame loss probability and stuffing jitter of 2048/8448kbit/s multiplexer.

The advantages of this kind of multiplex scheme are standardization, good generality, but the equipments are more complicated, the multiplex impair is more serious.

(2) Combinational Scheme of 8448/34 368kbit/s Multiplex Using Instructions to Control Justification Synchronization (Figure 11-21)

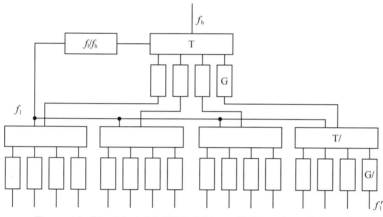

Figure 11-21   Model A 2/34Mbit/s multiplexer (sending end)

From Figure 11-21, compared with the previous combinational scheme, this kind of scheme only has one more plug-in unit of the tributary clock f_1 generated by f_h clock. Here, the tributary clock (f_1) of the high order group is just the multiplex clock (f'_h) of the lower order groups. It is completely consistent with the first plesiochronous/synchronous compatible scheme introduced in Chapter 8, at this time the high order group multiplexer is in synchronous working state. Then,

$$A_{jpp}=0$$

The overall stuffing frame loss probability and the stuffing jitter of this kind of combinational scheme are respectively:

$$P_{s\Sigma} = P_s + P'_s$$

$$A_{jpp\Sigma} = A'_{jpp}$$

Compared with the scheme (1), this kind of combinational scheme has one more circuit unit and eliminates the stuffing jitter of the high order group. This kind of combinational scheme is called model A 2/34Mbit/s multiplexer. Its complete configuration is shown in Figure 11-22 and Figure 11-23.

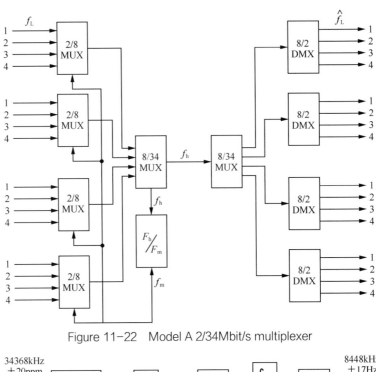

Figure 11-22   Model A 2/34Mbit/s multiplexer

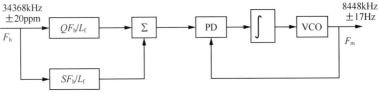

Figure 11-23 F_l/F_m chart

(3) Single Frame Structure Direct Multiplex Scheme (Figure 11-24)

This kind of multiplex scheme directly multiplexes 16 2048kbit/s tributary binary digits into 34 368kbit/s binary digits. If only the 2048kbit/s tributary plesiochronous multiplex is considered as the two previous schemes, it has 36 circuit units altogether. The stuffing frame loss probability and the stuffing jitter are respectively:

$$P_{s\Sigma} = P_s + P'_s$$

$$A_{jpp\Sigma} = A'_{jpp}$$

Obviously, the advantages of this kind of scheme are that the equipments are simple and the multiplex impair is smaller but the generality is poor. This kind of scheme is called model B 2/34Mbit/s multiplexer.

We will discuss the design of this kind of multiplexers detailed in the following:

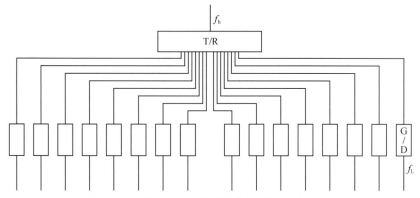

Figure 11-24　Model B 2/34Mbit/s multiplexer

11.4.2　Model B 2/34Mbit/s Multiplexer Design

(1) The Elementary Parameter Design

Given: f_h=34 368kbit/s±20ppm; f_1=2048kbit/s±50ppm; m=16, calculate according to the general formulas of the elementary parameters:

$$\left(\frac{1}{m}-\frac{f_1}{f_h}\right)L_s = K+S$$

$$L_{max} = 0.02m \left/ \left(\left|\frac{\Delta f_1}{f_1}\right|+\left|\frac{\Delta f_h}{f_h}\right|\right)\right.$$

$$L_{min} = 4 \left/ \left(\frac{1}{m}-\frac{f_1}{f_h}\right)\right.$$

$$2.909\ 684\times10^{-3}L_s = K+S$$

$$1374 \leqslant L_s \leqslant 4571$$

When the frame length is consistent with the frame length of the 8/34Mbit/s multiplexer in G.751,

$$L_s=1536$$
$$K+S=4.469\ 27$$
$$K=4$$
$$S=0.469\ 27$$

According to the variation range formulas of the tributary bit number and justification ratio:

$$Q = \frac{L_s}{m} - K$$

$$\Delta S \approx \left(\left|\frac{\Delta f_1}{f_1}\right| + \left|\frac{\Delta f_h}{f_h}\right| \right) Q$$

get:

$Q=92$
$\Delta S=0.006\ 44$
$S_{min}=0.462\ 83$
$S_{max}=0.475\ 71$

From Figure 11-1 $A_{jpp}=F\ (S\pm\Delta S)$ get:
$A_{jpp}=6.7\%UI$

(2) Frame Structure Design

Partition the primary frame into four subframes: primary cycle unit is 16 bits, the bits 1-10 of the first cycle of the first subframe are the frame alignment signal (1111010000), bit 11 is used for alarm of the other end, bits 12-16 are reserved; the stuffing indication uses 3 bits code which is placed in the first cycle of the 2,3,4 subframes; the stuffing position is placed in the second cycle of the fourth subframe. The detailed frame structure is shown in Table 11-2.

Table 11-2 2/34Mbit/s Frame Structure

Multiplex bit rate (kbit/s)		34 368
Tributary bit rate (kbit/s)		2048
Number of tributaries		16
Frame structure	Group number	Bit number
Frame alignment (1111010000)	I	1-10
Alarm indication		11
Reserved bits		12-16
Tributary bits		17-384
Justification control bits	II, III	1-16
Tributary bits		17-384
Justification bits	IV	1-16
Frame length		17-32
Tributary bits		33-384

续表

Frame length	1536
Number of bits per tributary	92
Maximum tributary justification rate(kbit/s)	22.375
Nominal justification rate	0.469

(3) Justification Transition Process

According to the general formula of the justification transition process, the read/write time difference when read out the xth tributary information code are:

$$\Delta t_x = \Delta t_0 + \left[\frac{mf_1}{f_h}(x+g) - x\right] \cdot UI$$

When $\Delta t_0 = 0$,

$$\Delta t_x = [0.953\,445(x+g) - x] \cdot UI$$

$$x = \frac{3}{4}Q + 1$$

$$g = 5$$

$$\Delta t_{max} = 1.51 \cdot UI$$

$$x = Q$$

$$g = 4$$

$$\Delta t_{min} = -0.47 \cdot UI$$

$$d = S \bigg/ \left(1 - \frac{f_1}{f_h}\right) \approx 10$$

The calculation curve is shown in Figure 11-25. Obviously, Δt_0 can be either $1 \cdot UI$ or $2 \cdot UI$.

(4) Jitter Distribution

According to the general calculation formulas:

$$\frac{S - S_0}{Q} \approx \frac{\Delta f_h}{f_h} - \frac{\Delta f_1}{f_1}$$

$$S \approx \left(\frac{\Delta f_h}{f_{h0}} - \frac{\Delta f_1}{f_{10}}\right)Q + S_0$$

Based on the given nominals f_{h0}, f_{10}, S_0 and the design parameter Q, only one group of offset Δf_h and Δf_l is selected, the corresponding S can be gotten. then according to Figure 11-1 curve $A_{jpp}=F(S)$ the corresponding stuffing jitter is gotten. The calculation result is shown in Figure 11-26. Obviously the maximum of stuffing jitter is 6.7%UI, the next peak is 5.9%UI, 5.3%UI.

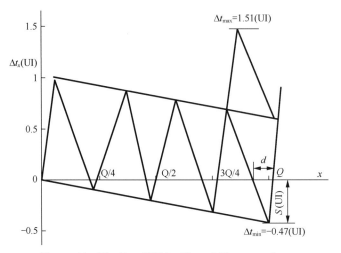

Figure 11-25　Read/Write Time Difference Curve

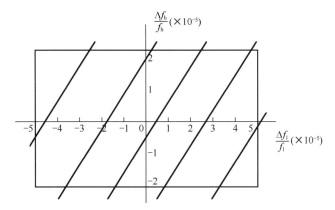

Figure 11-26　Model B 2/34Mbit/s Multiplexer Jitter Distribution

(5) Synchronous Compatible Design

This design uses instruction control justification synchronous compatible techniques introduced in Chapter 8. The configuration of a complete model B 2/34Mbit/s multiplexer is shown in Figure 11-27. Compare with a pure plesiochronous multiplex scheme, two additional circuits are added. One of

Chapter 11 Engineering Application Design

the plug-in unit is used to generate multiplex clock (f'_h) from the 2048kbit/s clock (f_1):

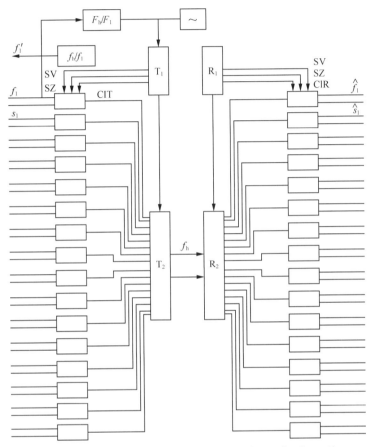

Figure 11-27 Model B 2/34Mbit/s Multiplexer Configuration

$$f'_h = \frac{L_s}{Q-S} \cdot f_1$$

The another is used to generate multiplex clock (f'_1) from the 34368kbit/s clock (f_h):

$$f'_1 = \frac{Q-S}{L_s} f_h$$

where, Q =02. b, L_s=1536. b. Let S=8/17, the tolerance of the external clock is not over 20ppm, then:

$$f_h' = 34\,368\,528\text{Hz}$$

$$\frac{\Delta f_h'}{f_h'} = \frac{\Delta f_1}{f_1} = 20 \times 10^{-6}$$

$$f_1' = 2\,047\,968\text{Hz}$$

$$\frac{\Delta f_1'}{f_1'} = \frac{\Delta f_h}{f_h} = 20 \times 10^{-6}$$

The logical diagram of generating f_h' and f_1' are shown in Figure 11-28 and Figure 11-29 respectively.

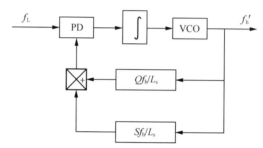

Figure 11-28 f_h'/f_1 Plug-in Unit Logical Diagram

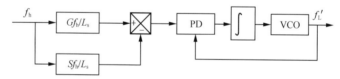

Figure 11-29 f_1'/f_h Plug-in Unit Logical Diagram

(6) Working States

The model B 2/34Mbit/s multiplexer has four working states (Table 11-3):

(a) Internal clock plesiochronous working state: a tributary and the multiplexer have their own independent clocks, f_1=2048kHz±50ppm, f_h=34 368kHz±20ppm;

(b) External clock plesiochronous working state: a tributary has an independent clock, f_1=2048kHz±50ppm, the standard clock of the network node provides the clock 2048kHz±20ppm for the multiplexer, the plug-in unit f_h'/f_1 generates multiplex clock f_h' =34 368kHz±20ppm;

(c) Internal clock synchronous working state: the standard clock of the network node provides the clock for the multiplexer and the tributaries: f_1' =2048kHz±20ppm, f_h' =34 368 528kHz±20ppm.

Table 11-3　　　Model B 2/34Mbit/s Multiplexer Working States

Working state		f_l (Hz)±$\Delta f/f$	f_h (Hz)±$\Delta f/f$	A_{jpp}	Number of plug-in units
Plesiochronous	Internal clock	2 048 000±50ppm	34 368 000±20ppm	6.7%UI	37
	External clock	2 048 000±20ppm	34 368 528+20ppm		
Synchronous	Internal clock	2 047 968±20ppm	30 368 000±20ppm	0	39
	External clock	2048000±20ppm	34 368 528±20ppm		

11.4.3　CCITT Recommendations

Now CCITT only gave one recommendation about the n- (n+2) asynchronous digital multiplex series, i.e. the digital multiplexer, which works at 139 264kbit/s and multiplexes sixteen 8448kbit/s tributaries, of the recommendation G.751.

The arrangement of such 8448/139 264kbit/s multiplex frame requires still keeping the 8448/34 368kbit/s frame structure and the 34 368/13 9264kbit/s frame structure. That is the 139 264kbit/s frame structure which is formed from sixteen 8448kbit/s is completely the same with that four 8448kbit/s form a 34 368kbit/s at first, then four 34 368kbit/s form a 139 264kbit/s frame structure.

The digital interfaces of 8448kbit/s and 139 264kbit/s should accord with the requirement of the recommendation G.730.

The jitter transmit characteristics require that the. jitter gain under 100Hz should not higher than 0.5dB, above 100Hz the attenuation is not less than 20dB per 10 times frequency. The peak-peak of the tributary output jitter without input jitter may not bigger than 0.35UI under 40kHz. The peak-peak with the probability 99.9% in 10 seconds may not beyond 0.05UI in the frequency band of 31kHz—400kHz. The peak-peak of the multiplex signal jitter may not beyond 0.05UI in the frequency band of 100kHz—3500Hz.

11.5　DESIGN OF BRANCH IN GROUP TRUNK TRANSMISSION

11.5.1　Introduction

The high order group trunk transmission often need group branches for inserting and extracting speech channels in the middle way. For example,

8448kbit/s branch in 34 368kbit/s trunk, 2048kbit/s branch in 8448bit/s trunk, 2048bit/s branch in 34 368kbit/s trunk, and etc. This is a general problem. CCITT calls this kind of branches an external access. This book calls this subject the branch problem in the group trunk transmission.

Using the standard multiplex CCITT recommended, such a branch, function should be provided. For example, to provide a 2048kbit/s branch in the 34 368kbit/s trunk, use two 8448/34 368kbit/s multiplexers according to the recommendation G.751 and two 2048/8448kbit/s multiplexers according to G.742, which are connected as shown in Figure 11-30.

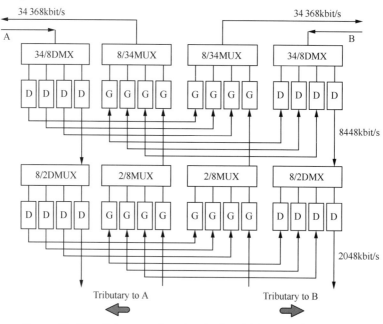

Figure 11-30 Scheme using standard multiplexer to implement 2048kbit/s branch in 34 368kbit/s trunk

Obviously from Figure 11-30 using such kind of scheme, all the trunk groups will be impaired. For example, 8448kbit/s trunk binary digits introduces a stuffing jitter of $A_{jpp}=14.3\%UI$ and 2048kbit/trunk binary digits introduces a stuffing jitter of $A_{jpp}=17.8\%UI$. If the main equipment of a standard multiplexer be an equipment of 12 standard circuits, there are 48 standard circuits altogether.

In a group trunk, especially in a dedicated trunk, there are often many branches each of them introduces such amount of stuffing jitters and need such number of equipments. This is not satisfied in both technique and economy. Hence it is necessary to find a proper trunk branch technique to reduce the impair as much as possible and to simplify the equipment as much as possible.

11.5.2 Trunk Branch Simplified Scheme One

If the entire 34 368kbit/s transmission system uses the multiplexer model B 2/34, then using two multiplexers of model B 2/34 may accomplish the trunk branch function. See Figure 11-31, its trunk stuffing jitter is A_{jpp}=6.7%UI which is 38% of the standard scheme, the basic equipments are 78 circuit units which are 162.5% of the standard equipment scheme. Obviously, this kind of scheme reduces the stuffing jitter but increases the number of units.

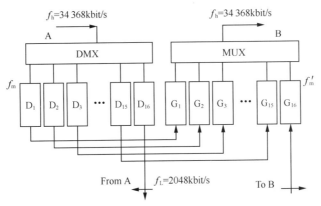

Figure 11-31 34 368kbit/s trunk with 2048kbit/s branch (scheme one)

11.5.3 Trunk Branch Simplified Scheme Two

Let us analyze the working process of Figure 11-31 at first: (a) 34 368kbit/s input binary digits is resolved into synchronous tributary binary digits (f_m) by synchronous demultiplexer (DMX); (b) through the recovery circuit unit (D), the synchronous tributary binary digits (f_m) are recovered into plesiochronous tributary binary digits (f_l); (c) through the justification circuit unit (G), the plesiochronous tributary binary digits (f_l) are justified into synchronous tributary binary digits (f_m); (d) through the synchronous multiplexer (MUX), the synchronous tributary binary digits (f_m) are multiplexed into 34 368kbit/s output binary digits.

Known from the principle of justification plesiochronous multiplex, in the structure of Figure 11-31, if the MUX frame sequence is locked by the DMX frame sequence, a relay tributary may omit the operations (b) and (c), i.e. after synchronous demultiplex, the synchronous tributary binary digits may be synchronously multiplexed. Such relay tributary no longer need the G unit and the D unit; of course, no stuffing jitter will be introduced. Figure 11-32 shows this kind of trunk branch simplified scheme two.

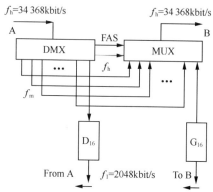

Figure 11-32　34Mbit/s trunk with 2Mbit/s branch (unidirection) (scheme two)

In this case, a relay tributary does not introduce stuffing jitter, the basic equipments are 18 circuit units which are 37.5% of the standard equipment scheme.

11.5.4　Trunk Branch Simplified Scheme Three

Figure 11-33 is the principle diagram of the simplified scheme two. In the following we will emphatically analyze the relay tributary synchronous demultiplex/multiplex process. When receive shift register just contains a primary cycle ($m=16$) tributary information code, by control of the signal LR, this cycle of tributary information code is written into the output tributary register; then it is shifted into input tributary register, again by control of the signal LT, it is written into send shift register.

If the prerequisite mentioned previously: "the MUX frame sequence is locked by the DMX sequence" particularly means that two frame alignment signals are completely aligned, the synchronous demultiplex/multiplex process described above must completed within a multiplex time slot. The particular time relationship is shown in Figure 11-34.

To analyze conveniently, move the synchronous demultiplex/multiplex part of Figure 11-33 into (1) of Figure 11-35. Here the receive shift register and the send shift register are separated, during a multiplex time slot, the trunk information code

Chapter 11 Engineering Application Design

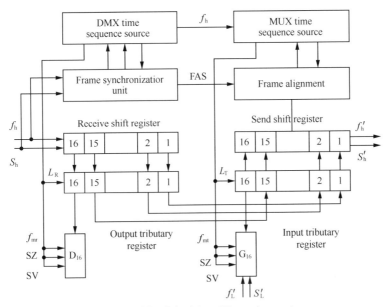

Figure 11-33 Principle of the scheme two

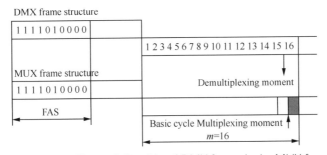

Figure 11-34 Time relationship of DMX frame locks MUX frame

in the receive shift register is moved into the send shift register in parallel. In this moving process, the content of the information code has no change, the time position has no change either, only the spatial position had changed. However, such a spatial movement is meaningless for transmission and multiplex. hence it is unnecessary. In this case two shift registers may be unified. See the logical diagram in Figure 11-35 (2), it may accomplish all the functions of Figure 11-35 (1): except being delayed the relay tributary information code passes through the shift register without any impair, under the control of LR and LT, the tributary information code completes a demultiplex and multiplex process within a multiplex time slot. In fact, the scheme of the Figure 11-35 (2) can be simplified further that as long as the control signals are designed properly, it is not necessary

−333−

for the shift register to take 16 steps, using the scheme of Figure 11-35 (3) can accomplish the above functions. But consider there may be many branches in a branch point and also consider the other supplementary functions being mentioned later, the scheme of Figure 11-35 (2) is still used. The principle of this kind of simplified scheme is shown in Figure 11-36. Compare with the simplified scheme two it saves 45 triggers which save about one standard circuit unit.

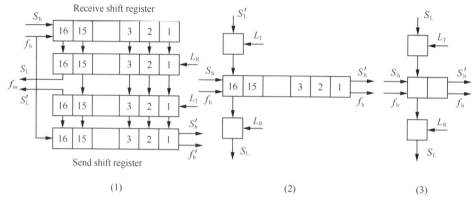

Figure 11-35 Synchronous Demultiplex/Multiplex diagram

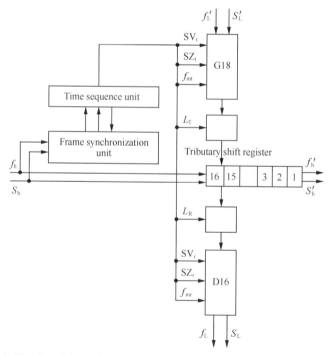

Figure 11-36 Trunk branch simplified scheme three (unidirection transmission)

Chapter 11 Engineering Application Design

11.5.5 Design of Branch Control Signals

We will design the branch control signals for trunk branch simplified scheme three bellow. These control signals are: justification position signal (SV_t), justification stuffing indication position signal (SZ_t), synchronous multiplex tributary clock (f_{mt}), multiplex control signal (LT), multiplex control signal (LR), synchronous demultiplex tributary clock (f_{mr}), recovery stuffing indicating position signal (SZ_r) and recovery stuffing position signal (SV_r). In a word, these are equivalent to all the control signals for justification and recovery of a general plesiochronous multiplexer. The principles to design these control signals are that in the prerequisite to satisfy the requirements of the simplified scheme, reduce the number of the control signals as much as possible and be consistent with the general multiplexer as much as possible, as well as accord with the common requirements of many branches.

The principle of the branch control part of the simplified scheme three is shown in Figure 11-37. After the information code of a primary cycle of S_h enters the specified position of the tributary shift register, the control signal LR writes the branch information code into the demultiplex register to complete the synchronous demultiplex; later then the control signal LT write the inserted information code into the idle slots of the tributary shift register to complete synchronous multiplex. It should be emphasized that such synchronous demultiplex/synchronous multiplex operation must be completed within a multiplex time slot. Hence, only two control signals LR and LT need to be specially designed, the other signals are no difference with the general multiplexer.

To design the control signals LR and LT, the strict relationship to the multiplex clock f_h are mainly considered. The time slot of 34 368kbit/s is 29ns, whereas with high speed TTL packages an NAND gate delays 3.5ns and a D flip-flop delays 6ns, with a ECL packages an NAND gate delays 3ns and a D flip-flop delays 4ns. When using TTL packages, the logical diagram of generating the control signals LR and LT is shown in 11-38 and the time relationship of the corresponding demultiplex/multiplex processes is shown in Figure 11-39. Obviously, both TTL packages and ECL packages may satisfy the design requirements.

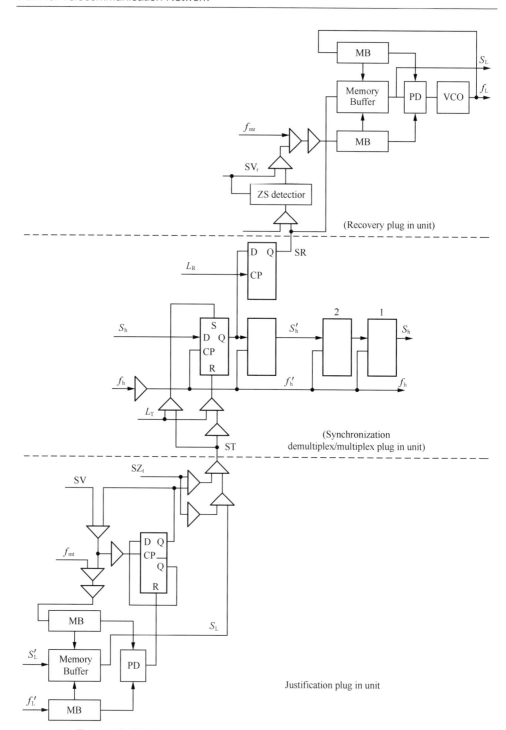

Figure 11-37 Branch control principle of trunk branch scheme three

Chapter 11 Engineering Application Design

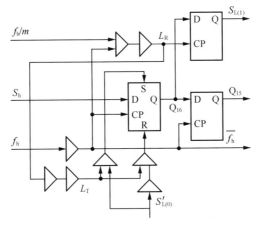

Figure 11-38 The logical diagram of generating the control signals L_R and L_T

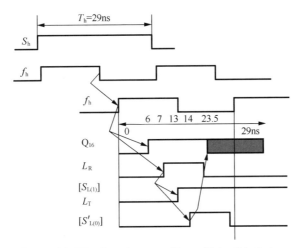

Figure 11-39 Synchronous Demultiplex/Multiplex time relationship (TTL packages)

11.5.6 Function Supplement

Through the above simplifications, the relay tributary information code no longer introduces stuffing jitter, the number of equipments is also reduced greatly. But, at the same time some advantages that the general equipments are used in branches are lost by the simplification. For example, when a general multiplexer is used in trunk branches, the clock and the frame alignment signal sent down are generated by the multiplexer at the branch point. Hence, the clock interrupts, frame

– 337 –

alignment signal errors and the frame losses of the upper reaches will not affect the work of lower reaches. However, the simplified scheme three does not have these advantages. Hence such useful functions must be supplemented to the simplified schemes.

The first has to be supplemented is the clock protection function. There are two schemes to be used. The first one is to use general subordinate clock system which may precisely remember the input clock within a quite long time and may precisely replace the upper reach clock in case it is interrupted, but the equipments are complicated. This kind of scheme may be used at a few of important nodes, however, it is undesirable in the case of many branches because of economic problem. The another scheme is to use a simple analog phase-locked loop. When the upper reach has clock it automatically locks on it. When the clock is interrupted the phase-locked loop automatically keeps its own central frequency and insures against the frequency exceeding the nominal tolerance. The later may be different a little bit from the general phase-lock. But it is not difficult to be realized that only the discriminator need to be designed reasonably. The particular circuit is shown in Figure 11-40.

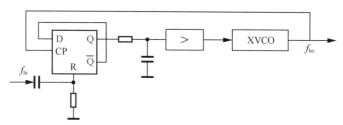

Figure 11-40 Clock Protection Phase-Locked Loop

The second is to install the frame alignment signal generation unit and forcedly set the frame alignment time slot of the multiplex binary digits, in order to eliminate the collecting errors of the frame alignment signals. Besides, the upper reach alarm test unit has to be installed. When it finds the frame loss at the upper reach, it controls the frame alignment logical unit that does not implement search operation. Thus the frame loss spread may be avoided, and automatically play a role of an end station to send binary digits of intact frame structure to the lower reaches. The trunk branch simplified scheme supplemented with the above three functions is shown in Figure 11-41.

Chapter 11 Engineering Application Design

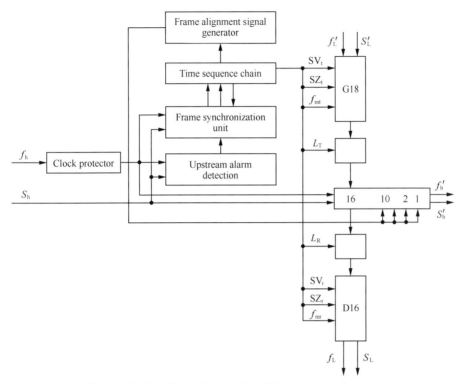

Figure 11-41 Trunk branch simplified scheme diagram

11.5.7 CCITT Recommendations

About the branch equipment of the group trunk transmission, now CCITT only passed one recommendation, Figure 11-42 gives a trunk/branch diagram. The trunk binary digits is 2948kbit/s, the equipment may insert and extract the bi-directional synchronous 64kbit/s binary digits, unidirectional synchronous 384kbit/s binary digits, and unidirectional asynchronous 384kbit/s binary digits.

The 2048kbit/s trunk signal frame structure accords with the recommendation G.704, i.e. the frame length is 256 bits, the frame alignment at interval of a frame is arranged (1-8bit):

Bit number Interval of a frame	1	2	3	4	5	6	7	8
Alignment frame	Si	0	0	1	1	0	1	1
Non-alignment frame	Si	1	A	Sn	Sn	Sn	Sn	Sn

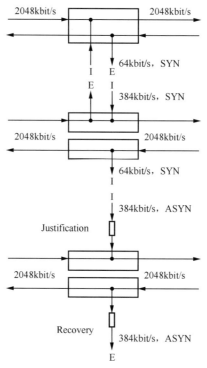

Figure 11-42 Trunk/Branch diagram

Where 0011011 is the frame alignment signal, S_i is reserved internationally, A is the alarm indication of the other end, S_n is reserved domestically.

The 64kbit/s branches give priority to select time slots: 6-22-14-30-2-18-10-26-4-20-12-28-24-5-21-13-29-1-17-9-25-3-19-11-27-7-23-15-31.

The time slots assigned for 384kbit/s branch parameters:

Inserting type	384kbit/s tributary				
	A	B	C	D	E
Synchronization	1-2-3 17-18-19	4-5-6 20-21-22	7-8-9 23-24-25	10-11-12 26-27-28	13-14-15 29-30-31
Asynchronization	1-7-11 17-23-27	3-9-15 19-25-31	4-8-12 20-24-28	5-10-13 21-26-29	2-6-14 18-22-30

The out of branch frame alignment and the frame alignment recovery specify that if the frame alignment errors are found in consecutive three frames or four frames, the out of frame alignment is believed. Recognize the frame alignment signal, find the second bit slot of 1 in the next frame, recognize again the frame alignment signal in the third frame distance again, then the frame alignment recovery is believed.

11.6 INTERNETWORK MULTIPLEX DESIGN

Now, there are two kinds of the digital networks internationally, i.e. the digital network which takes 2048kbit/s as the primary bit rate and has bit rate independent and the digital network which takes 1544kbit/s as the primary bit rate and has the minimum pulse density constraints. For present and future service requirements, it requires these two kinds of digital networks interconnected. Hence, CCITT recommends two multiplex methods of network interconnection: mixed multiplexes of 2048-6312kbit/s and 44 736-139 264kbit/s. But up to now, none of particular multiplex scheme is recommended. This section will discuss: the particular design of these two kinds of interconnection multiplexes.

11.6.1 2048-6312kbit/s Interconnection Multiplex Design

This design will use the same frame length of the recommendation G.743 (1544/6312kbit/s). Stuffing jitter calculation:

f_h=6312kbit/s±30ppm

f_l=2048kbit/s±30ppm

$m=3$

L_s=1176bits

$$\left(\frac{1}{m} - \frac{f_l}{f_h}\right) L_s = K + S$$

$K=10$

$S=0.42959$

$$Q = \frac{L_s}{m} - K$$

$Q = 382$

$$\Delta S = \left(\left|\frac{\Delta f_l}{f_l}\right| + \left|\frac{\Delta f_h}{f_h}\right|\right) Q$$

$\Delta S = 0.02292$

S_{max}=0.4525

S_{min}=0.4066

A_{jpp}=14.3%UI

The frame structure design is shown in Figure 11-43.

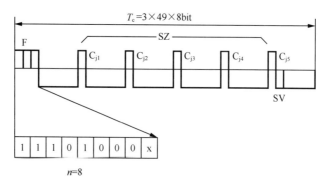

Figure 11-43 2048/6312kbit/s Frame Structure

The transition process calculation is shown in Figure 11-44:

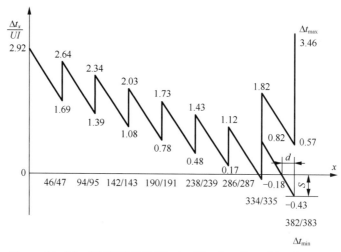

Figure 11-44 2048/6312kbit/s Justification Transition Process

$$\frac{\Delta t_x}{UI} = \frac{\Delta t_0}{UI} + \left[\frac{mf_1}{f_h}(x+g) - x\right]$$

$$g_0 = 3$$

When $\Delta t_0 = 0$,

$$\Delta t_{min} = -S \cdot UI = -0.43 UI$$

$$\Delta t_{max} = \left[\frac{mf_1}{f_h}(g_0+1) - S\right] \cdot UI = 3.46 UI$$

$$\Delta t_{pp} = \frac{mf_1}{f_h}(g_0+1) \cdot UI = 3.89 UI$$

$$d = \frac{S}{1-\dfrac{mf_1}{f_h}} = 16.14$$

The synchronous search design:

$$n=8$$
$$\eta=5$$
$$\alpha=2$$
$$\alpha = \frac{8}{0.3n} \approx 3$$
$$\beta = \frac{8}{3-\lg n} \approx 4$$

When $P_e = 1\times 10^{-3}$,

$$t_f = T_s /(nP_e)^\beta = 1.25\times 10^{11} \cdot T_s$$
$$t_r = \left(\alpha+\beta-\frac{1}{2}\right) = 6.5T_s$$
$$t_s = T_s / C_\eta^{\alpha+1} \cdot P_e^{\alpha+1} = 1\times 10^8 \cdot T_s$$

11.6.2 44 736-139 264kbit/s Interconnection Multiplex Design

In view of that the 34 368/139 264kbit/s frame structure of the CCITT recommendation G.751 is not a very good design scheme, hence the design here will not use the same frame length. The frame structure design requires: the frame length (L_s) is integer multiple of the primary cycle length (m); the none information bit number (K) of each the corresponding frame and the subframe number (h) keeps deterministic relationship of $K-h=2$, to insure that there are $3m$ bits time slot used for bunched frame alignment in the first subframe, there are m bits slot used for justification control and the other order-wires in the rest subframes. If the justification indication is at least 3 bits, then require $h \geqslant 4$, $K \geqslant 6$. Let the primary cycle number of each subframe be P, then the frame length expression is:

$$L_s = mph$$
$$K-h=2$$

∴ $$L_s = m(k-2)p$$

This kind of frame length selection should insure smaller stuffing jitter, hence the frame length selection should be constrained by justification design:

$$\left(\frac{1}{m} - \frac{f_1}{f_h}\right)L_s = K + S$$

Let $\alpha \equiv \left(\frac{1}{m} - \frac{f_1}{f_h}\right)$, the above formula may be rewritten:

$$\alpha m(K-2)p = K+S$$

In view of $S \leq 1$, then get the integer p expression:

$$p = \left[\frac{K+1}{\alpha m(K-2)}\right]$$

Take K as a variable, then get the corresponding p and L_s, as long as the corresponding S within a smaller A_{jpp} area, the value of K is a proper selection. The calculation data are as following:

$$f_h = 139\ 264 \text{kbit/s} \pm 15\text{ppm}$$
$$f_1 = 44\ 736 \text{kbit/s} \pm 20\text{ppm}$$
$$m = 3$$
$$\alpha = 0.012\ 101\ 4$$

K	6	7	9	10	11	12	13	14
P	48	44	39	37	36	35	35	34
Ls	576	660	819	888	972	1050	1155	1724
S	0.9704	0.9308	0.9110	0.7460	0.7625	0.7064	0.9771	0.8121
S'	0.0296	0.0692	0.0890	0.2539	0.2374	0.2936	0.0229	0.1879
Q	186	238	264	286	313	338	372	394
ΔS'	0.0202	0.0258	0.0287	0.0311	0.0340	0.0368	0.0404	0.0484
S'_{max}	0.0498	0.0950	0.1177	0.2850	0.2714	0.3304	0.0633	0.2363
S'_{min}	0.0094	0.0334	0.0603	0.2228	0.2034	0.2568	−0.0175	0.1395

The optimal scheme:

$$K = 8\text{bit}$$
$$P = 41$$
$$L_s = 738\text{bit}$$
$$Q = 238\text{bit}$$
$$h = 6$$
$$S = 0.9308$$
$$A_{jpp} = 9.1\%\text{UI}$$

The frame structure design is shown in Figure 11-45.

Chapter 11 Engineering Application Design

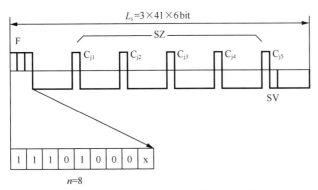

Figure 11-45　44 736/139 264kbit/s Frame Structure

The transition process calculation (Figure 11-46):

$$\frac{\Delta t_x}{UI} = \frac{\Delta t_0}{UI} + \left[\frac{mf_1}{f_h}(x+g) - x \right]$$

$$g_0 = 3$$

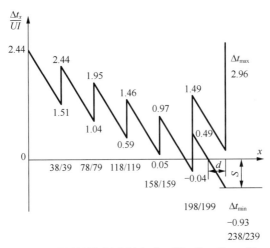

Figure 11-46　44 736/139 264kbit/s Justification Transition Process

When $\Delta t_0 = 0$,

$$\Delta t_{min} = -S \cdot UI = -0.931 UI$$

$$\Delta t_{max} = \left[\frac{mf_1}{f_h}(g_0+1) - S \right] \cdot UI = 2.934 \cdot UI$$

$$\Delta t_{pp} = \frac{mf_1}{f_h}(g_0+1) \cdot UI = 3.855 \cdot UI$$

$$d = \frac{S}{1-\frac{mf_1}{f_h}} = 25.65$$

The synchronous search design:

$$n = 8$$
$$\eta = 5$$
$$a = 2$$
$$\alpha = \frac{8}{0.3n} \approx 3$$
$$\beta = \frac{8}{3-\lg n} \approx 4$$

When $P_e = 1 \times 10^{-3}$,

$$t_f = T_s/(nP_e)^\beta = 1.25 \times 10^{11} \cdot T_s$$
$$t_r = \left(\alpha + \beta - \frac{1}{2}\right)T_s - 6.5T_s$$
$$t_s = T_s / C_\eta^{a+1} \cdot P_e^8 \cdot T_s$$

11.7 SUBGROUP MULTIPLEX DESIGN

The subgroup multiplex design contains two problems. which are related with each other, of subgroup rate selection and subgroup frame structure arrangement. Up to now, CCITT has not given any essential recommendation about them. This section will discuss the subgroup multiplex design techniques in view of these two problems.

11.7.1 Subgroup Rate Selection

To select the subgroup rate the following factors usually are considered:

(1) Transmission media. The transmission media suitable to subgroup transmission contain symmetrical cables, aerial conductors, channels over voice of mini-coaxial cables and quaded signal group, scatter radio stations, super short wave relay stations etc. Consider the characteristics of the transmission media, a radio station may change the transmission rate flexibly according to

the transmission distance and conditions required, whereas a wire system is difficult to change the regeneration distance and the transmission rate after the construction has completed. Consider the quantity of the transmission media, the symmetrical cable and mini-coaxial cable are used more than the others, but it may be different depending on the countries. The same wire channels may have very different practical characteristics depending on the level of design and construction and the level of maintenance. For example, the 0.008mm diameter local symmetrical cables of France, n,if its regeneration distance is 5km, the transmission rate is about 500kbit/s, whereas its regeneration distance is 10km, the transmission rate is about 250kbit/s. However the other countries are different.

(2) Service class. in a predict future the major services load bearing in the telecommunication networks are undoubtedly telephone service. Hence, the telecommunication network design is always based on the telephone services. But it should take account of the non telephone services increasing day by day. Now, the telephone signal coding in a fixed network are mainly of two kinds of PCM 64kbit/s and ΔM or DPCM 32kbit/s, the mobile networks mainly use ΔM 16kbit/s. Besides, the 384kbit/s of broadcast program transmission should also be considered. In addition to the above common points, each country uses various special rate domestically, for example, 19.2kbit/s.

(3) Digital multiplex series. Each country uses different primary rate, this will undoubtedly affect the selection of the subgroup rates, fortunately, the primary telephone channel rate are all 64kbit/s. May 1979, the CCITT study group 18 (Digital network) passed such a recommendation: in the rate series of 1544kbit/s and 2048kbit/s, only one unified subgroup rate may be specified in the future to communicate internationally. Unfortunately, after a decade, such a recommen-dation is still at the level of the recommendation of principle. However, after all it is a principle to cite as evidence.

(4) Multiplex techniques. After specify a kind of subgroup rate, naturally two kinds of equipments will appear, that is, the subgroup multiplexer between a telephone channel and the subgroup, and the primary multiplexer between a subgroup and the primary group. The telephone subscribers are usually centralized, hence a subgroup multiplexer usually need only the synchronous multiplex. Whereas the subgroup multiplexers may be centralized at one place or distributed in a wide area; they may use either a common clock source or their own

independent clock source. Hence, the primary multiplexer from a subgroup to the primary group may need synchronous/asynchronous compatible. In this case, it is suitable to use simple delta control synchronous/asynchronous compatible positive/negative/0 or positive/negative/0 justification techniques. Using such kind of techniques produces quite effects on the selection of the subgroup rate.

Summarize up, it is not difficult to see that the selection of subgroup rate should consider both the requirements for future use and the environment of the present telecommunication networks. However, the application requirements and the present environments of each country are different. This makes such a particular problem of the selection of subgroup rate complicated. See Table 11-4, each country and each telecommunication network specifies their own subgroup rate. These number are seemingly not far different, but it is hard to make a common recommendation.

Table 11-4 Examples of international subgroup rate selections

Country or system	Telephone channel rate	Subgroup rate (kbit/s)								
US army strategy network	32/16	56		128			256				
US army tactical network	16						256				
UK army grouse	16				144			288			
France army RITA	48										576
Canada network	64							256			
France network	64										528
Sweden network	64										704
China network	64/32							256		512		1024
China dedicated network	19.2		115.2				230.4			460.8		
CCITT broadcast program	64×2								384			
CCITT ISDN	64					160 192						

We can see from the Table that up to now there are only two kinds of CCITT recommendations which are the 384kbit/s using in broadcast program transmission and 160kbit/s using in bi-directional digital transmission for analog user loop. These two precise rates are all assigned to specific cases. All the other numbers are specially selected by each country and each specific network according to their particular cases and requirements.

11.7.2 Subframe Structure Arrangement

Because the subgroup rates have not an unified selection, the subgroup frame structure is hard to be unified arranged. This section will discuss generally the arrangement method of the subgroup frame structure.

The subgroup frame structure arrangement contains subgroup multiplex frame structure arrangement and the relevant primary multiplex frame structure arrangement. The subgroup multiplex and the primary multiplex contains channel multiplex and group multiplex. Both the channel multiplex and the group multiplex may use synchronous multiplex or asynchronous multiplex. The group synchronous multiplex may use the general group synchronous multiplex which only requires the bit slot aligned, may also use the compatible group synchronous multiplex which requires frame aligned. The asynchronous multiplex may use slipping asynchronous multiplex, may also use justification asynchronous multiplex. The above subgroup digital multiplex methods are summarized in the table. The seven kinds of subgroup multiplex methods listed in the table may all be used, however their frame structures are not complete the same, among them the synchronous multiplex is the most elementary. Now let us discuss respectively as the following.

Subgroup multiplex	Channel multiplex	Synchronous channel multiplex	
		Asynchronous channel multiplex	Slipping asynchronous channel multiplex
			Justification asynchronous channel multiplex
	Group multiplex	Synchronous group multiplex	General Synchronous group multiplex
			Compatible synchronous group multiplex
		Asynchronous group multiplex	Slipping asynchronous group multiplex
			Justification asynchronous group multiplex

(1) Synchronous Channel Multiplex Frame Structure

A synchronous channel multiplex frame should contain frame alignment signal (FAS), tributary information code (CTS), signaling (S), multiframe alignment signal (MFAS) and necessary service digits (V).

Signaling (S): The CCITT recommendation G.732 requires to provide a speech channel four signaling channels (a, b, c, d) with capacity of 500bit/s. Generally believing, the capacity of channel-associated signaling channel provided to each speech channel is at least two 500bit/s. Hence, the multiframe frame frequency of the synchronous channel multiplex frame should be 500Hz.

Frame Alignment: The basic principles of the frame alignment design are that the average synchronous search time should be as small as possible and the search circuits should be as simple as possible. the frame alignment signal recognition of the synchronous search unit may use either parallel recognition or serial recognition. The calculation formula of its synchronous search time (T_a) is:

$$T_a \approx \begin{cases} \dfrac{1}{2}(1+2P_1+L_s P_y)T_s \cdots \text{parallel-recognition} \\ \dfrac{L_s}{2}(1+2P_1+P_y)T_s \cdots \text{serial-recognition} \end{cases}$$

where, virtual recognition probability $P_y \approx \left(\dfrac{1}{2}\right)^n$;

leak recognition probability $P_1 \approx nP_e$;

n—frame alignment signal length;

P_e—receiving binary digits average error ratio;

T_s—frame cycle;

L_s—frame length.

According to distribution of frame alignment slots within a frame, it may be classified into the bunched frame alignment and distributed frame alignment. Consider the corresponding recognition methods, the practical meaningful may be classified into: bunched frame alignment/parallel recognition, distributed frame alignment/parallel recognition and distributed frame alignment/serial recognition. The shift register width required by the synchronous search unit in the corresponding scheme respectively is:

$$W = \begin{cases} n\bullet\bullet\text{bunched\_farme\_alignment/parallel\_recognition} \\ (n-1)L_s/k + 1\bullet\bullet\text{distributed\_frame\_alignment/parallel\_recognition} \\ n\bullet\bullet\text{distributedl\_frame\_alignment/serial\_recogniyion} \end{cases}$$

where k is the primary cycle number in a frame. The synchronous search mechanism usually uses bunched frame alignment/parallel recognition as much as possible. But in the subgroup multiplex, it often uses distributed frame alignment.

The most basic component of a multiplex frame is the primary cycle. The general expression of the frame structure is:

$$L_s = (M+m)k$$

where, M—number of multiplex tributaries participating;

m—number of channels occupied by non-tributary information bits;

k—number of primary cycles in a frame:

$$M+m = f_h/f_1$$
$$k = f_1/F_s$$

where, f_h—multiplex rate;

f_1—tributary rate;

F_s—frame frequency.

Consider the practical conditions of the subgroup design, the general expression of the frame structure usually uses the following form:

$$L_s = (jM + jm)\frac{k}{j}$$
$$j = 2^i$$
$$i = 0, 1, 2, \cdots, k$$

The maximum relative width offset of the corresponding tributary multiplex/demultiplex code element:

$$\frac{\Delta t_1}{t_1} = \frac{m}{M+m}(j-1)$$

When the code elements of the demultiplex tributary are equal-widen through buffer, the read/write time difference of the buffer when the xth tributary code element is read out is:

$$\Delta t = \left[\frac{Mf_1}{f_h}\left(x + \frac{jm}{M}\right) - x\right]t_1$$

From it the maximum read/write time difference is:

$$\Delta t_{max} = j \cdot \frac{m f_1}{f_h} \cdot t_1$$

When $j=1$,

$$L_s = (M+m)k$$

$$\frac{\Delta t_1}{t_1} = 0$$

Obviously, the demultiplex code elements are equal-wide, hence buffer is not necessary, but only the distributed frame alignment may be used.

When $j = \frac{n}{m}$,

$$L_s = \left(\frac{n}{m}M + n\right)\frac{m}{n}k$$

$$\frac{\Delta t_1}{t_1} = \frac{n-m}{M+m}$$

$$\Delta t_{max} = \frac{n}{M+m} \cdot t_1$$

Obviously, the demultiplex code elements ale not equal-wide, hence buffers have to be set up, and its size may not be less than $\frac{n}{M+m}$. However, the bunched frame alignment with octet length of n may be just used.

When $j = \frac{n}{m} = 1$,

$$L_s = (M+n)k$$

$$\frac{\Delta t_1}{t_1} = 0$$

Obviously, this kind of scheme may use the bunched frame alignment without tributary buffer. Theoretically it is the optimal scheme. But if the $m=n$ is bigger, the multiplex efficiency may be too low, whereas the $m=n$ is smaller, the frame alignment signal length may be too short. Hence, it is usually not used in engineering. Of course, such kind of optimal scheme may be used in some specific cases.

Chapter 11 Engineering Application Design

(2) Slipping Asynchronous Channel Multiplex Frame Structure

The slipping asynchronous channel multiplex may use the same frame structure of the synchronous channel multiplex, but the tolerance of the multiplex clock must be as small as possible. The particular number requirements depend on the limitation of the slipping frequency.

(3) Justification Asynchronous Channel Multiplex Frame Structure

The justification asynchronous channel multiplex may uses the same frame structure of the synchronous channel multiplex, but the justification frame must be formed additionally. When the relationship between the multiplex frequency (f_h) and the tributary frequency (f_l) is simply integer multiple, the positive/0/negative justification is suitable. The justification frame length depends only on the positive/0/negative justification technical design. Usually, the justification frame length is far bigger than that of multiframes. In this case, although the multiframe structure may be still kept, however, the multiframe synchronous mechanism is not necessary to set up. Such multiframe synchronous functions may be considered by the justification frame synchronous mechanism.

(4) General synchronous Group Multiplex Frame Structure

This kind of multiplex frames only suits to transmission multiplex. It may not in common use with exchange frames, and it is unnecessary to be compatible with the channel multiplex frames. Hence, the design of such kind of frame structure is relatively simple without any specific limitation. The following gives design of two typical schemes:

Design of the distributed flame alignment:

$$L_s = (M+1)k$$

When the relationship between the multiplex frequency (f_h) and the tributary frequency (f_l) is simply integer times, at least one channel slot is used for the frame alignment signals and the other service bits. Usually speaking, the efficiency of such arrangement is relative low.

Design of bunched frame alignment:

$$L_s = (nM + n)\frac{k}{n}$$

This design may use bunched frame alignment, hence may get better frame synchronization characteristics. But the tributary buffer must be set up in this case.

(5) Compatible Synchronous Group Multiplex Frame Structure

A compatible synchronous group multiplex is that as long as the tributary groups are multiplexed into the primary group, this kind of frame structures are then disappeared, the positions of all the channels in the primary group are the same arrangement completely as that the tributaries are multiplexed directly into the primary group. The compatible synchronous group multiplex process is: at the tributary entry of the multiplexer, firstly recognize the frame alignment signal of the tributary binary digits, then delayed justify each tributary frame to align the tributary frame alignment signals and multiplex frame alignment signal; finally write each tributary information bits into the specified slots of the multiplex frame.

It is not difficult to see from the above introduction that some special requirements to implement the compatible synchronous multiplex are: firstly, all the tributary frames multiplexed must have the same cycle with the multiplex frame, that is, using equal frame period (or equal frame rate) design; secondly, each tributary frame is required to have the same frame alignment signal configuration with the multiplex frame, for example, all may use bunched frame alignment; finally, each tributary entry is required to set up frame justifier to complete frame delayed justification. Obviously, the equipment to implement compatible synchronous multiplex is relative complicated. But this kind of multiplex has obvious ad vantages: firstly, this kind of multiplex has good compatibility, for example, it may be compatible with exchange frames. The time division exchange units of some digital exchanges are actually compatible synchronous group multiplexers; secondly, the multiplex efficiency is relative high. For example, while multiplexing, all the tributary frame alignment signals may not enter into the multiplex binary digits. Hence, these idle time slots may be used to transmit the other information codes.

(6) Slipping Asynchronous Group Multiplex Frame Structure

The slipping asynchronous group multiplex may use the same frame structure with the compatible synchronous group multiplex, actually it may use the same equipment, because the tributary frame justifier has also the synchronous justification function and the frame delayed justification function. The only difference is that it does not make slip when used in synchronous group multiplex and the slip frame will be made when used in asynchronous group multiplex.

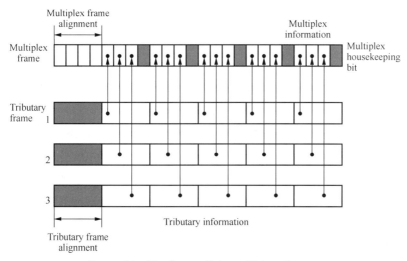

Figure 11-47 Compatible multiples diagram

(7) Justification asynchronous Group Multiplex Frame Structure

The justification technique has only the function justifying the rate to be synchronized, but no function of frame delayed-justification. Hence, it may not unify the frame structure with the compatible synchronous group multiplex, but may be uniform with the frame structure of the general synchronous multiplex. When the multiplex rate and the tributary rate are in a relationship of simple integer multiple, the positive/0/negative justification techniques is suitable. In this case the justification frame must be formed based on the general synchronous group multiplex frame structure.

11.7.3 Examples of Subgroup Frame Structure Design

According to the above principles of the subgroup frame structure design, this section will take 32/256/2048kbit/s subgroup as examples to design frame structure.

(1) 32/256kbit/s the Simplest Equipment Channel Synchronous Multiplex Design

256kbit/s and 32kbit/s are in the relationship of simple integer multiple, hence either using synchronous channel multiplex or asynchronous multiplex, non-information bits will occupy a 32kbit/s tributary. That is, $M=7$, $m=1$. Consider the capacity requirement of the non-information bits (frame alignment signal, signaling and necessary service digits), the primary frame length is at least $L_s=128$, the multiframe rate must be $F_{MS}=500Hz$. Then get the number (k) of the primary cycle within the primary frame and the number (y) of the primary frames in the multiframe:

$$k = \frac{L_s}{M+m} = 16$$
$$y = F_s / F_{MS} = f_h / L_s \cdot F_{MS} = 4$$

The frame structure arrangement is shown in Figure 11-48. For this kind of the frame structure, equipments are simple that suit to be used together with the compatible 256/2048kbit/s group synchronous multiplex, but only the distributed frame alignment may be arranged.

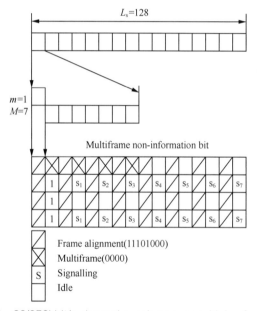

Figure 11-48 32/256kbit/s channel synchronous multiplex frame structure

(2) 256/2048kbit/s Compatible Group Synchronous Multiplex Design

The frame rate of the 256/2048kbit compatible multiframe must be equal to the frame rate of the relevant tributaries, i.e. using $F_s=2kHz$, then get the frame length $L_s=1024bit$; the primary cycle must also be completely compatible with the tributary frames, i.e. $m=1$, $M=7$, and the slot arrangement of the information bits and the non-information bits must also be the same. The multiplex frame structure is shown in Figure 11-49.

Chapter 11 Engineering Application Design

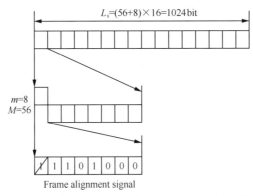

Figure 11-49 256/2048kbit/s compatible group multiplex frame structure

The multiplex frame consists of 16 primary cycles, each of them contains 64 bit slots that are 8 non-information bits and 56 information bits.

(3) 256kbit/s and 2048kbit/s General Multiplex Frame Structure Design

Implementation of either general synchronous multiplex or justification asynchronous multiplex, 32/256kbit/s multiplex and 256/2048kbit/s multiplex may use common frame structure, and even use the same main multiplex equipment. In this case, these two kinds of multiplex frames must be designed in equal frame length. Consider CCITT already recommended that the 2048kbit/s multiplex frame length is 256bit, hence the general frame length should be $L_s=256$ bit. Then we may get the corresponding parameters of two kinds of multiplex frames:

f_1 (kbit/s)	f_k (kbit/s)	F_s (kHz)	L_s (bit)	M (bit)	m (bit)
256	2048	8	256	7	1
32	256	1	256	7	1

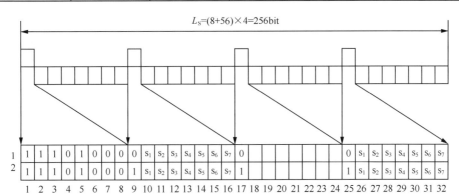

Figure 11-50 256/2048kbit/s general frame structure

We can see from the general frame structure diagram that the frame length $L_s = 256$, each frame consists of 4 primary cycles which contains 64 bit slots that are 8 non-information bits and 56 information bits. The non-information bits of the entire frame are partitioned into 4 groups, each group has 8bit, altogether are 32bit. Because the non-information bits are bunched, it is easy to arrange the bunched frame alignment (11101000). The first group of the non-information bits carry the frame alignment signal, whereas the rest three groups carry the multitrame alignment, signaling and service digits. Here a multiframe contains two primary frames, the multiframe rate is 500Hz.

(4) 256kbit/s Frame and 512kbit/s Frame Compatible Design

The equal frame frequency design must be taken to let 256kbit/s frame and the 512kbit/s frame compatible. In this case, partitioning every slot of the 256kbit/s frame structure into two parts may get the 512kbit/s frame structure. see Figure 11-51.

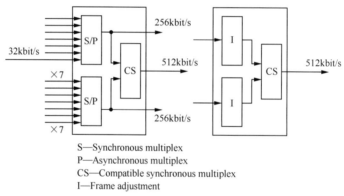

S—Synchronous multiplex
P—Asynchronous multiplex
CS—Compatible synchronous multiplex
I—Frame adjustment

Figure 11-51　256kbit/s frame and 512kbit/s frame compatible design

This kind of 512kbit/s multiplexer may multiplex fourteen 32kbit/s tributary binary digits, and may also multiplex two 256kbit/s binary digits. The design must insure that the non-information bit arrangement in the 512kbit/s multiplex frame and that two 256kbit/s frames must be strictly aligned, such that a 512kbit/s frame may look like the combination of two 256kbit/s frames.

(5) 32/256/2048kbit/s General Frame Structure Multiplex System

32/256/2048kbit/s general frame structure multiplex system diagram is shown Figure 11-52. All the three multiplexers use the frame structure shown in Figure 11-50. The advantages of this system are good standardization, common use of most the plug-in units, flexible to use, equipment compatible with functions of synchronous and asynchronous multiplex. The shortages are low efficiency and only forty nine 32kbit/s tributaries.

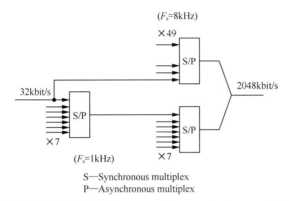

Figure 11-52 32/256/2048kbit/s general frame structure multiplex system

(6) 32/256/2048kbit/s Compatible Synchronous Multiplex System

The diagram of the 32/256/2048kbit/s compatible synchronous multiplex system is shown in Figure 11-53. The 256kbit/s multiplex frame uses the frame structure shown in Figure 11-48; the 2048kbit/s multiplex frame uses the frame structure shown in Figure 11-49. The advantages of the system are that compatibility is good that may be compatible with exchange frame, efficiency is high that may contain 56 32kbit/s tributaries. The shortages is that the equipments are relative complicated and that the 256kbit/s tributary frame. justifier must be set up.

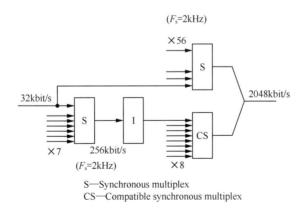

Figure 11-53 32/256/2048kbit/s compatible synchronous multiplex system

11.8 ISDN USERS/NETWORK INTERFACE MULTIPLEX DESIGN

11.8.1 2B+D Interface Multiplex Design

There are bi-directional digital transmission between the user terminal equipment (TE) and the network terminations (NT), the rate is 192kbit/s, the frame length is 48bit; but the frame structures in two transmission directions are different (see Figure 11-51).

From the network (NT) to the terminal (TE) direction:

Bit position	Contents
1	Frame alignment signal
2,48	DC balance bits
3-10,27-34	B_1 channel bits
11,24,35,46	E-returned channel bits
12,25,36,47	D channel bits
13	A bits used to set working state
14	Auxiliary frame alignment bit FA
15	Auxiliary frame alignment bits
16-23,38-45	B_2 channel bits
26	S_1 reserved bits
37	S_2 reserved bits

From the terminal (TE) to the network (NT) direction:

Bit position	Contents
1	Frame alignment signal
2,11,13,15,24,26,35,37,46,48	DC balance bits
3-10, 27-34	B_1 channel bits
12,25,36,47	D channel bits
14	Auxiliary frame alignment Signal
16-23,38-45	B_2 channel bits

The network terminations (NT) get the timing signals from the network clock, the user terminals (TE) extract the signals of bit synchronization, octet synchronization and frame synchronization from the NT→TE signals received, and use these timing signals to form the frame structure of the TE→NT signals. The TE→NT signal frame are delayed 2UI relative to the NT→TE signal frame.

Within the network termination (NT), the contents of the D channel received from the TE→NT signals are delayed 10UI and inserted into the E slot of the NT→TE signals, returned back to the original user terminal (TE).

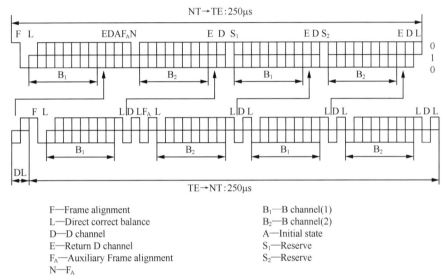

F—Frame alignment
L—Direct correct balance
D—D channel
E—Return D channel
F_A—Auxiliary Frame alignment
N—F_A

B_1—B channel(1)
B_2—B channel(2)
A—Initial state
S_1—Reserve
S_2—Reserve

Figure 11-54 2B+D Bi-directional transmission frame structure

Both the line codes of the two transmission directions use pseudo ternary code of 100% duty cycle. The coding principle is as following: the line code "0" level stands for the binary code 1: the positive pulse or the negative pulse of the line code stand for "0". The polarity of the first binary code "0" following the frame alignment balance bit is the same with the polarity of the frame alignment balance bit. Since then, the polarity of the binary code "0" varies alternatively. If the number of binary code "0" following the previous balance bit is odd, the next balance bit is a binary code "0"; If the number of binary code "0" following the previous balance bit is even, the next balance bit is a binary code "1".

According to the above line coding principle: at the first binary code "0" bit after the frame alignment balance bit a AMI violence point win be produced, which is used to recognize the frame alignment. In order to carry out the frame alignment more reliable, set up a pair of auxiliary frame alignment bits F_A and N in the direction of NT→TE; Set up the auxiliary frame alignment bit F_A and its balance bit L in the direction of TE→NT. In the direction of NT→TE, bit F_A or $N=\overline{F_A}$ is 0; whereas in the direction of TE→NT, F_A=L is 0 bit always. Hence, there is always a AMI violence point every 14 bits or less than 14 bits from the frame alignment F bit.

The frame alignment searching process in the direction NT→TE: Passing through two 48 bits cycles, if no AMI violence point is detected within≤14 bits interval, a frame loss is believed; if the AMI violence points are found three times consecutively within≤14 bits interval, a frame alignment completed is believed.

The frame alignment searching process in the direction TE→NT: Passing through two 48 bits cycles, if no AMI violence point is detected within≤13 bits interval, a frame loss is believed; if the AMI violence points are found three times consecutively within≤13 bits interval, a frame alignment completed is believed.

11.8.2 2048kbit/s Primary Interface Multiplex Design

The frame rate is 8kHz, the frame length is 256 bits which are partitioned into 32 time slots (0-31) that each slot is 8 bits (1-8).

The 0th slot in even frames are is used to transmit the frame alignment signal at interval of a frame.

Interval of a frame \ Bit number	1	2	3	4	5	6	7	8
Alignment frame contained	Si	0	0	1	1	0	1	1
Alignment frame not contained	Si	1	A	Sn	Sn	Sn	Sn	Sn

The usage of Si, A, Sn are undefined.

The 16th slot is used as D channel, E channel or the other applications.

The slots 1-15 and 17-31 are used as B, H0, H12 channels. Where B=64kbit/s, H0=384kbit/s, H12=1929kbit/s.

Timing extraction specification: NT extracts the timing signal from the network clock; TE extracts the timing signals of bit synchronization, octet synchronization and frame synchronization from the NT→TE signal, and uses these timing signals to form the frame structure of TE→NT signals.

Frame alignment searching process recommended: a frame loss is believed if no correct frame alignment signal is recognized at the consecutive 3-4 specified positions. The frame alignment is recovered if a frame alignment signal is found, bit 1 is found at the 2nd bit of the slot 0 of the next frame, and the frame alignment signal is found the second time at the next two frame.

11.8.3 Multiplex Design of Entering 64kbit/s Channel

8kbit/s, 16kbit/s and 32kbit/s binary digits may directly be multiplexed into 64kbit/s B channel.

There are two multiplex methods for these three rates of binary digits entering 64kbit/s, i.e. a fixed multiplex method and a varied multiplex method.

The fixed multiplex method specification:

(1) 8kbit/s binary digits may occupy any position within a 64kbit/s eight bit octet; 16kbit/s binary digits may occupy (1,2), (3, 4) or (5, 6) bit slots; 32kbit/s binary digits may occupy (1,2,3,4) or (5,6,7,8) bit slots.

(2) In the eight bits of each consecutive B channel, subrate binary digits occupy the same bit positions.

(3) The bit transmission sequence of each subrate binary digits are the same before the multiplex and after the demultiplex.

(4) All the unoccupied bit slots are set to binary code 1.

The varied multiplex method specification:

(1) The subrate binary digits occupy the same bit positions in the eight bit octet of each consecutive B channel.

(2) A newly multiplexed subrate binary digits should occupy the most front idle bit slots in the eight bit octet of the B channel.

(3) All the idle bit slots are set to 1.

Except 8kbit/s, 16kbit/s and 32kbit/s, all the other low rate binary digits must be multiplexed firstly into one of these three kinds of rates and then multiplexed again into B channel.

11.8.4 Multiplex Design of X_1 Rate Entering 8/16kbit/s

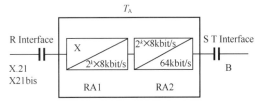

Figure 11-55 X_1 Rate Multiplexed into B Channel

600bit/s, 2400bit/s and 4800bit/s are multiplexed firstly into 8kbit/s; 9600bit/s is multiplexed firstly into 16kbit/s. The multiplex frame lengths are all 40 bits. Two kinds of 5×8 bit multiframes are recommended, i.e. odd multiframe and even multiframe whose particular arrangements are shown in the following table.

Octet number \ Bit number		1	2	3	4	5	6	7	8
0	Odd frame	0	0	0	0	0	0	0	0
	Even frame	1	E_1	E_2	E_3	E_4	E_5	E_6	E_7
1		1	P_2	P_2	P_3	P_4	P_5	P_6	SQ
2		1	P_7	P_8	Q_1	Q_2	Q_3	Q_4	X
3		1	Q_5	Q_6	Q_7	Q_8	R_1	R_2	SR
4		1	R_3	R_4	R_5	R_6	R_7	R_8	SP

An odd frame plus an even frame form a multiframe of 80 bits long. Within an 80 bits multiframe, the 8 zeros of the first octet of an odd frame plus the beginning 1 of the following 9 octet form the multiframe alignment signal: 8 zeros and 9 ones so altogether 17 bits long.

The SP, SQ, SR in the table are state bits; E is supplementary signaling bit; P, Q, R are data bits.

To shorten the frame synchronization and to reduce the transmission delay, CCITT recommends to use user-bit-repeat method. In this case, the frame arrangements of different rates are listed respectively as following:

(1) 600bit/s to 8kbit/s frame arrangements:

Octet number \ Bit number	1	2	3	4	5	6	7	8
0	0	0	0	0	0	0	0	0
1	1	P_1						SP
2	1			P_2				X
3	1					P_3		SP
4	1							SP
5	1	1	0	0	E_4	E_5	E_6	1
6	1	P_4						SP
7	1			P_5				X
8	1					P_6		SP

续表

Bit number\Octet number	1	2	3	4	5	6	7	8
9	1							SP
10	0	0	0	0	0	0	0	0
11	1	P_7						SP
12	1			P_8				X
13	1					Q_1		SQ
14	1							SQ
15	1	1	0	0	E_4	E_5	E_6	1
16	1	Q_2						SQ
17	1			Q_3				X
18	1					Q_4		SQ
19	1							SQ
20	0	0	0	0	0	0	0	00
21	1	Q_5						SQ
22	1			Q_6				X
23	1					Q_7		SQ
24	1							SQ
25	1	1	0	0	E_4	E_5	E_6	1
26	1	Q_8						SR
27	1			R_1				X
28	1					R_2		SR
29	1							SR
30	0	0	0	0	0	0	0	0
31	1	R_3						SR
32	1			R_4				X
33	1					R_5		SR
34	1							SR
35	1	1	0	0	E_4	E_5	E_6	0
36	1	R_6						SR
37	1			R_7				X
38	1					R_8		SR
39	1							SP

(2) 2400bit/s to 8kbit/s frame arrangements:

Octet number \ Bit number	1	2	3	4	5	6	7	8
0	0	0	0	0	0	0	0	0
1	1	P_1		P_2		P_3		SP
2	1	P_4		P_5		P_6		X
3	1	P_7		P_8		Q_1		SQ
4	1	Q_2		Q_3		Q_4		SQ
5	1	1	1	0	E_4	E_5	E_6	E_7
6	1	Q_5		Q_6		Q_7		SR
7	1	Q_8		Q_1		Q_2		X
8	1	R_3		Q_4		Q_5		SR
9	1	R_6		R_7		Q_8		SP

(3) 4800bit/s to 8kbit/s frame arrangements:

Octet number \ Bit number	1	2	3	4	5	6	7	8
0	0	0	0	0	0	0	0	0
1	1	P_1	P_2	P_3	P_4	P_5	P_6	SQ
2	1	P_7	P_8	Q_1	Q_2	Q_3	Q_4	X
3	1	Q_5	Q_6	Q_7	Q_8	R_1	R_2	SR
4	1	R_3	R_4	R_5	R_6	R_7	R_8	SP
5	1	0	1	1	E_4	E_5	E_6	E_7
6	1	P_1	P_2	P_3	P_4	P_5	P_6	SR
7	1	P_7	P_8	Q_1	Q_2	Q_3	Q_4	X
8	1	Q_5	Q_6	Q_7	Q_8	R_1	R_2	SR
9	1	R_3	R_4	R_5	R_6	R_7	R_8	SP

(4) 9600bit/s to 16kbit frame arrangements:

The same as 4899bit/s to 8kbit/s frame arrangements.

11.8.5 Design of V Rate Multiplexed into Middle Rate

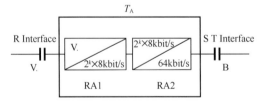

Figure 11-56 V Rate Multiplexed into B Channel

The V rate is multiplexed into a middle rate (8kbit/s, 16kbit/s, 32kbit/s) firstly, then it is multiplexed into B channel rate (64kbit/s). The rates adaptation is specified as following:

V rate (bit/s)	Intermediate rate		
	8kbit/s	16kbit/s	32kbit/s
600	√		
1200	√		
2400	√		
4800	√		
7200		√	
9600		√	
12 000			√
14 400			√
19 200			√

Octet number \ Bit number	1	2	3	4	5	6	7	8
0	0	0	0	0	0	0	0	0
1	1	D_1	D_2	D_3	D_4	D_5	D_6	S_1
2	1	D_7	D_8	D_9	D_{10}	D_{11}	D_{12}	X
3	1	D_{13}	D_{14}	D_{15}	D_{16}	D_{17}	D_{18}	S_3
4	1	D_{19}	D_{20}	D_{21}	D_{22}	D_{23}	D_{24}	S_4
5	1	E_1	E_2	E_3	E_4	E_5	E_6	S_7
6	1	D_{25}	D_{26}	0_{27}	D_{28}	D_{29}	D_{30}	S_6
7	1	D_{31}	D_{32}	D_{33}	D_{34}	D_{35}	D_{36}	X
8	1	D_{37}	D_{38}	D_{39}	D_{40}	D_{41}	D_{42}	S_8
9	1	D_{43}	D_{44}	D_{45}	D_{46}	D_{47}	D_{48}	S_9

The frame structure: a middle rate frame length is 80 bits which contains ten octets numbered 0-9. The octet 0 is an eight 0: the first bit of all the octets 1-9 is 1; the rest 7 bits of the octet 5 is E (signaling) bits; The last bit of the octet 1-4 and 6-9 is state bit (S and X.); The rest slots contain data (D) bits. The details are shown in the following table:

Frame alignment signal: The eight 0 bits of the multiframe octet 0 and the nine 1s of the first bit of the octet 1-9 form the multiframe alignment signal of octet length in 17.

The bit assignment:

(1) 600bit/s to 8kbit/s frame arrangements

Octet number \ Bit number	1	2	3	4	5	6	7	8
0	0	0	0	0	0	0	0	0
1	1	D_1						S_1
2	1			D_2				X
3	1					D_3		S_3
4	1							S_4
5	1	1	1	0	E_4	E_5	E_6	E_7
6	1	D_4						S_6
7	1			D_5				X
8	1					D_6		S_8
9	1							S_9

(2) 1200bit/s to 8kbit/s frame arrangements

Octet number \ Bit number	1	2	3	4	5	6	7	8
0	0	0	0	0	0	0	0	0
1	1	D_1	D_1	D_1	D_1	D_2	D_2	S_1
2	1	D_2	D_2	D_3	D_3	D_3	D_3	X
3	1	D_4	D_4	D_4	D_4	D_5	D_5	S_3
4	1	D_5	D_5	D_6	D_6	D_7	D_6	S_4
5	1	0	1	0	E_4	E_5	E_6	E_7
6	1	D_7	D_7	D_7	D_7	D_8	D_8	S_6
7	1	D_8	D_8	D_9	D_9	D_9	D_9	X
8	1	D_{10}	D_{10}	D_{10}	D_{10}	D_{11}	D_{11}	S_8
9	1	D_{11}	D_{11}	D_{12}	D_{12}	D_{12}	D_{12}	S_9

(3) 2400bit/s to 8kbit/s frame arrangements

Octet number \ Bit number	1	2	3	4	5	6	7	8
0	0	0	0	0	0	0	0	0
1	1	D_1	D_1	D_2	D_2	D_3	D_3	S_1
2	1	D_4	D_4	D_5	D_5	D_6	D_6	X

续表

Octet number \ Bit number	1	2	3	4	5	6	7	8
3	1	D_7	D_7	D_8	D_8	D_9	D_9	S_3
4	1	D_{10}	D_{10}	D_{11}	D_{11}	D_{12}	D_{13}	S_4
5	1	1	1	0	E_4	E_5	E_6	E_7
6	1	D_{13}	D_{13}	D_{14}	D_{14}	D_{15}	D_{15}	S_6
7	1	D_{16}	D_{16}	D_{17}	D_{17}	D_{18}	D_{18}	X
8	1	D_{19}	D_{19}	D_{20}	D_{20}	D_{21}	D_{21}	S_8
9	1	D_{22}	D_{22}	D_{23}	D_{23}	D_{24}	D_{24}	S_9

(4) 7200bit/s to 16kbit/s frame arrangements

Octet number \ Bit number	1	2	3	4	5	6	7	8
0	0	0	0	0	0	0	0	0
1	1	D_1	D_2	D_3	D_4	D_5	D_6	S_1
2	1	D_7	D_8	D_9	D_{10}	F	F	X
3	1	D_{11}	D_{12}	F	F	D_{13}	D_{14}	S_3
4	1	F	F	D_{15}	D_{16}	D_{17}	D_{18}	S_4
5	1	1	0	0	E_4	E_5	E_6	E_7
6	1	D_{19}	D_{20}	D_{21}	D_{22}	D_{23}	D_{24}	S_6
7	1	D_{25}	D_{26}	D_{27}	D_{28}	F	F	X
8	1	D_{29}	D_{30}	F	F	D_{31}	D_{32}	S_8
9	1	F	F	D_{33}	D_{34}	D_{35}	D_{36}	S_9

F—Stuffing bit

(5) 14400bit/s to 32kbit/s frame arrangement

The same with 7200bit/s to 16kbit/s frame arrangements.

(6) 4800bit/s to 8kbit/s frame arrangements

Octet number \ Bit number	1	2	3	4	5	6	7	8
1	0	0	0	0	0	0	0	0
2	1	D_1	D_2	D_3	D_4	D_5	D_6	S_1
3	1	D_7	D_8	D_9	D_{10}	D_{11}	D_{12}	X
4	1	D_{13}	D_{14}	D_{15}	D_{16}	D_{17}	D_{18}	S_3
5	1	D_{19}	D_{20}	D_{21}	D_{22}	D_{23}	D_{24}	S_4

Octet number \ Bit number	1	2	3	4	5	6	7	8
6	1	0	1	1	E_4	E_5	E_6	E_7
7	1	D_{25}	D_{26}	D_{27}	D_{28}	D_{29}	D_{30}	S_6
8	1	D_{31}	D_{32}	D_{33}	D_{34}	D_{35}	D_{36}	X
9	1	D_{37}	D_{38}	D_{39}	O_{40}	D_{41}	D_{42}	S_8
10	1	D_{43}	D_{44}	D_{45}	D_{46}	D_{47}	D_{48}	S_9

(7) 9600bit/s to 16kbit/s frame arrangements

The same with 4800bit/s to 8kbit/s frame arrangements.

(8) 19200bit/s to 32kbit/s frame arrangements

The same with 4800bit/s to 8kbit/s frame arrangements.

(9) 1200bit/s to 32kbit/s frame arrangements

No CCITT recommendation up to now.

(10) 48kbit/s to 64kbit/s frame arrangement

Octet number \ Bit number	1	2	3	4	5	6	7	8
1	1	D_1	D_2	D_3	D_4	D_5	D_6	S_1
2	0	D_7	D_8	D_9	D_{10}	D_{11}	D_{12}	X
3	1	D_{13}	D_{14}	D_{15}	D_{16}	D_{17}	D_{18}	S_3
4	1	D_{19}	D_{20}	D_{21}	D_{22}	D_{23}	D_{24}	S_4

(11) 56kbit/s to 64kbit/s frame arrangements

Octet number \ Bit number	1	2	3	4	5	6	7	8
0	D_1	D_2	D_3	D_4	D_5	D_6	D_7	1
1	D_8	D_9	D_{10}	D_{11}	D_{12}	D_{13}	D_{14}	1
2	D_{15}	D_{16}	D_{17}	D_{18}	D_{19}	D_{20}	D_{21}	1
3	D_{22}	D_{23}	D_{24}	D_{25}	D_{26}	D_{27}	D_{28}	1
4	D_{29}	D_{30}	D_{31}	D_{32}	D_{33}	D_{34}	D_{35}	1
5	D_{36}	D_{37}	D_{38}	D_{39}	D_{40}	D_{41}	D_{42}	1
6	D_{43}	D_{44}	D_{45}	D_{46}	D_{47}	D_{48}	D_{49}	1
7	D_{50}	D_{51}	D_{52}	D_{53}	D_{54}	D_{55}	D_{56}	1

11.9 DESIGN OF MULTIPLEX SYSTEM MAINTENANCE

11.9.1 Introduction

The digital multiplexer is a basic network node equipment. The channel signals and the group signals are multiplexed, demultiplexed and relayed through these equipments. These kinds of equipments are usually used in great amount repeatedly which are always the case in digital networks no matter what terminals or nodes. Besides, the external interfaces of these kinds of equipments are relative many, a multiplex interface contains signal interface and clock interface; The tributary interfaces contain multiple amount of signaling interfaces in addition to the signal interfaces and the clock interfaces. The rate ranges of these interfaces are quite wide, the lowest may be 50bit/s, whereas a higher one already reaches at 622 080kbit/s now.

Obviously, it is essential to ensure the normal operation of the multiplex system. In order to guarantee the normal operation of the system, the maintenance work will be very heavy. In the maintenance of the multiplexing system, it is very difficulty to use some effective normal maintenance methods (e.g. switchover of the integrated equipment), due to lots of high rate interfaces in the multiplexer. Therefore, the important on maintenance design for the multiplexing system needs to be carefully considered. Firstly, this maintenance design should accord with the basic maintenance principle recommended by CCITT. Secondly, the specific system plan should be worked out according to the basic features of the multiplexing system. Finally, a proper technical method should be selected to realize the maintenance design target.

11.9.2 Maintenance Principles

The aim of maintenance is to let the maintainers effectively recognize the faults of the equipments, recover services and repair the equipments. Hence we have to specify maintaining entity, maintenance and service alarm indication and faults location methods. Thus it may insure the equipment working satisfactorily, provide a definite fault location, and prevent from unnecessary actions.

In order to describe and execute the maintenance principles conveniently, partition a network equipment into several maintenance entities. See Figure 11-57, an equipment between two neighbor distribution frames (or other equivalent objects) are defined as a maintenance entity. Known from the Figure, this maintenance entity may be digital exchange, digital multiplexer, digital wired section and digital wireless section.

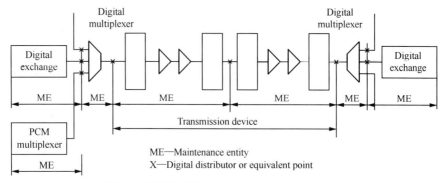

Figure 11-57 Maintenance Entity Concepts

Maintenance Principles:

- When faults happen to the network, Require that at the maintenacce entity where the faults happens maintenance alarm indication appears; If this requirement may not be realized, require that at the maintenance entity near as much as possible such an alarm indication appears.
- Require that any maintenance indication at a maintenance entity will not cause the other maintenance entities to appear the relevant alarm indication; If this requirement may not be realized, require clearly pointing out that the fault appears at the upstream, but not at the maintenance entity which displays such an indication.

Alarm Indication and Corresponding Steps:

- Immediate Maintenance Alarm Indication: When a maintenance entity varies into unusable state, immediately send out such an indication. Seeing such an indication, the maintainers start the maintenance activities immediately: withdraw the maintenance entity out of the service, the operating equipments recover the services and repair the fault entity.
- Delayed Maintenance Alarm Indication: When a maintenance entity varies into a poor state, send out such an indication. Seeing such an indication, the maintainers do have to start the maintenance activities immediately until a proper time.
- Service Alarm Indication: When a performance index is lower than the specified standard of a special service, send out the indication at the maintenance

entity where the service starts or ends. Such an indication marks that the special service is unusable.

- Alarm Indication Signal (AIS): When a maintenance entity varies into unusable state, send out the alarm indication immediately at the entity, at the same time send out a signal to affected direction (the downstreams). Tell the maintenance entities in the downstreams: the fault entity is found, and forbid the other maintenance alarms caused by this fault. AIS uses the code of all "1"s to replace the normal transmission signals.
- Upstream Fault Indication: send out the indication from a normal maintenance entity. It points out that input signals reaching at the maintenance entity have faults, that is, the upstream appear faults, this entity does not do any unnecessary maintenance activities.
- Alarm Information to Remote Ends: In the case of maintenance alarm, the entity of the digital signal source produces the information and transmit the information to the digital ends through the specified slots of the multiplex frames. The aim is to make some evaluation at the link ends.

11.9.3 Multiplexer Maintenance Requirements

The maintenance requirements of three typical multiplexers are summarized in the following:

The maintenance requirements of the group multiplexer:

Equipment position	Defect/failure condition	Consequent actions				
		Immediate maintenance alarm indication	Send AIS to remote end	AIS applied		
				All	Multiplex	Relevant
Multiplex/demultiplex	Power faults	yes		yes	yes	
Multiplex	Lost of one tributary input signal	yes				yes
Demultiplex	Lost of input multiplex signal	yes	yes	yes		
	Out of frame	yes	yes	yes		
	Receive AIS from remote end					

The requirements of the group multiplexer are the most basic requirements in multiplex maintenance. When faults appear at equipment, the alarm indication appears at the local equipment; at the same time, send out the alarm indication signals to the opposite ends or the downstreams. The maintenance requirements of other two kinds of multiplexers are based on that of the group multiplexers.

The maintenance requirements of PCM multiplexers:

Equipment position	Defect/failure condition	Consequent actions					
		Service alarm indication	Immediate maintenance alarm indication	Send AIS to remote end	Forbid audio output	AIS applied on output of ributary	AIS applied in relevant slots
Multiplex/ demultiplex	Power faults	yes	yes	yes	yes	yes	yes
	Encoder/decoder faults	yes	yes	yes	yes		
Multiplex	Lost of tributary input signal		yes				yes
Demultiplex	Lost of multiplex signal	yes	yes				
	Out of frame	yes	yes	yes	yes	yes	
	Frame alignment 1×10^{-3}	yes	yes	yes	yes	yes	
	Receive AIS from remote end	yes					

The difference of the maintenance requirements of the PCM multiplexers from that of the group multiplex is: Encoder/decoder faults and frame alignment error rate exceeds unusable threshold (1×10^{-3}) indication. Both of these two have relationship with the tributary used and final service quality. When the equipment is faulted and the service is poor to an unusable state, the tributary voice output is stopped.

The maintenance requirements of the group multiplexers with switching over the back up power:

Equipment position	Defect/failure condition	Consequent actions			
		Immediate maintenance alarm indication	Delayed maintenance alarm indication	Send AIS to remote end	Automatically power switch over
Multiplex/ demultiplex	Power faults	no	yes		yes

Chapter 11 Engineering Application Design

续表

Equipment position	Defect/failure condition	Consequent actions			
		Immediate maintenance alarm indication	Delayed maintenance alarm indication	Send AIS to remote end	Automatically power switch over
Multiplex	Lost or poor of tributary input signal	yes			
Demultiplex	Lost or poor of multiplex signal		yes	yes	no
	Receive AIS from remote end		yes		

11.9.4 System Consideration of Maintenance Design

Summarize the above maintenance principles and the particular requirements, may get the relationship of receiving signal status, maintenance entity equipment status, corresponding maintenance indication and maintenance operations as the following:

Equipment position		Equipment state		Maintenance indication	Maintenance operation
Normal		Normal		No	No
Abnormal	AIS	Normal		UFI	No
	Unusable				
Normal		Fault	Poor	Delayed maintenance alarm	Delayed maintenance
			Unusable	Immediate maintenance alarm	Immediate maintenance

The final aim of the maintenance design is to use simple maintenance indication to guide the correct effective maintenance operations. Hence, the decisions to determine the maintenance indication are the key point. The basic contents of the fault decisions is to decide if the receiving signals abnormal (i.e. the upstream maintenance unit fault) or the local unit fault, and the abnormal type of the receiving signals and the fault type of the equipment.

Discrimination of equipment normal/abnormal: The most important and also the easiest are to recognize AIS signals, hence set up detecting circuits of consecutive "1"s; detect unusable ($P_e \leqslant 1\times10^{-3}$) receiving signals, set up error test unit, usually use frame alignment signal error rate measurement, and set up unusable error threshold, send out indication when the threshold is exceeded.

Equipment normal/fault detection: usually use loop back method to measure the equipment normal/abnormal.For a maintenance entity, when it inputs a specified input signal, it may output a specified output signal, it is decided be normal, or else abnormal. For multiple maintenance entities, to form many independent loops, i.e. decide any one of the many entities is normal or faulted, it is similar with solving algebra equations.

For a particular multiplexer, consider reliability and economy, use specific integrated circuits as much as possible.In this case, a multiplexer is a relatively simple circuit unit. When these simple multiplexers are used greatly and repeatedly, a quite complicated system is formed. For such a system, the test of various input signals and the operation of equipment self loop back a quite complicated control operations. Now, these maintenance control operations are usually implemented by microcomputers.

The primary channels of the multiplex systems are high speed channels, but many auxiliary operations are low speed.except the input signal recognition and self loop controls, there are some operations are also low speed. For example, forming/sending signaling and receiving/discriminating signaling, discriminations, transforms and displays of the maintenance and the alarm signals, and etc. All of these auxiliary operations are done by microcomputers.

The primary channels use the special integrated circuits as much as possible: auxiliary controls are done by microcomputers. This is the basic idea of design of multiplex systems recently.

11.9.5 Examples of Maintenance Design
(1) Structure of Multiplex Systems

In the multiplex system, the tributary rates are 16/32kbit/s, the multiplex rate is 512kbit/s, the maximum number of tributaries is 15, the frame length is 32 bits, where the first slot transmits frame alignment signals, the second slot transmits signaling, the 3rd to the 32nd slots transmit information codes. See figure 11-58, the system constitution units and their functions are described bellow:

Chapter 11 Engineering Application Design

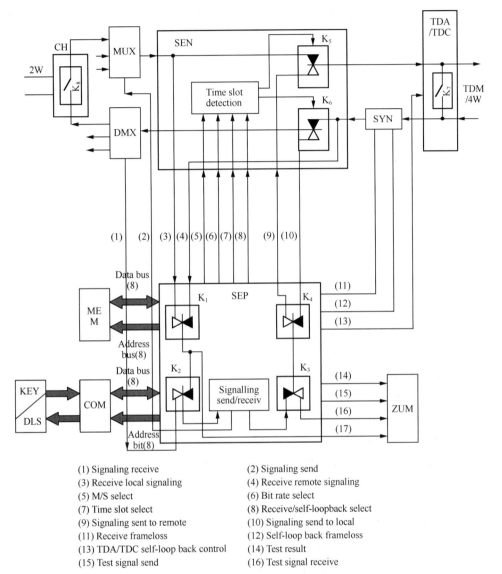

Figure 11-58 Structure of multiplex systems

Channel Units (CH): CHA, CHB CHT are used to connect to different user terminal equipments. CHA connects to analog telephones, CHB connects to digital telephones, CHT connects to Fax;

Multiplex/Demultiplex Units (MUX/DMX): Multiplex/demultiplex 15 information codes and signaling, addressing operations are controlled by SEN;

Central Logical Unit (SEN): Produce the frame synchronous pulses to control MUX/DMX work correctly;

Synchronous Unit (SYN): Frame synchronous search and keep;

Circuit Interface Units (TDA/TDC) Interface standard Code transforms;

Central Processing Unit (SEP): Clock selection control, TDA/TDC interface selection control, channel unit adaptation programming, signaling sending/receiving, channel units testing, interface unit testing, operators and users communication controls, channel unit model data access and display;

Memory Units (MEM): SEP controls program access, working process data access, power lost protection access;

Panel Communication Units (COM): KEY/DIS and SEP interfaces;

Keyboard/Display Units (KEY/DLS): Man-machine interactions, maintenance alarm indications;

Dial Tune Units (ZUM): Generate testing signals and decide the testing signals received if they are correct.

(2) System Maintenance Control

See from the Figure 11-59, the entire multiplex system may be partitioned into channel part and control part. The channel part contains CH, MUX/DMX, SEN, and TDA/TDC which form the information channels; the control part contains SEP, ZUM, MEM, COM and KEY/DIS which form the channel control system. In the control part, all the control operations are executed by SEP. The control functions are channel state control, signaling sending/receiving control, system unit self loop back testing control and unit working state display. where, the later two functions belong to system maintenance control. The typical self-loop testing control and state display are described below:

TDA/TDC unit self-loop back testing:

SEP controls SEN to transfer into TDA/TDC self-loop state through link (8); ZUM sends out testing signals through link (16), through K_3, K_4, K_5 to TDA/TDC units, through K_8 to SEN unit, through K_1 and through link (17) again back to ZUM: after decision ZUM reports the testing results to SEP through link (14); the testing results are sent by SEP to DLS to tell the operator.

SYN Working State Detection:

The SYN unit may be in frame synchronous state or out of frame alignment state even though its equipment is completely good. Hence, it must be monitored all the time. When SYN is in normal receiving state, SEP monitors it through link (11) if there is frame loss; when SYN is in self-loop back working state, SEP monitors it through link (15) if there is frame loss. the particular results are displayed through DLS.

CH unit self-loop Testing:

SEP controls SEN to transfer into self-loop state through link (8), at the same time indicates through link (5) that it is channel self-loop testing; ZUM sends out testing signals through link (16), which enter the DMX unit through K_3, K_4, K_6; then enter channel units; go back to ZUM through K_8, MUX unit, again through link (8), K_1 and link (17); after decision ZUM reports the results to SEP through link (14); and SEP sends the testing results of CH unit to DLS unit to display. There are 15 CH channel units altogether, SEP selects any specific CH channels through link (7), the corresponding self-loop back testing is completed.

(3) SEP Hardware System Structure

See Figure 11-59, the SEP hardware system consists of CPU, address lock decoder, DMA control unit, interrupt control unit, slot activation testing unit, signaling sending unit, signaling receiving unit and input/output collection units. The signaling sending circuits contain three 8 bits parallel input/serial output cycle shift registers, one of them is used for the slot 2 to send signaling to the remote ends; the second is used to send information codes to selected (local or remote) channel units; the third is used for the selected channel to send invalid code in the opposite direction. The signaling receiving circuits are an 8 bits serial input/parallel output shift register which is used to receive signaling. The collection input and control output circuits consist of channel number register, state signal input register and output signal register, which completes the functions of bit rate selection, channel loop testing, interface selection, clock selection and interface testing and etc.

(4) Control Keyboard Configuration

Control keyboard has 16 keys altogether which consist of 10 digital keys of 0-9, and 6 functional keys:

CLEAR key: clear the contents of the display screen;

LIST key: display channel unit model and internal/external clock interface model;

PROG key: programming the channel models, internal/external clock interface model;

OPR key:operators and channel unit users communications;

TEST key: test channel units and interface units;

ENTER key: start to process the above functions and display the results.

PDH for Telecommunication Network

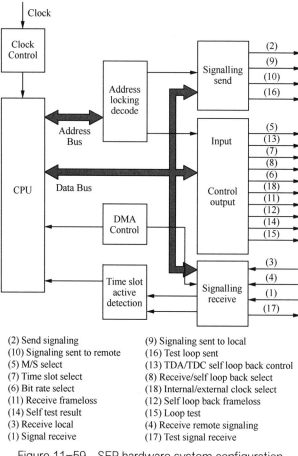

(2) Send signaling
(10) Signaling sent to remote
(5) M/S select
(7) Time slot select
(6) Bit rate select
(11) Receive frameloss
(14) Self test result
(3) Receive local
(1) Signal receive
(9) Signaling sent to local
(16) Test loop sent
(13) TDA/TDC self loop back control
(8) Receive/self loop back select
(18) Internal/external clock select
(12) Self loop back frameloss
(15) Loop test
(4) Receive remote signaling
(17) Test signal receive

Figure 11-59 SEP hardware system configuration

(5) Control Software Structure

The control software is shown in Figure 11-60, which has 6000 lines of assembly programs.

The initial program: sets the program parameters, channel unit state model, output control register state and interrupt vectors;

Interrupt programs: implements the controls of M_1, M_2, M_3 in turn;

M_1 function module: makes specified responses to the signaling from channel units or slot 2, selects suitable signaling from the module M_2 and sends to the destinations, implements communications to channel units and slot 2;

M_2 function module: calls the lower modules, processes data from the keyboard and the control actions, and sends signaling to channel units and slot 2.

M_2 has 9 kinds of lower modules. Where, the module $M_{2,4}$ executes testing of channel units; $M_{2,7}$ executes testing of interfaces.

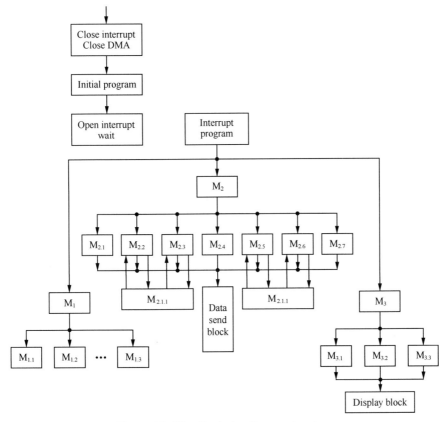

Figure 11-60 Control software structure

M_3 function module: manages keyboard and display screen.

Bibliography

[1] CCITT 6th Plenary Assembly Vol. III-2 Recommendation G.702 Vocabulary of Pulse code modulation and digital transmission terms pp369-388.

[2] Maurizio Decina: Planning of a Digital Hierarchy IEEE Trans. on Communication Technology, February 1972 Vol. COM-20, No.pp60-64.

[3] CCITT Study Group XV III.8-17, April 1980, Temporary docament No. 76 Proposed Reply to Question 8/XV III-PCM and Digital Multiplexing for Telephony and Other Signals.

[4] CCITT 6th Plenary Assembly Vol. III-2, Recommendation G.732、G.734、G.742、G.744、G.746、G.751.

[5] CCITT Study Group XV III 8-17 April 1980, Temporary document No.52 Report of Group XV III/3-Line Transmission, Recommendation G.922.

[6] H. HABERLE: Frame Synchronizing PCM Systems, ITT Electrical Communication 1969 Vol.44, No.4, pp 280-287.

[7] 孙玉：延迟锁定环最佳工程设计，<无线电通信技术>1972年第一期.

[8] 董利民：同步复接和同步分离设计，<无线电通信技术>1978年第五期.

[9] N.Kuroyanagi H Saito: Multiplexer-Demultiplexer for PCM-16 System Reuiew of the Electrical Communication Laboratory 1969. Vol. 17, No.5-6.

[10] 孙玉：正码速调整异步复接技术，<国外电子技术>1977年第三期.

[11] F.J.Witt: An Experimental 224 MBit/s Digital Multiplexer-Demultiplexer Using Pulse Stuffing Synchronization. BSTJ No.9 1965 pp 1843-1885.

[12] 袁斌：码速调整设计，<无线电通信技术>1978年第五期.

[13] 法国：TNM1-2 二次群数字复接器，Cables ET Transmission 29 Annee, DEC 1975<数字通信>1977年第1-2期.

[14] 孙玉、于绍颖：码速恢复设计，<无线电通信技术>1978年第五期.

[15] S. KOZUKA: Phase controlled Oscillator for Pulse Stuffing Synchroniyation System, Reuiew of the Electrical Communication Laboratory 1969 Vol. 17, No.5-6.

[16] CCITT Study Group XV III 8-17 April 1980, Temporary document No. 76 Proposed Reply to Question 8/XV III-PCM and Digital Multiplexing for Telephony and Other Signal.

[17] CCITT Study Group XV III 8-17 April 1980, Temporary document No. 78, Proposed Reply on Ouestion 6/XV III.

[18] 孙玉：正码速调整的塞入抖动测量，<无线电通信技术>1978年第五期.

[19] CCITT Period 1977-1980 COMXV III-No. 305(CHINA) The Reduction of

Stuffing Jitter in the Digital Multiplex Equipment Based on 2048 kbit/s and Using Positiue Justification.

[20] 孙玉：正/负码速调整异步复接技术，<国外电子技术>1978 年第一期.
[21] 孙玉：采用正/负码速调整的二次群异步复接器的复接分接设计，<数字通信>1979 年第二期.
[22] 孙玉：采用正/负码速调整的二次群异步复接器的码速调整初步试验，<数字通信>1979 年第二期.
[23] CCITT Period 1973-1976 COM Sp. D-No. 312: Comparsion of Digital Transmission Systems With Positiue Justification and With Positiue-Negatiue Jushification.
[24] CCITT 6th Plenary Assembly Vol. III-2, Recommendation G.745.
[25] CCITT Period 1973-1976 COM Sp. D-No. 167，澳大利亚：数字复接码速调整引起的抖动测试结果（1975 年 1 月）.
[26] Karl-Heinz Stolp: Digitale Multiplexgerate des PCM30D-Netzes.
[27] CCITT Period 1977—1980 COMXVIII-No. 75, Multiplexing of 30 Signals at 64 Kibit/s Using Positiue/Zero/Negatiue Justification: The PCM 30D Multiplexer.
[28] 孙玉、袁斌：正码速调整复接误码测量，<无线电通信技术>1978 年第五期.
[29] 孙玉：CCITT 第 XVIII 研究组（数字网）1977—1980 研究期研究进展，<通信学报>1980 年第一期.
[30] 董利民、袁斌、孙玉：采用正码速调整的二次群异步复接器，<无线电通信技术>1978 年第五期.
[31] 法国 CIT-ALCATEL：TNM2-4 数字复接器及其检测设备，<C&T> 32^{e}A, n°2, avril 1978.
[32] CCITT Period 1977-1980 COMXVIII-No.24; FRENCH: n-$(n+2)$ Multiplex Equipments and Transfer.
[33] CCITT: Period 1977-1980 COMXV COMX III-No. 183; ITALIAN: Remarks on the Specification for type n-$(n+2)$ Multiplexing Equipments.
[34] CCITT: Period 1981-1984 COMXV III-No.; CHINA: Design for 2048/34368 kbit/s MULDEX.
[35] CCITT: Period 1981-1984 COMXV III-No.; CHINA: A simple 2048/34 368 kbit/s Digital MULDEX.
[36] 孙玉：2048/34 368kbit/s 数字复接设计，通信学会：1980 年九月昆明会议文件.
[37] 刘国存：2048/34 368kbit/s 三次群数字复接器，<无线电通信技术>1978 年第五期.
[38] 高广明：三次群数字复接器简化工程设计，<无线电通信技术>1978 年第五期.
[39] 孙玉：非标速率码流在标准速率通道中的传输<电子信息技术>1980 年第一期.
[40] 苏联：码速调整的改进方法，<数字通信>1975 年第三期.
[41] J.A.Bylstra and R. Coxhill: A Hand-book for the 8448kbit/s Digital Multiplex, Jun 17, 1976. Telecom Australia.
[42] 董利民译：时分交换系统中相位同步存储器设备设计<交换研究>：SE74-22-36(1974-08).

[43] 董利民、吴戎云：准同步帧调整设备，<无线电通信技术>1980年5月.
[44] 吴戎云：帧调整存储器的一种现实方法，<无线电通信技术>1980,No.11-12.
[45] 刘国存、安建亭：一种用于气象卫星数据传输的数字复接设备，<无线电通信技术>1980, No.11-12.
[46] CCITT SGXVIII 8-17 April 1980, Temporary document No. 76.
[47] CCITT Period 1977-1980 COMXVIII-No.248.
[48] CCITT Period 1977-1980 COMXVIII-No.207.
[49] CCITT SGXVIII 1980 April Delayed Comtribution DI.
[50] 汪家顾：1980年CMTT中期会议报告，声音节目数字化现状，<无线电通信技术>1981.
[51] 高广明、李国华：正/0/负码速调整数字复接技术，<无线电通信技术>1981.
[52] 高广明：正/0/负码速调整数字复接技术，<无线电通信技术>1981 No.11-12.
[53] 孙玉、谢冬蓉、毛凤莲：晶体压控振荡器工程设计，<无线电通信技术>1980, 5.
[54] CCITT 6th Plenary Assembly Vol. III-2, Recommendation G.702.
[55] CCITT 关于脉码通信的建议<数字通信>1977年专刊.
[56] 清华大学：增量调制数字电话终端机，人民邮电出版社1977年版.
[57] 北京大学：脉码调制复接设备，人民邮电出版社1981年版.
[58] 孙玉：数字网传输损伤，人民邮电出版社1985年版.
[59] 孙玉：数字网专用技术，人民邮电出版社1988年版.
[60] 孙玉：数字网中的子群帧结构<电子信息技术>1982.4.
[61] 冀克平：抗衰落帧同步<无线电通信技术>1988.4.
[62] 王俊芳：正/负码速调整准同步复接实验研究石家庄通信测控技术研究所1988研究生论文.
[63] CCITT 建议 G.702（1984—1988）数字分级比特速率.
[64] CCITT 建议 G.731（1984—1988）音频用基群PCM复用设备.
[65] CCITT 建议 G.732（1984—1988）工作在2048kbit/s的基群PCM复用设备的特性.
[66] CCITT 建议 G.735（1984—1988）工作在2048kbit/s并提供384kbit/s和64kbit/s分支的基群PCM复用设备的特性.
[671] CCITT 建议 G.736（1984—1988）工作在2048kbit/s的同步数字复接器的特性.
[68] CCITT 建议 G.737（1984—1988）工作在2048kbit/s并提供384kbit/s和64kbit/s同步分支的分支设备的特性.
[69] CCITT 建议 G.741（1984—1988）二次群复接器的一般考虑.
[70] CCITT 建议 G.742（1984—1988）工作在8448kbit/s并采用正码速调整的二次群复接器.
[71] CCITT 建议 G.744（1984—1988）工作在8448kbit/s的二次群PCM复用设备.
[72] CCITT 建议 G.751（1984—1988）工作在34368kbit/s 三次群比特速率和13926kbit/s 四次群比特速率，并采用正码速调整的数字复接器.
[73] CCITT Study GroupXVIII Geneua, 6-17 Jone 1988 TD — 172 Draft Recommendation G.707. Synchronous digital Hierarchy Bit Rates.
[74] CCITT Study GroupXVIII Geneua, 6-17 Jone 1988 TD — 173 Draft

Recommendation G.708. Network Node Interface for the Synchronous Digital Hierarchy.

[75] CCITT Study GroupXVIII Geneua, 6-17 Jone 1988 TD — 174 Draft Recommendation G.709. Synchronous Multiplexing Structure.
[76] CCITT Period 1985—1988 COMXVIII-138: Proposed Reuisions to Draft Recommendetion G.70X、G.70Y、G.70Z.
[77] CCITT Period 1985—1988 COMXVIII-142: Mapping of Asynchronous 139264kbit/s into VC-4 Container.
[78] CCITT Period 1985—1988 D.1103/XVIII: Synchronous Optical Network Standaudization in the United States.
[79] CCITT Period 1985—1988 D.1122/XVIII: U.S.A.Use of 32kbit/s at Network Nodes.
[80] CCITT Period 1985—1988 D.1143/XVIII:United States of America: Clarification of 64kbit/s Access in the New Synchronous Hierarchy.
[81] Annex-3 Draft Recommendation G.704. Synchronous Frame Structures Used at Primary and Secondary Hierarchical leuels.
[82] CCITT Period 1985—1988 D.1643/XVIII: Proposal for 34368kbit/s and 8448kbit/s mapping respectiuely in VC-31 and VC-22 of the STM-1.
[83] CCITT 建议（1984—1988）I.430. 基本用户—网络接口—第一层规范.
[84] CCITT 建议（1984—1988）I.431. 基群速率用户—网络接口—第一层规范.
[85] CCITT 建议（1984—1988）I.460. 复接、速率适配和现有接口的支持.
[86] CCITT 建议（1984—1988）I.461. ISDN 对 X.21 和 X.21bis 基本数据终端（DTE）的支持.
[87] CCITT 建议（1984—1988）I.463. ISDN 对 V 系列接口数据终端设备（DTE）的支持.
[88] CCITT 建议（1984—1988）G.803. 数字网的维护.
[89] CCITT 建议（1984—1988）G.743. 工作在 6312kbit/s 并采用正码速调整的二次群复接器.
[90] CCITT 建议（1984—1988）G.802. 采用不同技术的数字通道的互连.
[91] 裴军：微机控制的增量调制复接器控制体系分析，石家庄通信测控技术研究所 1989 研究生论文.
[92] CCITT 建议（1988）：G.73、G.74、G.75.
[93] CCITT SGXVIII（1988）：TD.173、TD.174.
[94] 孙玉、冀克平：抗衰落帧同步潜在技术性能，<无线电通信技术>1991.2.

全集出版后记

衷心感谢人民邮电出版社为我出版这套全集。

这套全集与人民邮电出版社有几十年的缘分。因此，我想用我为人民邮电出版社成立60周年纪念册《历程》的题词，作为全集出版的后记。

> 我作为人民邮电出版社50年的读者和作者，以十分敬佩和感恩的心情祝贺人民邮电出版社成立60周年。
>
> 从1962年起，我就是人民邮电出版社受益丰厚的读者；从1983年出版专著《数字复接技术》起，直到2007年出版专著《电信网络总体概念讨论》，又成了人民邮电出版社备受关照的作者。可以说，在这50年间，我与人民邮电出版社结下了不解之缘；与那些敬业奉献的编辑们，从白发苍苍的长辈到风华正茂的后生，建立了深厚的感情。我从内心感谢他们，敬佩他们。
>
> 可以确切地说，人民邮电出版社为我国电信技术发展建立了实实在在的不朽功勋。祝愿人民邮电出版社繁荣昌盛。
>
> 中国电子科技集团公司
> 第54研究所研究员
> 中国工程院院士 孙玉
> 2013年7月1日

我作为人民邮电出版社50年的读者和作者，以十分敬佩和感恩的心情祝贺人民邮电出版社成立60周年。从1962年起，我就是人民邮电出版社受益丰厚的读者；从1983年出版专著《数字复接技术》起，直到2007年出版专著《电信网络总体概念讨论》，又成了人民邮电出版社备受关照的作者。可以说，在这五十年间，我与人民邮电出版社结下了不解之缘；与那些敬业奉献的编辑们，从白发苍苍的长辈到风华正茂的后生，建立了深厚的感情。我从内心感谢他们，敬佩他们。确切地说，人民邮电出版社为我国电信技术的发展建立了实实在在的不朽功勋。祝愿人民邮电出版社繁荣昌盛！

感谢人民邮电出版社对于我国电信技术发展的支持和贡献！敬佩沈肇熙先生、李树岭编辑、梁凝编辑、杨凌编辑四代编辑的敬业精神和专业水平！感谢邬贺铨院士为我的全集作序！